Fundamentals and Applications of Chemical Sensors

ACS SYMPOSIUM SERIES **309**

Fundamentals and Applications of Chemical Sensors

Dennis Schuetzle, EDITOR
Ford Motor Company

Robert Hammerle, EDITOR
Ford Motor Company

James W. Butler, ASSOCIATE EDITOR
Ford Motor Company

American Chemical Society, Washington, DC 1986

Library of Congress Cataloging-in-Publication Data

International Chemical Congress of Pacific Basin
   Societies (1984: Honolulu, Hawaii)

   Fundamentals and applications of chemical sensors.
   (ACS symposium series, ISSN 0097-6156; 309)

   Includes bibliographies and indexes.

   1. Chemical detectors—Congresses.

   I. Schuetzle, Dennis, 1942–   . II. Hammerle,
Robert, 1940–   . III. Series.

TP159.C46156   1984   660.2′8′00287   86-3518
ISBN 0-8412-0973-1

ACS Symposium Series

M. Joan Comstock, *Series Editor*

Advisory Board

FOREWORD

CONTENTS

PREFACE

Sensors can be used with microprocessors to control automatically and more effectively a wide variety of devices. For example, sensors are used in manufacturing machines and consumer items, such as cars and ovens, as well as in gas safety alarms and medical drug delivery devices for diseases such as diabetes. The growing interest in these sensors stems from the enormous benefits anticipated from the widespread use of sensors and microprocessors for control. Although microprocessors have been extensively developed, sensors have received relatively little attention. As a result, sensor development is expanding rapidly.

To reach their full potential, sensors developed for control applications must be small and inexpensive because the space and money available for improved control are generally limited. In manufacturing, for example, one machine tool may need 5–10 sensors to provide automated control. These sensors must fit on the machine without interfering with transfer of parts or access by humans. This requirement is especially demanding when the sensors monitor a small part of the machine such as a bearing or drill. In addition, the sensors may be exposed to cutting fluid spray and vibration and have limited maintenance, all of which may lead to frequent failures. Therefore, the sensors must be easy to fix or replace. Installing a new sensor and throwing away the old one is usually easier than repairing a larger, more expensive device. Finally, the cost of instrumenting an entire plant may be prohibitive if the sensors cost $5000, whereas it may be very attractive if the sensors cost less than $100. Similar considerations limit the size and cost of the other control sensors mentioned earlier.

In addition to size and cost concerns, control sensors also must be accurate to reach their full potential. They are often used when control is critical. For example, if a position sensor on a milling machine is inaccurate, the parts produced on that machine may not meet specifications. Not only are the time and materials used to make those parts wasted, but also an entire assembly operation could be shut down for lack of parts.

The size, cost, and accuracy requirements for successful sensors place a substantial burden on the processes used to manufacture them. Typically, today's sensors are manufactured by processes that are specific to one sensor or, at best, to a limited class of sensors. This process specificity has occurred because each type of sensor is usually made of unique materials in unique configurations that best convert the quantities to be measured into electrical signals. For example, pressure transducers may use piezoelectric or piezoresistive materials on thin diaphragms, whereas ion-selective electrodes use ion-conductive glasses or polymers around electrodes. Unfortunately, this situation implies large developmental costs.

Many people envision that sensor design and production can be greatly simplified by making sensors of silicon. In some cases, silicon devices can generate electrical signals directly from the quantity to be measured, but, in most cases, other materials must be added. In addition, silicon must be formed in configurations that are unusual for integrated circuits but that are necessary for sensors. The principal advantages are (1) that sensors made of silicon can be made with integrated-circuit manufacturing techniques that naturally allow for miniaturization, low process costs, and high reproducibility, and (2) that many of the same manufacturing techniques can be used for a large variety of sensors.

Chemical sensors, those that measure the presence or concentration of chemical species, are the subject of this book. Until recently, they received even less attention than other sensors; in general, they are not as well developed. They have the same need to be small, inexpensive, and accurate as other sensors. However, accomplishing these requirements for chemical sensors is often more difficult than for other sensors because chemical sensors are noted for interferences. For example, a chloride sensor may be sensitive to other halides. One popular way to counter this limitation is to use an array of somewhat different sensors, each responsive to the same set of related compounds but with different sensitivity. The output of the sensor array can be processed by a computer to give greater accuracy than a single sensor for the concentration of one compound. Unfortunately, this approach tends to gain better accuracy at the expense of increased size and cost.

This book discusses a wide variety of chemical sensor subjects ranging from the development of new sensor concepts to improvements in sensors that have been mass produced for several years. It is not always clear whether the new sensor concepts or the improvements described will ever be commercialized. However, the aim of this book is to present these new ideas to help the sensor community describe where further development may be worthwhile.

We wish to acknowledge the efforts of G. G. Guilbault and T. Seiyama, who organized the symposium from which this book was developed, and E. Poziomek, who organized the symposium, "Microsensors for Chemical Detection, Identification, and Analysis," at the ACS Middle Atlantic Regional Meeting held April 6–8, 1983, where most of these papers were first presented.

DENNIS SCHUETZLE
ROBERT HAMMERLE
Ford Motor Company
Dearborn, MI 48121

March 1986

GENERAL DISCUSSIONS

1

Recent Advances in Chemically Sensitive Electronic Devices

Jay N. Zemel, Jan Van der Spiegel, Thomas Fare, and Jein C. Young

Center for Chemical Electronics, Department of Electrical Engineering, University of Pennsylvania, Philadelphia, PA 19104

Significant advances have occurred in microfabricated ion sensitive and Pd gated field effect devices and fiber optic, chemically sensitive elements. These elements are beginning to find their way into commercial development. Recent advances in these devices are discussed and compared. Pyroelectric sensor devices developed here are reviewed. A discussion of the utility of these devices is presented.

An important goal'of the research presented in this volume is the development of instantaneous or near instantaneous data acquisition elements which, with the assistance of suitable control algorithms, could define a system's state in real time. The need for process control information has become even more urgent as high speed, low cost, single chip microcomputers become generally available. The development of new real time sensors based on contemporary materials and phenomena knowledge requires a more systematic approach than has been the case to date. While the need for new measurement tools to advance scientific and technological inquiry is certainly well appreciated, sensor and sensor related research is frequently viewed as "widget making" by the scientific and technological communities. By and large, the development of new sensors has been a byproduct of the normal course of materials or phenomena research. It is only in the last few years that the scientific and technological community has taken note of the fact that sensor research is becoming a distinct and important topic.

In the past decade, microfabrication methods developed in the microelectronic industry have led to new opportunities for device research and development involving chemically sensitive electronic structures. In 1980, this subject was reviewed in depth at a NATO Advanced Study Institute (1). Over the last five years, there have been three international conferences (2-4), devoted to sensors with a strong emphasis on chemical sensors as well as a number of national and specialized meetings on the subject (5,6). In this paper, some recent developments that will have long term consequences on the study of chemically sensitive electronic devices will be reviewed. To simplify the discussion, the topics are divided into the following categories:

    a. Pd based metal-oxide-semiconductor hydrogen sensors.
    b. Integrated electrochemical sensors.
    c. Fiber optic chemical sensors (FOCS) using fluorescence and
       absorptive behavior.
    d. Thermally based chemical sensors employing pyroelectric
       elements.

Connecting these disparate elements is their mode of manufacture. All employ the same basic procedures developed by the microelectronic industry to manufacture its elements, i.e. planar photolithography. The one class that has not yet been microfabricated with planar photolithography is the FOCS. In general, the choice of phenomena and materials for a particular type of measurement requirement is determined by their compatibility with planar photolithography. In other words, if the material is not suitable for planar processing, it will not be used. With the exception of FOCS, if the phenomena on which the device operation depends requires a structure that cannot be microfabricated with planar technology, it too will not be considered. As it turns out, these restrictions are not too severe and the flexibility of planar processing, combined with its potential for low cost batch processing, makes it a very attractive fabrication methodology.

Pd MOS STRUCTURES: The Pd MOS device (capacitor and field effect transistor) has been extensively studied as a model chemical sensor system and as a practical element for the detection of hydrogen molecules in a gas. There have been two outstanding reviews of the status of the Pd MOS sensor with primary emphasis on the reactions at the surface (7,8). In this section, the use of the device as a model chemical sensor will be emphasized. As will be seen, the results are applicable not only to the Pd based devices, they also shed light on the operation of chemfet type systems as well. Because of its simplicity and the control that can be exercised in its fabrication, the discussion will focus on the study of the Pd-MOSCAP structure exclusively. The insights gained from these studies are immediately applicable to the more useful Pd-MOSFET.

In a typical experimental study, a Pd-MOSCAP is kept at a fixed temperature in a controlled ambient atmosphere of oxygen or an inert gas like pure nitrogen. The gas is exchanged with a dilute mixture of hydrogen in nitrogen, typically 10-1,000 ppm of H_2 in N_2. A reaction occurs at the Pd-gas interface between the residual adsorbed oxygen and the hydrogen (9). When enough oxygen is removed as water vapor

from the Pd surface, the gaseous hydrogen adsorbs and dissociates
into atomic hydrogen. This atomic hydrogen then diffuses into the Pd
(H_{Pd}), eventually reaching the $Pd-SiO_2$ interface. The diffusion time
through a typical 100 nm thick Pd film is of the order of a mil-
lisecond (7). The H_{Pd} generates a dipole layer, illustrated in Figure
1a), at this interface that lowers the electronic barrier height. The
resulting displacement current changes the space charge density in
the silicon part of the Pd-MOS structure. The magnitude of this
change can be measured by observing the displacement of the admit-
tance versus voltage curve of Pd-MOSCAP as indicated in Figure 1b).

One of the standing problems in this $Pd-SiO_2-Si$ system is the
uncertainty about the boundary conditions relative to the motion of
H_{Pd} into the SiO_2 and from there, on to the SiO_2-Si interface or the
Si itself. Problems with long term drift of the steady state flat
band voltage have raised the possibility of H_{Pd} diffusing further
into the oxide. Early work by Hofstein had raised the possibility
that protons could be injected into the oxide (10). However, an extensive set of studies eventually demonstrated that it was most
likely Na^+ rather than H^+ that was responsible for Hofstein's observations (11,12). One of the more significant observations was that
Na^+ apparently enhances the diffusion of H in the SiO_2 (9). These
results imply that hydrogen does not go into bulk SiO_2 as a charged
species, but could be injected as H^+ at the Pd-oxide or electrolyte-
oxide interface.

Lundstrom has suggested that most of the H^+ near the $Pd-SiO_2$
interface may be neutralized prior to diffusing into the bulk oxide,
but that enough remain near the interface to account for the long
term instabilities (13). The neutral species (H) is not constrained
by the image charge at the Pd gate and is able to move to another
hydrogen active site. In their investigation of the hydrogen induced
drift (HID), Nylander et al. found that Na^+ contaminated samples
demonstrated an enhanced HID (14,15). This is consistent with the
observations of Holmberg et al. (12).

The mechanism for the $H-SiO_2$ and $H-SiO_2-Na^+$ interactions was
treated by Doremus in a model for gel formation at the surface of
glass electrodes (17). The incorporation of water in SiO_2 has been
reviewed by Nicollian and Brews (16). The water molecule may diffuse
into the oxide without interacting with the bulk or it may take part
in an exchange reaction with SiOH already present. This requires that
H_2O transforms an existing Si-O-Si bond to a pair of silinol groups
in a manner analagous to the Doremus model for low temperature
hydrogen transport (See Figure 2a). In other words, hydrogen is in-
terstitially transported by water or by a complex Si-OH-HO-Si
exchange with water. Neutral water is the diffusant here and SiH
species do not take part in the exchange reaction. The general
mechanism is shown in Figure 2 where the motion of O^{18} is
illustrated. The same principles would apply for the motion of H attached to the water molecule. When sodium is added to the oxide, it
serves as a site for Na-OH dissociation. This model was expanded to
include SiO^-, SiOH and $SiOH_2^+$ sites in order to explain surface adsorption at the electrolyte-oxide interface (18,19). The
incorporation of H in SiO_2 is almost impossible to avoid under normal
conditions. The density and nature of the defects associated with
thermally grown oxides not only varies from sample to sample, but

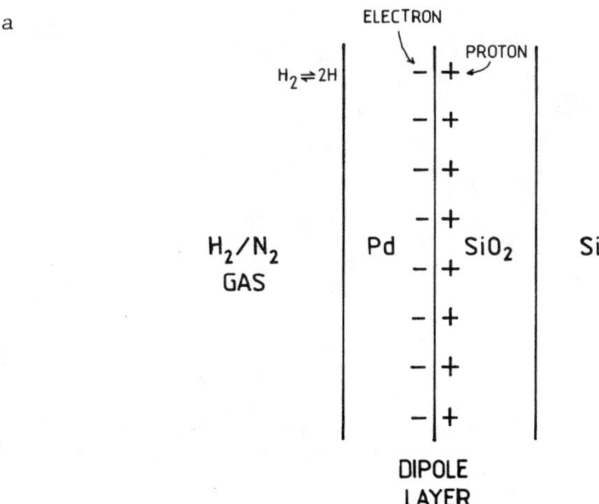

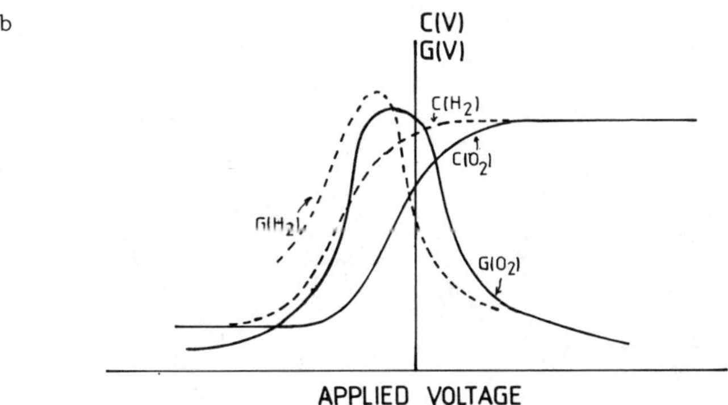

Figure 1. a. Illustration of the dipole layer formed at the palladium–silicon dioxide interface as a result of H interactions; b. Schematic capacitance versus applied voltage, (C–V), and parallel conductance at frequency δ versus applied voltage, (G–V), for an n–type silicon based Pd MOSCAP in the presence of oxygen (solid line) and hydrogen (dashed line).

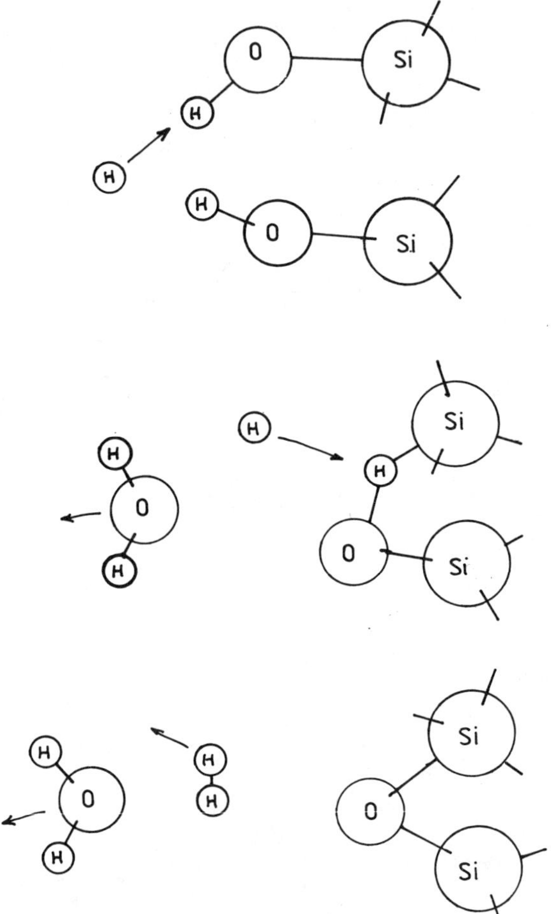

Figure 2. Mechanism for the diffusion of hydrogen in SiO_2:
a. Hydrogen atom approaches two neighboring Si-OH bonds; b.
Formation of water molecule from the Si-OH structure and the
formation of a hydroxyl bridge between the two Si atoms.
Approach of another hydrogen atom while the water molecule
diffuses away; c. Combination of another free H with the
bound H to form molecular hydrogen. Both the H_2 and the H_2O
then diffuse until they find a site where they can react
with a SiO_2 molecule and reverse the process.

also strongly depends on the method of preparation. Thus the preparation of the oxides will play a major role in the degree to which devices will exhibit various types of hydrogen induced behavior.

The study of bulk defects in SiO_2 has relied heavily on ESR measurements. Johnson et al. employed a combination of deuterium plasma and ESR to follow the behavior of both the so-called P_b signal corresponding to a paramagnetic defect and the interface states at $E_v + 0.3$ eV and $E_c - 0.25$ eV, characteristic of unannealed samples, while the sample is subjected to a standard post-oxidation thermal anneal (20,21). In the presence of the deuterium (D) plasma, the P_b signal correlated with the decrease in the interfacial state density. SIMS measurements correlated the increase of D at the $Si-SiO_2$ interface with the decrease in P_b and the surface states, in agreement with other types of studies (22-24). In one study, it was shown that in freshly oxidized Si specimens, there was no sign of H contamination. However, after a period of time in the air, hydrogen signals are observed (25). This correlates with observation of Wen and Zemel on exposing oxides to acid-base solutions (26). Other studies showed that Cl additives tended to increase the hydrogen concentration spontaneously (27). These studies make it evident that hydrogen and hydrogen bearing compounds interact strongly with SiO_2. As a result, it is difficult to obtain unambiguous information on the interaction of hydrogen with the Pd-MOSCAP system because the oxide is not a passive component.

A further conclusion to be drawn from investigations of hydrogen interacting with bulk and thin film SiO_2 is that H-Pd-MOS device operation will be influenced by the type of oxide preparation used. In fact, the first use of a Pd-MOSCAP was in a study of hydrogen annealing of interfacial states (28).

Most recently, Fare et al. demonstrated that the oxide preparation can strongly influence the behavior of the Pd-MOSCAP (29-31). In particular, their measurements demonstrated that H could be isothermally injected not only to the Si-oxide interface, but that the hydrogen will also reversibly enter and leave the silicon substrate (28,29). The data showed that the introduction of H by the Pd gate would cause variations in the recombination cross section and density of the interfacial states (29,31). The oxides used in this study were quite thin (11 nm) and were not subjected to any interfacial state reduction processing. These studies provided general confirmation that atomic hydrogen is injected from the $Pd-SiO_2$ interface, through the oxide, and on into the $Si-SiO_2$ interface where they form interfacial states at $E_v + 0.3$ eV and $E_c - 0.25$ eV as originally proposed by Keramati and Zemel (32,33). However, the Fare et al. studies also provide some clues as to why the evidence of different investigators on the influence of hydrogen injected by a Pd gate varied so much (34-36). Most of these studies employed wet oxides that had been well annealed, generally with a hydrogen bearing ambient gas to reduce the interfacial state density. In the absence of the strained bonds or chemical defects, the interaction of H with the interface is likely to be difficult to observe. This data also points out that chemical shielding of all chemically sensitive electronic elements will be a critical step in device processing.

The conclusions to be drawn from these studies are that the composition and past history of both the Pd-SiO$_2$ and Si-SiO$_2$ interfaces, the method of producing the oxide and the post-growth annealing steps play key roles in determining the nature of the Pd-MOS response to hydrogen. As a model system for a chemical sensor, it points out the important role that catalytic materials like Pd can play in future sensor designs.

INTEGRATED SILICON-BASED ELECTROCHEMICAL SENSORS: The idea of marrying integrated circuit fabrication technology with membrane science to generate new classes of chemically sensitive electronics devices began in the early 70's (37-39). The rationale for this effort was the expectation that planar microfabrication technology would produce the same dramatic impact on information sensing systems as it had on information processing systems. In particular, it was hoped that these new sensor systems would embody the following advantages:

1. small, rugged solid-state structures
2. low impedance outputs
3. high dimensional precision
4. low cost batch fabrication
5. prospect of multi-species sensors with on-chip electronics

These prospects have stimulated a significant amount of research. Despite these efforts, progress has been slow to date. It is only recently that a commercial single species chemically sensitive field effect transistor (chemfet) has become available (40). Among the reasons for the slow development are: the lack of planar technologies for micromachining the chemically sensitive membrane that are still compatible with standard microelectronic technology and use of unshielded field effect based devices where ionic motion creates almost insurmountable packaging problems. In the chemfet, there is a membrane in the gate region which is in direct contact with the fluid to be analyzed. As a result, the electronically active region of the chemfet is directly exposed to the environment. Any ionic leakage paths from the solution, around the chemically sensitive membrane and along the surface of the gate-dielectric will discharge the potential generated in the chemically sensitive membrane by the ions entering from the solution. This is illustrated in Figure 3. Another major problem produced by inadequate packaging of the chemically sensitive ionic device is ionic drift similar to that shown in Figure 3 which produces an unacceptable cross talk between neighbouring sensors. Also, because the chemfet is a capacitive structure it is unforgiving to even very small leakage currents in the dielectric medium consisting of the gate insulator and the chemically sensitive membrane.

The ion controlled diode was an initial attempt to isolate the active electronics from the chemical solution by producing a metallic-like via that allows the isolation of the chemically sensitive region from an area where electronic components could be deposited (41,42). However, the limited precision of the non-standard microfabrication techniques made this process difficult and costly. Since this device is still essentially a capacitive membrane-insulator-semiconductor structure like the chemfet, the same problems of hermetic isolation of the gate remain.

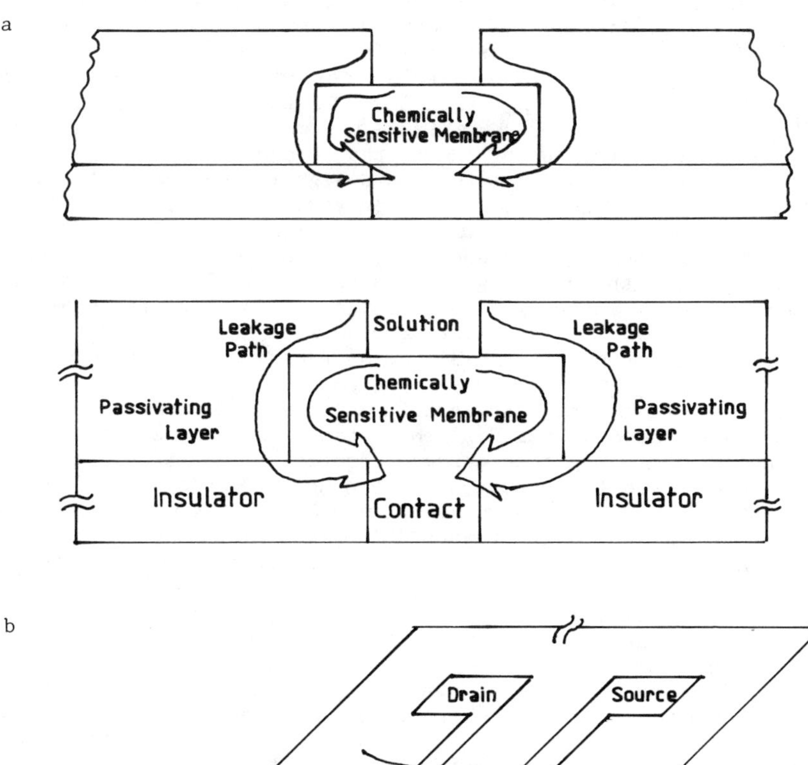

Figure 3.   Ionic leakage paths in chemfet structures:
a.Schematic illustration of ionic leakage paths around the
chemically sensitive membrane. Leakage through the membrane
also occurs but is not illustrated; b. Schematic illustration of leakage at the surface of a standard ion sensitive
field effect transistor.

A new approach to semiconductor based sensors, developed in our laboratories, is the integrated shield, electrochemical sensor. The goal of this research was to develop multispecies sensors which could be completely fabricated by wafer level processing, while at the same time completely isolating the electronically active zone from the chemically sensitive layers which are in contact with the fluids. The structure is schematically shown in Figure 4 (43). The device consists of two sections. The first part is the chemically sensitive layer in direct contact to the shielded signal line which forms a complete electrochemical electrode to the second part, the input of a high impedance buffer. Central to this design is the insulated shield around the the line and sensors. This provides the necessary electrical isolation and chemical passivation to reduce the cross-talk between different sensors. In addition, the shield is bootstrapped in order to minimize the coupling between the shield and line. In fact, this structure is an integrated monolithic version of the classical ion selective membrane electrode contacted via a wire to an electrometer. The sensors are designed as faradaic elements.

One of the most important aspects of the integrated, shielded electrode structure is that it lends itself to large scale chemical sensor integration, i.e. four or more chemical sensors on a chip. A wafer is processed to produce a large number of uncoated sensors by the same standard microfabrication technology used to manufacture integrated circuits. The chemically sensitive membranes are then added following the testing of the electronic circuitry. This procedure makes it quite feasible to produce an array of sensors, each responding to a different chemical quantity. By separating the preparation of the sensor materials and silicon device structures, the sensor scientist can obtain wafers containing his design of the silicon structure at the wafer level through a silicon foundry. Use of silicon foundries has become standard in the custom IC design industry because it has been amply demonstrated that quite complex silicon integrated circuits can be fabricated there at relatively low cost and with a high degree of reliability. Not only can the sensor circuitry be prepared this way, it is also possible to integrate on-chip signal processing circuits which improve and condition the sensor signal. This is particularly important because the sensor signals are often weak and susceptible to noise during transmission. The deposition of the chemically sensitive layers by appropriate planar processing steps is the last phase prior to packaging (43,44).

An important development in chemfet research was the discovery that the protonic conductor IrO_x, prepared by DC reactive sputtering, is an excellent faradaic pH sensitive material (45-47). When properly prepared, these materials show no significant redox interferences (47). Because the materials are faradaic and highly conducting, the outer potential of the IrO_x obey the Nernst relation over the range ~0<pH<~12. These materials are also Nernstian over the temperature range 0^0 C<T<150^0 C. Other inorganic thin film membranes have demonstrated interesting characteristics suitable for detecting different chemical species. Esashi and Matsuo employed CVD sodium aluminosilicate and borosilicate glasses of the gate of their chemfets to detect Na^+ and K^+ ions but their published data is not especially promising as yet (48). Deposited films of LaF_3 (49), AgCl (43,50) and AgBr (51) have been used to monitor F^-, Cl^-, and Br^-, respectively. Evaporated tantalum layers have been deposited on the gate region of a chemfet, anodized to form an oxide and then employed to measure pH (52). CVD Ta_2O_5 has been successfuly deposited also and used as a pH sensitive material (48). Various inorganic and polymeric membranes have been investigated as pH sensor materials for chemfets and a selected group of these are listed in Table I with appropriate references.

A first version of the integrated shielded electrode structure was a four element sensor, containing chemically sensitive layers of LaF_3, AgCl, IrO_2 and Ag (43). In these early devices, the buffer consisted of a simple source follower. An improvement is obtained by replacing the source followers by high impedance operational amplifiers. An eight channel multivariate sensor chip has been fabricated with on-chip CMOS operational amplifiers (44). A picture of the chip lay-out is given in Figure 5. This chip has been a collaborative effort between three different institutions. The processing of the silicon circuitry up to the deposition of the chemically active layers was done at the ESAT Laboratory of the University of Leuven. A standard 5 μm, p-well CMOS technology was employed for the fabrication of the wafers. The deposition of the chemically sensitive membranes was carried out co-jointly between the University of Pennsylvania's Center for Chemical Electronics and Integrated Ionics, Inc. after the operational characteristics of the circuits were tested.

Because the procedure for the deposition of the chemically sensitive layer is not that well known, we will describe the procedures developed here. The first step in depositing the chemically sensitive membrane involves coating the electronic circuits with a passivating layer everywhere. A window is then opened in this layer in the portion over the contact area of the signal line by means of standard photolithographic procedures. Different chemically sensitive layers can then be sequentially deposited and patterned using modified photolithographic procedures (53). The various sensors produced included: pH, pCl^-, pBr^-, pI^-, pAg^+, pCu^{++}, and pCa^{++}. The results obtained with these devices can be found in reference 44.

The process compatibility and the isolation of the ionics section of the chip from the electronics makes it relatively safe to increase the complexity of the chip. An eight channel structure which has a multiplexer and an A-to-D converter in addition to the buffers has been reported (54). Another example of a chip lay-out which contains CMOS circuitry is shown in Figure 6. This chip contains operational amplifiers, a pulse edge modulator and an RF transmitter (55).

Other authors have produced multi-ion chemfets. Matsuo and Esashi have demonstrated that multi-ion chemfet structures can be planar processed in a two-ion needle shaped form (48). Their ingenious needle shaped design permits them to incorporate two or possibly three sensitive elements at the tip. However, the demands of electrical and chemical isolation become increasingly difficult as the number of ion sensitive elements increases in these devices. Pace has demonstrated that a pH sensitive element, combined with a multilayer composite structure, can be used to detect a wide range of chemical species (56). This concept is quite versatile and has been proposed as the means to extend the capability of the pH sensitive

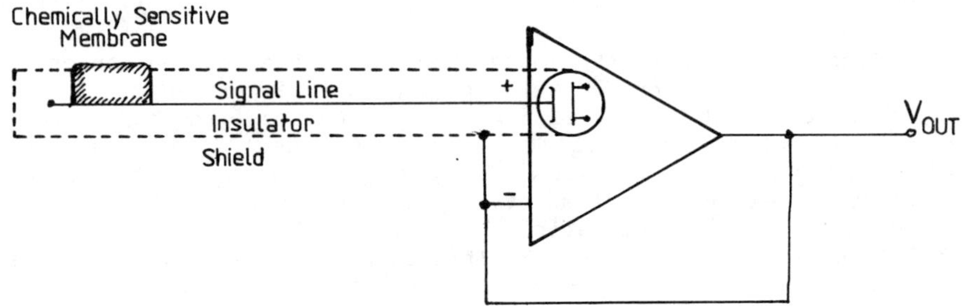

Figure 4. Schematic diagram of an integrated shielded
electrochemical sensor.

TABLE I

pH Sensitive Materials for integrated silicon based sensors
Applications

Inorganic Layers

Material	Preparation	References
IrO_x	Reactive Sputtering	49,50,67
Si_3N_4	Reactive Growth	38,97,98,99
SiO_2	Thermal Oxidation 37, 41, 99, 100, 101, 102	
Al_2O_3	CVD	98
Corning 0150	Screen Print	103,104
Ta_2O_5	Evaporation/Anodization, CVD	48,52

Polymeric Materials

Materials	References
Stearic Acid/Methyl-n-Octyl Ammonium Stearate	105
Tri-n-Dodecylamiane	106,107

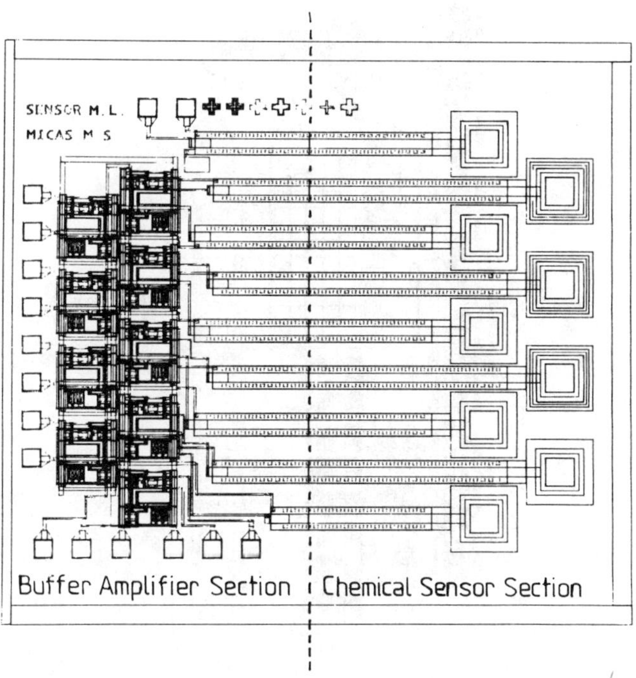

Figure 5. Layout of an eight element electrochemical sensor with on-chip CMOS operational amplifiers as buffers.

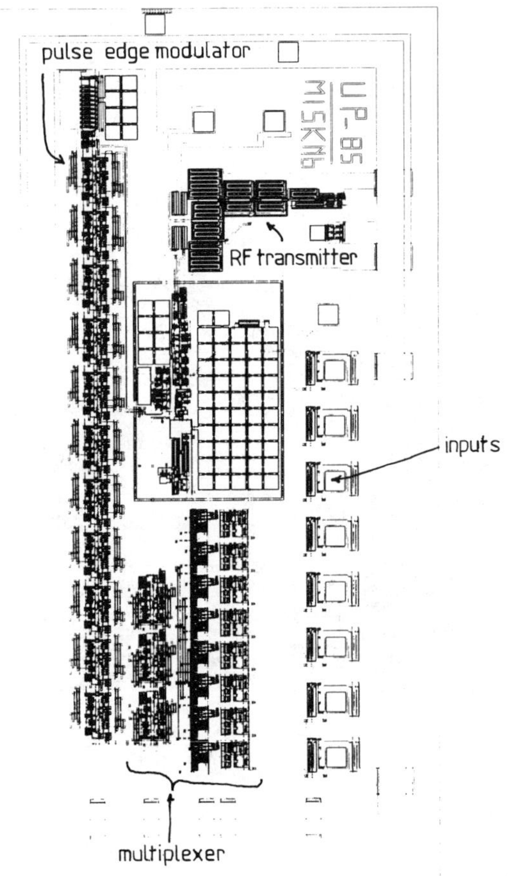

Figure 6. Layout of a sensor data acquisition CMOS chip consisting of: (a) eight buffer amplifiers; (b) a multiplexer; (c) a pulse edge modulator; and (d) a RF transmitter.

chemfet to measure a variety of important biological and medical chemical constituents (57).

While the chemfet device fabrication has advanced in a substantive fashion, it is interesting to note that there has been equally important advances in packaging of the structures. Perhaps the most important paper on this subject is the work of Ho et al. who demonstrated that Kapton pressure sensitive tapes can be used for packaging (58). Dietz and Zemel demonstrated that polyimide layers can act as effective high$_o$temperature silicon encapsulants for temperatures as high as 200 C (59). The growing availability of working ISFET structures and systems makes it important to give more attention to packaging. This will become essential if chemically sensitive electronic devices are to have broad application.

FIBER OPTIC CHEMICAL SENSORS, (FOCS): One of the oldest methods of chemical detection employs the optical properties of the analyte as a means of determining the types of ligands present and their concentration. Among the advantages of optical chemical detection methods are their selectivity and sensitivity. In some instances, particularly for gaseous molecules, the vibrational "finger" bands can serve as highly specific markers. The magnitude of the absorption is directly proportional to the oscillator strength of the absorbers. This, in turn, is the quantitative measure of the number of absorber molecules in the optical path length of the measuring light beam. In the case of an analyte in solution, the absorption may arise from an interaction with a dye molecule instead of a direct absorption with the analyte ligand. The absorption spectra of dyes like bromothymol blue are much broader than those of gas molecules but they still can be used to obtain quantitative information on pH or other activities. A related measurement method employs the interaction of an analyte with a fluorophor or phosphor, either enhancing or quenching the fluorescence or phosphorescence. This technique has been exceptionally useful in the study of biological processes inside of living cells.

By incorporating these three material phenomena for chemical measurements with fiber optic structures creates a basis for a fiber optical chemical sensor or FOCS. Two classes of FOCS have been reported in the literature:

1. An absorption based sensor.
2. A fluorescence or luminescence based sensor.

Each structure has its own area of utility but all FOCS share certain advantages. These include:

a. The light source used to probe the analyte zone is far removed from the chemical bath. Any analysis of the light beam is also conducted well away from the chemical bath.
b. Because of the substantial advances in fiber optic technology, optical losses between the source and the detector, aside from those in the interaction region, are negligible. Further, the cost of the optical fiber materials is minimal.

c. The fiber materials tend to be quite inert chemically
though it is possible to degrade the fiber as a result of
exposing it to harsh environments.
d. The small diameter of single mode fibers and their inherent mechanical strength and flexibility are positive
factors in their use in remote locations.

The basis of FOCS was established by Harrick in his pioneering
studies of the absorption of chemical species on germanium and
silicon total internal reflection probes (60). The structure devised
by Harrick is shown schematically in Figure 7. A light beam is
reflected internally so that the light emerges as shown into a
spectrometer. The light beam interacts with the surrounding medium by
means of an evanescent wave coupling to adsorbed ligands. The interaction mechanism is as follows: The total internally
reflected wave has an imaginary propagation vector outside of the
germanium or silicon probe. This causes the Poynting vector to decay
exponentially with distance from the surface into the measurement
medium. When a ligand adsorbs on the surface, the extinction coefficient corresponding to the complex portion of propagation coefficient
will be large in some part of the spectral region. If the evanescent
wave from the probe overlaps the ligand, the two imaginary terms
couple (the interaction is a square law process), allowing energy to
be transferred to the ligand. This loss of energy to the adsorbed
ligand is observed as an absorption process in the optical spectrum.
The evanescent wave coupling process could be an important mechanism
for the operation of a FOCS but, curiously, has not been so far.

The most commonly used method consists of injecting light into
a cavity containing a suitable optically active medium by means of an
optical fiber and measuring the intensity scattered by this medium
into another optical fiber (61-65). A schematic of this type of
structure is shown in Figure 8. The optically active medium is
usually an indicator dye immobilized in some appropriate fashion in
an analyte permeable container. This immobilization might be in a
solution or a polymeric matrix. In either case, the rate limiting
process was the diffusion of the analyte through the permeable
membrane and its gradual equilibration with the volume containing the
indicator dye. As a result, response times of the order of minutes
were typical.

Recent work here at Pennsylvania has demonstrated that another
form of FOCS has interesting attributes (66-68). By using a twisted
pair of optical fibers, one structure behaves like a radiator and the
other acts as an antenna (68). The reason for this is that the curvature associated with the twist gradually converts every guided mode
into a radiated mode. By complementarity, every radiated mode entering the twisted antenna, becomes a guided mode. The result is that
this pair behaves like an "inside-out" couvette, with the absorption
of the radiated modes occurring in the medium between the radiator
and anntenna elements of the FOCS. The experimental setup is illustrated in Figure 9. The advantage of this structure is that it
relies on the optical properties of the analyte directly and does not
require an intermediate such as a chemically sensitive dye (68).
However, a chemically sensitive dye can be used in combination with
these structures (68).

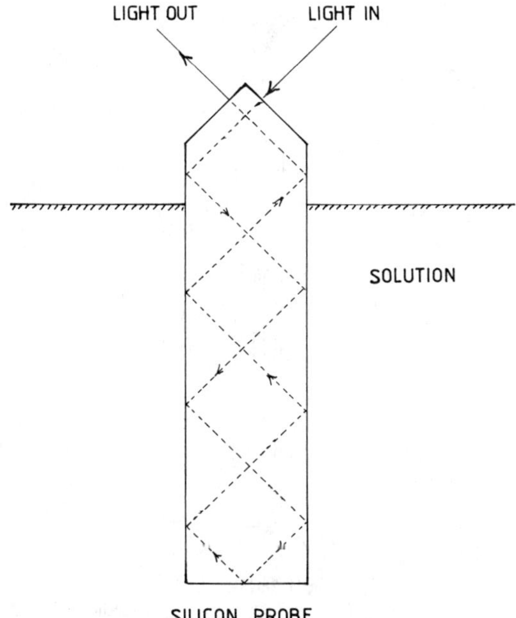

Figure 7. Schematic drawing of a silicon internal reflection probe for monitoring surface absorption processes. After N. J. Harrick (60).

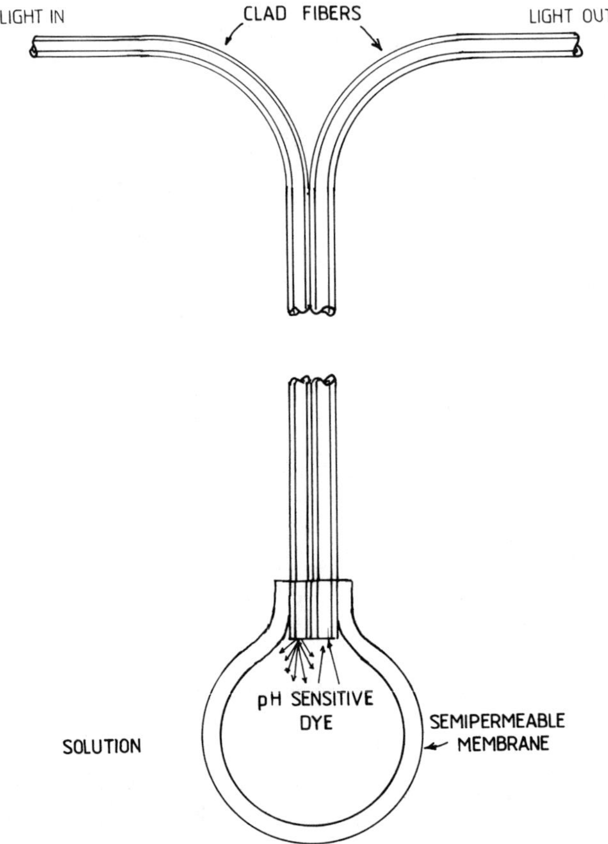

Figure 8. Fiber Optic Chemical Sensor: This structure is
the equivalent of a standard couvette with dye titratation
taking place through the permeable membrane.

Chemically sensitive optical structures can be generated that employ a fluorphor. A description of the current status of optical fluorescence sensors can be found in the recent review paper by Lubbers and Opitz (69). A recent example of the type of fluorescent optical element that can be used in biotechnical reactors is presented by Kroneis and Marsoner (70). Recent work by Hirshfeld and co-workers have demonstrated that FOCS can be devised using the fluorescence principle (71). They have successfully demonstrated that pH and pCl can be measured with a fluorescent based FOCS or optode. It should be pointed out that two different terms are employed in the literature for this type of structure. Lubbers and Opitz refer to their structures as optodes, while Hirshfeld calls his devices optrodes. Both devices employ the same general physiochemical principles but the optical configurations are different.Illustrative examples of the optode structures are presented in Figures 10a) and b). The structure in Figure 10a) is drawn from the work of Lubbers and Opitz (69). While hardly a microfabricated element, it illustrates the operational principles of the device rather nicely. The extension of the optode principle to include a fiber optic element is shown in Figure 10b). This is taken from the work of Hirshfeld and coworkers (71). A variation on this theme which combines the permeable chamber of the Peterson type structure with the optode fluorescence method is described by Schulz (72). In this measurement scheme, the fluorescent material is bound to the wall of the container. The fluorescent dye is released when the desired analyte diffuses into the chamber and substitutes for it. The fluorescent dye then enter into the solution which is irradiated by light coming from the optical fiber. The fluorescence from the dye in the main body of the container is detected with a fiber and transmitted to the optical detector.

Considering the variety in the FOCS structures, a corresponding variety also exists in their modes of operation. The coupling of the evanescent wave to the surrounding medium probably provides the most compact structure while still being capable of providing conoidorable sensitivity. This was illustrated in a theoretical calculation (73,74). There has not been a direct experimental verification of the utility of the evanescent wave coupling approach outlined by Wolfe and Zemel et al. However, work by Johnson et al. has shown that it is possible to impose a periodic grating on a fiber element (75). More recently, Giuliani et al. presented a very simple structure based on the evanescent wave interaction of a FOCS with adsorbed material (76). Further work on this configuration is clearly warranted.

The last type of FOCS to be discussed here is the work of Freeman and Seitz who employ a chemiluminescent process for generating light (77,78). While this measurement did not involve the use of a fiber optical element, it is pointed out by Seitz that this would not be too difficult (78). The measurement employed 1,1',3,3'-tetraethyl-$\Delta^{2,2'}$-bi(imidazolidene), (EIA) reacting with O_2 to produce a chemiluminescent reaction. The light was detected with a photomultiplier tube and the response measured as a function of various conditions. The general conclusion of this work was that the rate of reaction was limited by the diffusion of O_2 through the teflon membrane separating the EIA solution from the gas stream. This diffusion was temperature dependent, leading to a temperature dependence in the intensity of the chemiluminescence. However, the 90% of full

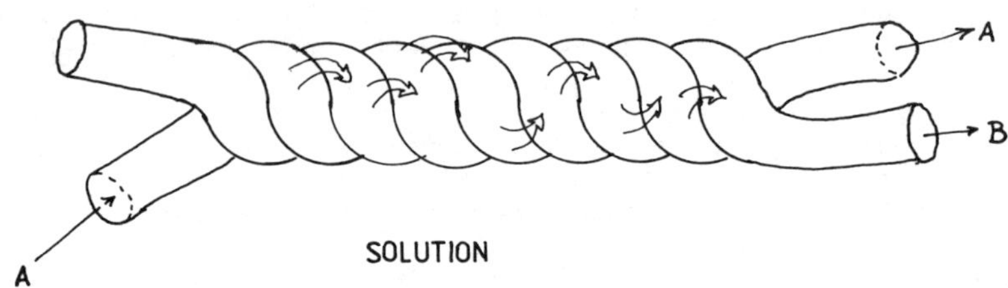

Figure 9. Fiber Optic Chemical Sensor: This structure is the equivalent of an inside-out couvette which probes the external solution.

a

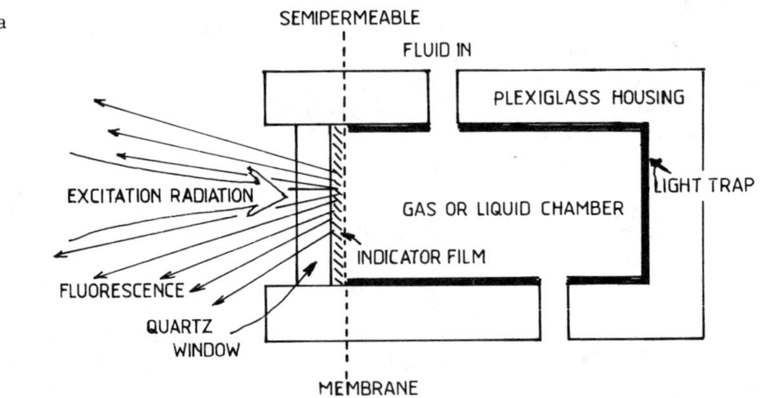

b

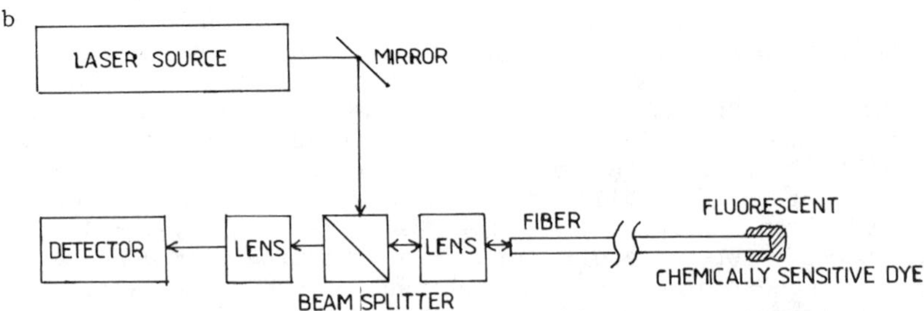

Figure 10. Basic elements for flourescent fiber optic chemical sensors: a. Optode Structure: Design after Lubbers and Opitz (69); b. Optode Structure: Design after Hirshfeld, et al. (71).

scale response to 20% O_2 was well under 30 seconds, a highly respectable response time. Furthermore, the authors estimate that the detection limits could be as low as 1 ppm (v/v) of O_2.

These structures have promise for serving particular measurement niches. By far, the most promising method appears to be the optode technology, specifically for monitoring a variety of biochemical species. While there already is a substantial amount of research on these devices, much more remains to be done to insure rapid advances in these areas.

PYROELECTRIC BASED THERMAL SENSORS: Thermal means of monitoring chemical processes are a mature, well established technology (79,80). The principle involved is simple and straightforward. If the mass of the system is M and its heat capacity is C_t, then a change ΔH in the thermal energy of the system will produce a change ΔT in the temperature T of the system. The relationship between the change in temperature, the change in the thermal energy and the mean temperature of the system provides the basic information for a number of different type of sensors. One way of assessing the value of a thermal sensor material is the minimum energy that it can measure. On this basis, pyroelectrics like $LiTaO_3$ are exceptionally useful sensor materials. Because pyroelectricity is not a commonly used phenomena, we will describe it briefly here.

Thermal bolometers have been produced from these materials for a number of years and have largely replaced thermopiles and bolometers for infrared measurements (81-83). A typical noise equivalent power for a $LiTaO_3$ device is 10^{-9} watts/cm^2 corresponding to a thermal input of 2.4×10^{-12} Kcal/cm sec or 6×10^{-10} eV/cm^{2}sec. This suggests that thermal processes involving less than a picomole of material could be detected providing the principle can be reduced to practice. Another application of the pyroelectric phenomena is the detection of fluid flow with heat lost from a Joule heated sample immersed in the fluid. This type of thermal anomometry is presented in other papers (84-86). Lang has described a calorimetric study which indicates that computer controlled calorimetry can be conducted with pyroelectric sensors (87). The reduction of this process to a microstructure using photolithographic processing of the pyroelectric wafer was conducted by Young (88). In this paper, some recent enthalpimetric and calorimetric studies using pyroelectric elements will be discussed (89,90).

Pyroelectricity is an intrinsic property of certain classes of non-cubic, non-centrosymmetric materials (91,92). While many pyroelectrics are crystalline in form, one important material, polyvinylidene fluoride, PVF_2, is a plastic that has been subjected to an inelastic stretching deformation that induces a quasi-crystalline character to the polymer (93). When these materials are subjected to a change in their thermal energy, sublattices of positive and negative charges move with respect to each other and induces a net change in the bulk polarization (91). In other words, in pyroelectric materials, there is a permanent temperature dependent polarization vector. Typical values for the surface polarization induced in these materials are of the order of 10^{-10} -10^{-8} coul/cm K (87). Under appropriate conditions, extremely small temperature changes, down to less than 10^{-6} ^{0}C can be measured. In the simplest case, there is only one direction that has a temperature dependent polarization of any magnitude. The net surface charge arising from a temperature dependent polarization can be written as

$$Q = CV - p(T)A\{<T> - T_o\} \qquad 1)$$

where C is the capacitance of the pyroelectric element, defined as

$$C = \epsilon\epsilon_o/d \qquad 2)$$

where ϵ is the dielectric constant for the material, ϵ_o is the dielectric permittivity of free space and d is the thickness of the pyroelectric. V in Equation 1 is the applied voltage across the capacitor C, p(T) is the temperature dependent pyroelectric constant in the direction normal to the electrodes of area A, T_o is the reference temperature where the net surface charge vanishes and $<T>$ is the temperature of the pyroelectric element that is volume averaged over the entire pyroelectric material

$$<T(t)> = \frac{1}{v}\int_V T(x,y,z,t)dv \qquad 3)$$

The current induced in an external circuit by changing the temperature of the pyroelectric is

$$i = -\dot{Q} = -C\frac{dV}{dt} + \frac{d}{dt}\{p(T)A<T>\} \qquad 4)$$

This expression is the basic description for the use of the pyroelectric effect in a host of sensor applications including the well known optical detection devices (82,83). A particularly useful way of describing this type of system is with an equivalent circuit where the pyroelectric current generator drives the pyroelectric impedance and the measuring amplifier circuit as shown in Figure 11.

The origin of the pyroelectric effect, particularly in crystalline materials, is due to the relative motions of oppositely charged ions in the unit cell of the crystal as the temperature is varied. The phase transformation of the crystal from a ferroelectric state to a paraelectric state involves what is called a "soft phonon" mode (94). In effect, the excursions of the ions in the unit cell increase as the temperature of the material approaches the phase transition temperature or Curie temperature, T_c. The Curie temperature for the material used here, LiTaO$_3$, is 618 C (95). The properties of a large number of different pyroelectric materials is available through reference 87. For the types of studies envisaged here, it is preferable to use a pyroelectric material whose pyroelectric coefficient, p(T), is as weakly temperature dependent as possible. The reason for this is that if p(T) is independent of temperature, then the induced current in the associated electronic circuit will be independent of ambient temperature and will be a function only of the time rate of change of the pyroelectric element temperature. To see this, suppose p(T) is replaced by p_o. Then Equation 4 becomes

$$i = -C\frac{dV}{dt} + p_oA<\frac{\partial T}{\partial t}> \qquad 5)$$

The time dependence of the temperature of the pyroelectric material can be related to the spatial dependence of the temperature by means of Fourier's equation for heat

$$\frac{\partial T}{\partial t} = D_t \nabla^2 T(x,y,z,t) = D_t \{\frac{\partial^2 T}{\partial x^2} + \frac{\partial^2 T}{\partial y^2} + \frac{\partial^2 T}{\partial z^2}\} \qquad 6)$$

where D_t is the thermal diffusivity (96). If the spatial average of the time rate of change of the pyroelectric's temperature is taken by inserting Equation 6 into Equation 3, then

$$\langle\frac{\partial T}{\partial t}\rangle = \frac{1}{v}\int_v D_{th} \nabla^2 T(x,y,z,t)dv = \frac{1}{v}\int_v D_{th}\{\frac{\partial^2 T}{\partial x^2} + \frac{\partial^2 T}{\partial y^2} + \frac{\partial^2 T}{\partial z^2}\}dv \qquad 7)$$

If the indicated integrations are performed, the result is

$$\langle\frac{\partial T}{\partial t}\rangle = \frac{1}{C_v v}\{[\Delta H_x A_x] + [\Delta H_y A_y] + [\Delta H_z A_z]\} \qquad 8)$$

where the ΔH_j and the A_j are the net rate of heat flow through the j surfaces of the pyroelectric. This is a formal statement of the fact that the time rate of change of the average temperature of a solid is due to the net energy flowing though its boundaries. When combined with Equation 5, it demonstrates that the current induced by temperature changes in the pyroelectric element is a measure of the change in the total thermal energy stored in the pyroelectric material.

It can be shown from Figure 11 and Equation 4 that the rise time of the signal will be controlled by the RC time constant of the circuit loading the pyroelectric current generator. The simplest case is where the heat flow into the pyroelectric is one dimensional as shown in Figure 12. Then the signal across the parallel load resistor of the pyroelectric element is

$$V = \frac{p_0 A_e r_p}{C_v d} \exp(-t/r_p C) \int_0^t \Delta H_e(t')\exp(t'/r_p C)dt' \qquad 9)$$

where $\Delta H_e(t)$ is the rate at which heat enters or leaves the pyroelectric. For times much smaller than $r_p C = \tau$, the induced voltage will be directly proportional to the rate of heat flow into the pyroelectric. However, the signal is proportional also to t/τ at short time intervals. In this short time regime, $\frac{dV}{dt}$ will be proportional to the heat input only. This is illustrated with data from our laboratory using a joule heat input as shown in Figure 13.

Two specific and related applications that have been investigated here are:

1. an integrated differential scanning microcalorimeter (with sensitivities in the sub-microcalorie range.
2. A high sensitivity microenthalpimeter for monitoring catalytic processes.

It is possible to obtain chips that are 50 microns or less in thickness. Because $LiTaO_3$ is an oxide, it tends to be resistant to most chemical treatments. Experience both here and at other laboratories indicates that the wafer will etch slowly in hot HF

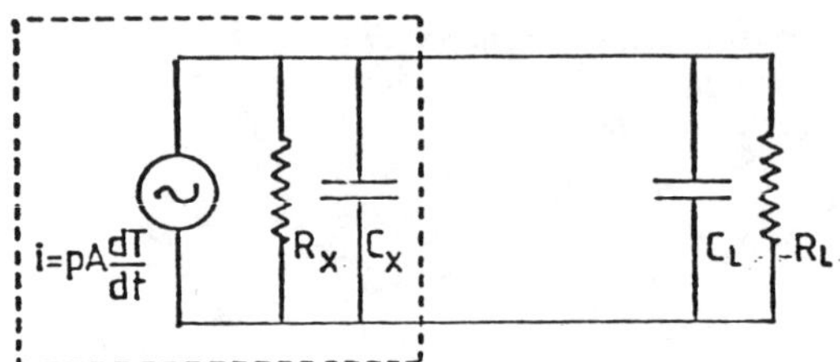

Figure 11. Equivalent circuit for the pyroelectric sensor
and the attached amplifier.

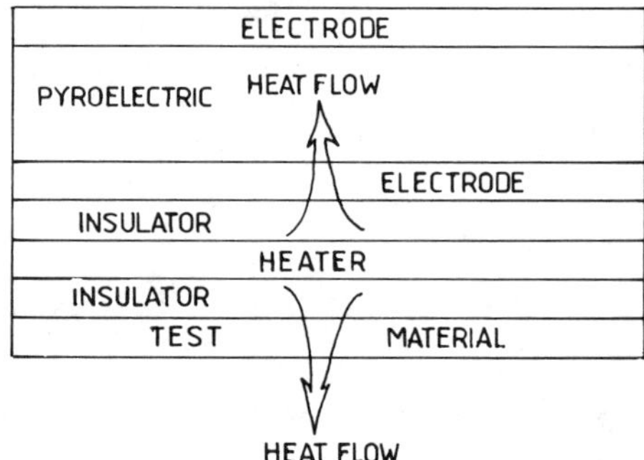

Figure 12 One dimensional heat flow in a pyroelectric
microcalorimeter.

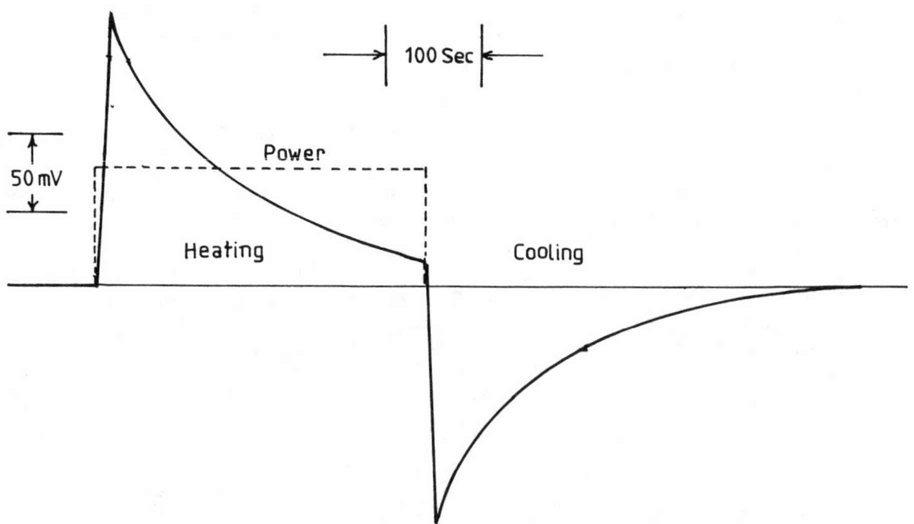

Figure 13. Variation of output voltage from a pyroelectric
microcalorimeter arising from the joule heating of the
integrated heater. Power in a 100 mW, the load resistor is
22 M$_\Delta$ and the pyroelectric capacitance across the measuring
electrode is μF.

acid. Both studies employ similar pyroelectric structural elements. Manufacture of microcalorimeters and microenthalpimeters employs standard microfabrication technology. There is a heater element integrated directly onto the $LiTaO_3$ pyroelectric chip. We have found that the NiCr films are well suited for use as the heater element on these chips. A schematic of the pyroelectric element is shown in Figure 14a). The entire assembly is mounted on a ceramic substrate as depicted in Figure 14b). By placing these structures in a vacuum system, the thermal losses to the environment are kept to a minimum.

Because the pyroelectric calorimeter structure has an extremely small mass and a heater element can be integrated directly onto the pyroelectric chip itself, some of the problems arising in scanning calorimetry can be mitigated. In conventional scanning differential calorimetry, the calorimeter and the sample are placed in a furnace and the system is allowed to heat up in a controlled fashion. The furnace mass has to be appreciable to avoid spurious fluctuations. As a rule, the large mass also imposes a limitation on the heating rate. This is not the case for the pyroelectric scanning calorimeter (PSC) since the heater element is integrated directly on the device. An interesting aspect of this structure is that extremely small quantities of thermal energy can be injected into it by means of controlled pulses of electrical power to the heater film (76). As a result, the average temperature of the PSC can be incremented during the measurement run in predetermined amounts. If a suitable algorithm for the heating cycle is available, it becomes possible to vary the rate of temperature increase; thereby allowing the measurement to scan through thermal peaks in a much more controlled and predetermined fashion. The principle for this measurement has been demonstrated using the integrated scanning pyroelectric microcalorimeter depicted in Figure 14a) and the data is presented in Figures 15 and 16. In Figure 15, representative response data is shown when the integrated heater is pulsed with a 100 msec current. These signals are distorted when some of the heat is used to initiate a phase transition (desorption, melting or a chemical reaction) on the measuring electrode. If these signals are integrated by a summation method (using a fast analog to digital converter to generate digital information on the magnitudes of the pulsed signals), then the data shown in Figure 16 is obtained. Curve a) is the response after a high temperature run has removed all adsorbed gases from an activated charcoal adsorber. Curve b) shows the response to an effective heating rate of about 100°C/min when the specimen has been exposed to a partial pressure of butyl phthalate. The corresponding response when the effective heating rate is halved by decreasing the energy per pulse is shown in curve c). Note that there is no substantial variation in the number of peaks in curves b) and c) but the peaks are far better resolved in curve c). While the data is quite encouraging, a detailed theoretical model that could be used as a control algorithm for the heating cycle has not been developed as yet. This will be another important step in the study of this type of sensor system and has the promise for providing a good means of defining thermal kinetic behavior in very small samples.

To have the PSC under real time control of the algorithm, a computer controlled measurement system must be used. The one employed by Young consisted of: an S-100 based data logging system with twelve bit analog-to-digital and digital-to-analog converter boards; a power

a

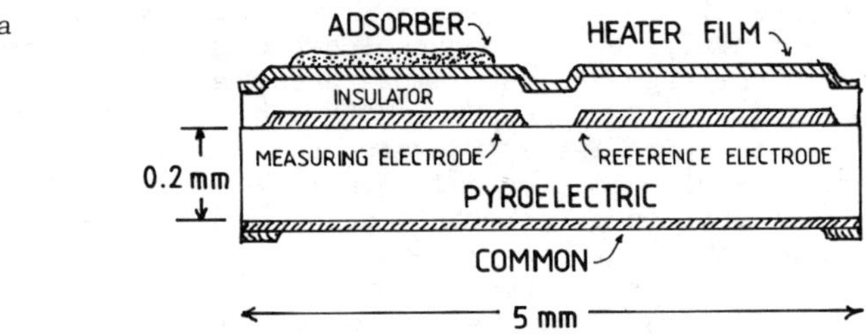

b

Figure 14.   Integrated Scanning Pyroelectric Microcalorimeters;
a. schematic drawing of the integrated scanning pyroelectric
microcalorimeter; b. photograph of an assemble integrated
scanning pyroelectric calorimeter on its ceramic mount.

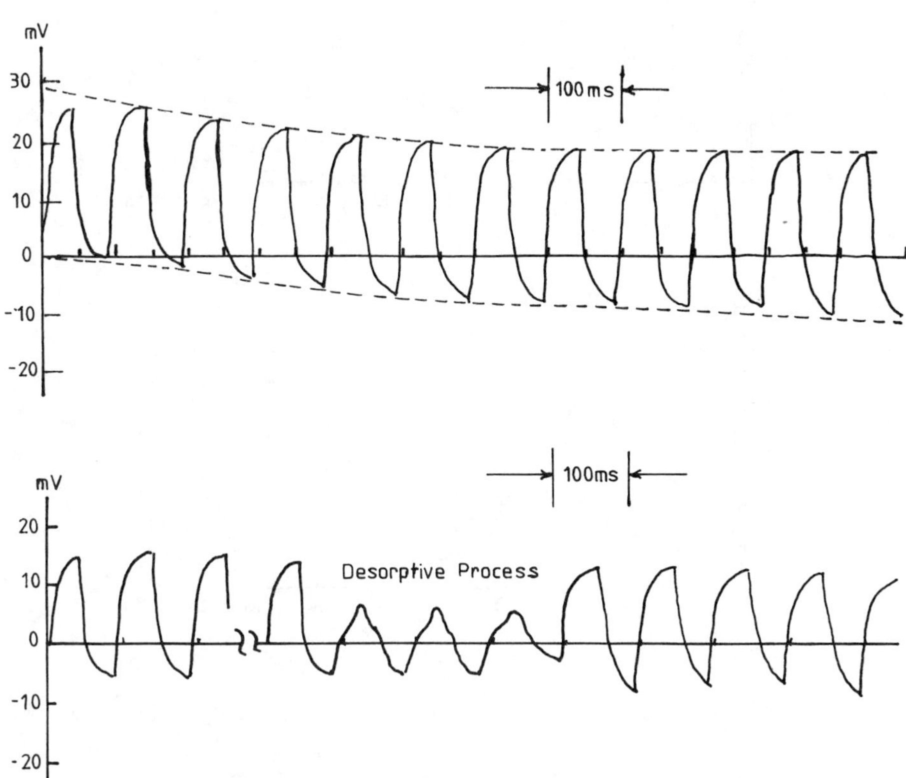

Figure 15. Response of the Integrated Scanning Pyroelectric Microcalorimeter to a Pulsed Electrical Heat Input: a.Response of a clean activated charcoal coated scanning microcalorimeter in a standard oil pumped vacuum system at a pressure of 10^{-4} Pa; b. Response of an activated charcoal coated scanning microcalorimeter to equilibrium adsorbed gasses in a standard oil pumped vacuum system at a pressure of 10^{-4} Pa.

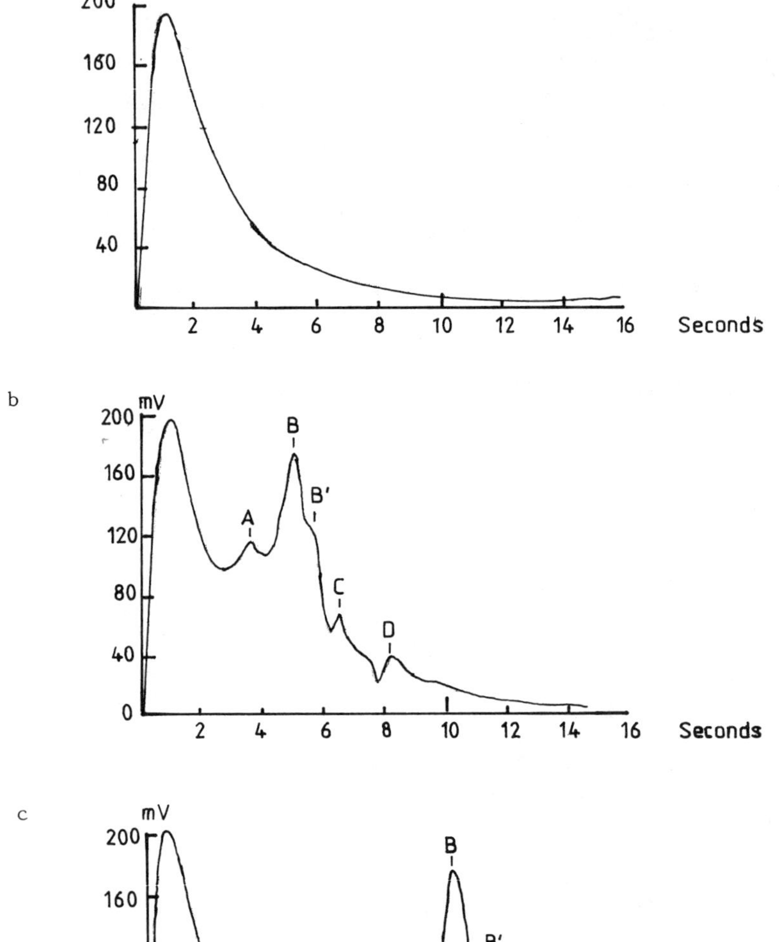

Figure 16.   Integrated response of a pulsed scanning
pyroelectric microcalorimeter. Activated charcoal adsorber
is exposed to a butyl phthalate partial pressure in a vacuum
system at 10^{-4} Pa: a. Response after removal of adsorbed
materials; b. Response to an effective heating rate of
100°C/min.; c. Response to an effective heating rate of
500°C/min.

op-amp capable of supplying the needed electrical power for the heater element of the PSC; a high input impedance preamplifier for the electronic measuring circuit. The schematic diagram for the measuring system is shown in Figure 17 (88). The software allows various time dependent power level functions such as: a step voltage of constant amplitude, an initial step voltage followed by a steadily increasing voltage, and a pulsed voltage with the pulse height and interval (88).

In another set of experiments, a much simpler pyroelectric calorimeter was employed as shown in Figure 18a) and b). These structures consisted of a common gold electrode and two measuring electrodes; one Pd and the other gold. Some typical data obtained with the structure in Figure 18a) is shown in Figure 19 (89). Typical data taken at elevated temperatures is shown in Figure 20 where the pyroelectric response is compared to the transient current response of a Pd-MOS capacitor subjected to exposures to hydrogen and oxygen as noted (90). The significance of these measurements is that they demonstate that thermal measurements on thin films can be related to chemical effects on the electronic properties of sensor devices. Here too, additional research is needed to obtain a better picture of the value of this class of sensors.

The general conclusion to be drawn from these studies is that the use of small pyroelectric elements as heat flow sensors in chemical investigations holds some promise. The early stage of the studies makes it difficult to assess the extent of their utility. New adsorber materials are an essential requirement if these structures are to fulfil their promise.

DISCUSSION: There are two significant observations to be made from the body of research reviewed. First is the variety of materials and phenomena that can be utilized for converting chemical information into electrical signals. The choice of optimum device principle will depend on a large number of factors not the least of which will be obtaining the lowest overall system cost. One definite conclusion to be drawn from this review is that the integrated silicon based sensors now appear to be close to commercial development. In other words, the integrated silicon based sensors have demonstrated capabilities, but for the many and varied information acquisition needs that are arising, other elements like the FOCS may prove even more versatile. The second observation is that the microfabrication manufacturing capability established by the electronics industry is just now beginning to impact on other areas of science and technology such as sensors. It also permits us to employ new materials and structures. In order to fully exploit the potential advantages of the small size and electronic sophistication that can be incorporated using microfabrication technology, our understanding of the interaction of chemical species in the vicinity of various solids has to advance substantially before chemically sensitive electronic devices will make the contribution they should to both science and technology.

As just pointed out, the development of integrated silicon based sensors structures is far and away the most advanced aspect of microfabricated chemical sensor research and development. It is now possible to visualize complex analytical chemistries being performed

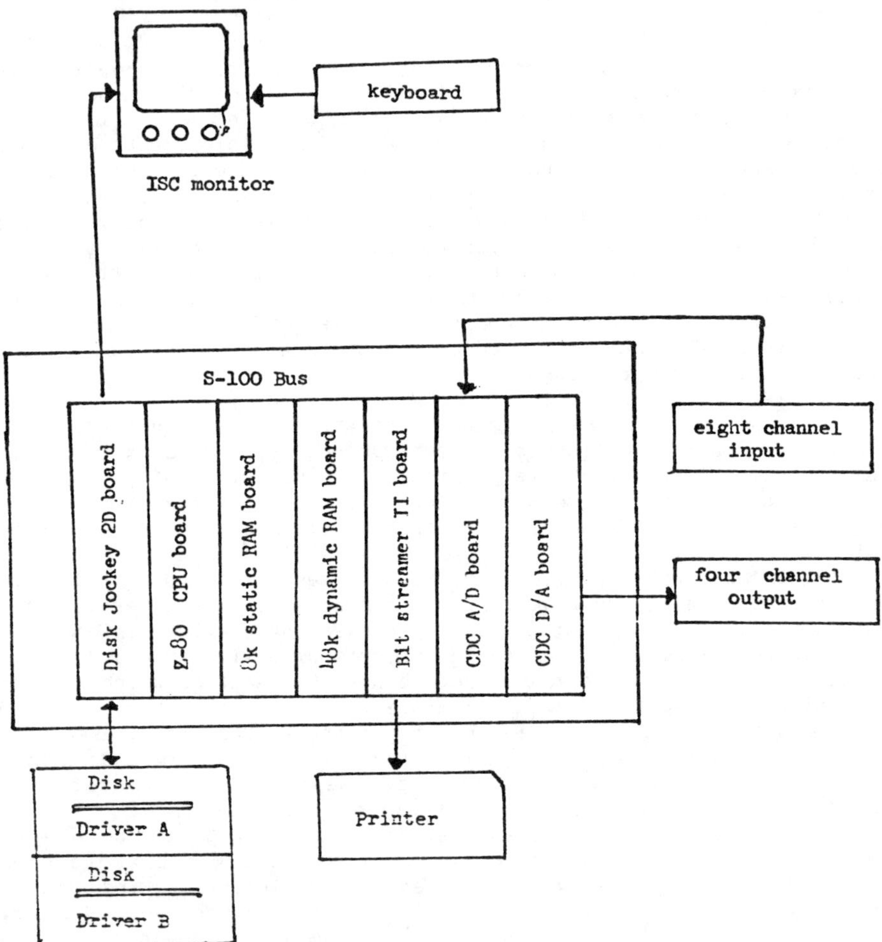

Figure 17. Schematic diagram of the computer assisted measuring system used to control the integrated scanningpyroelectric microcalorimeter.

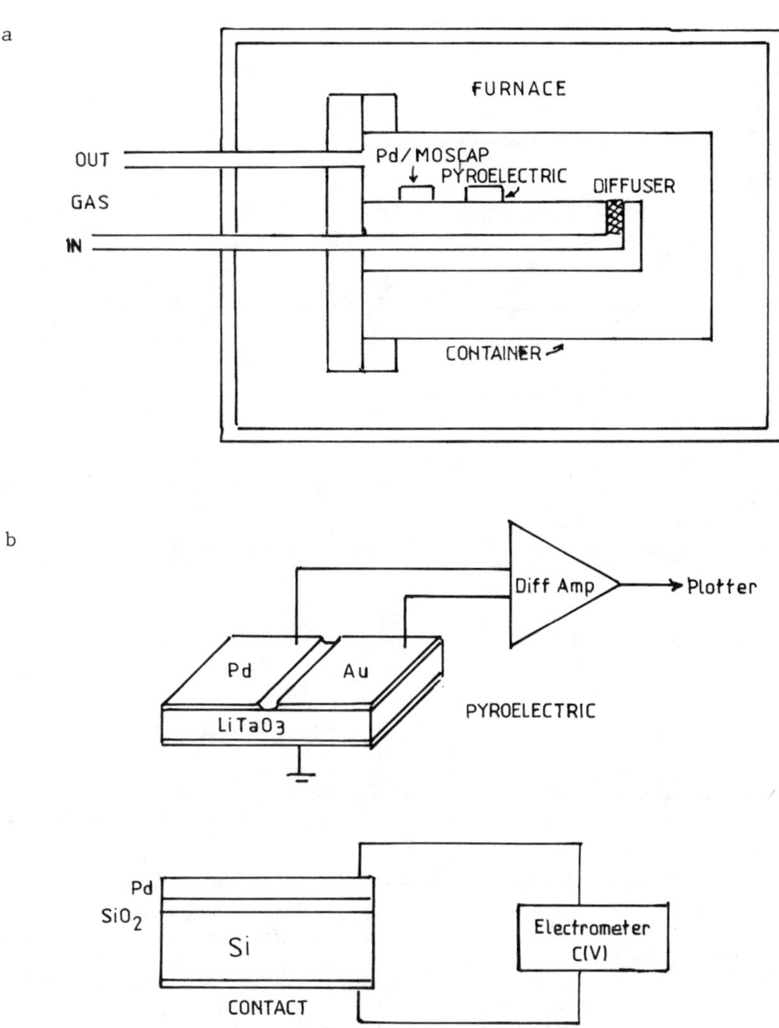

Figure 18. Pyroelectric Enthalpimetric Structures: a. Schematic of the pyroelectric enthalpimeter and Pd-MOSCAP structures and circuits used to study the reaction of hydrogen-oxygen on palladium; b. Layout of the apparatus used to monitor the chemical and electronic response of hydrogen-oxygen reactions on palladium.

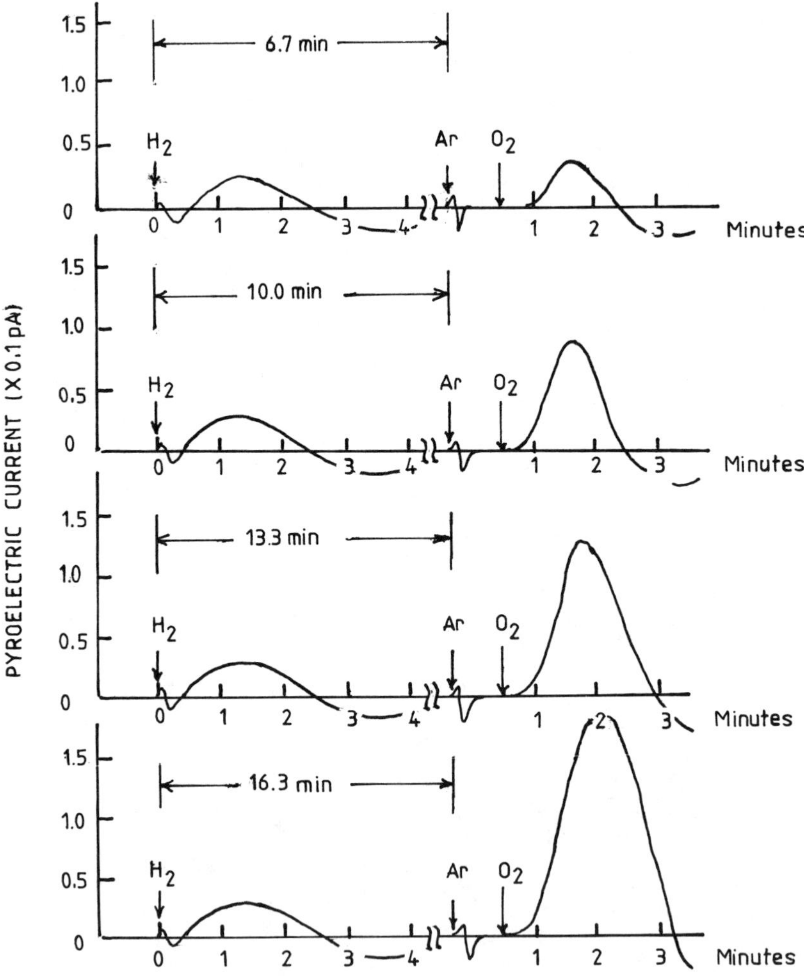

Figure 19. Oxygen response of a pyroelectric enthalpimeter of Figure 18.a. to increasing the exposure times to 1% hydrogen in nitrogen.

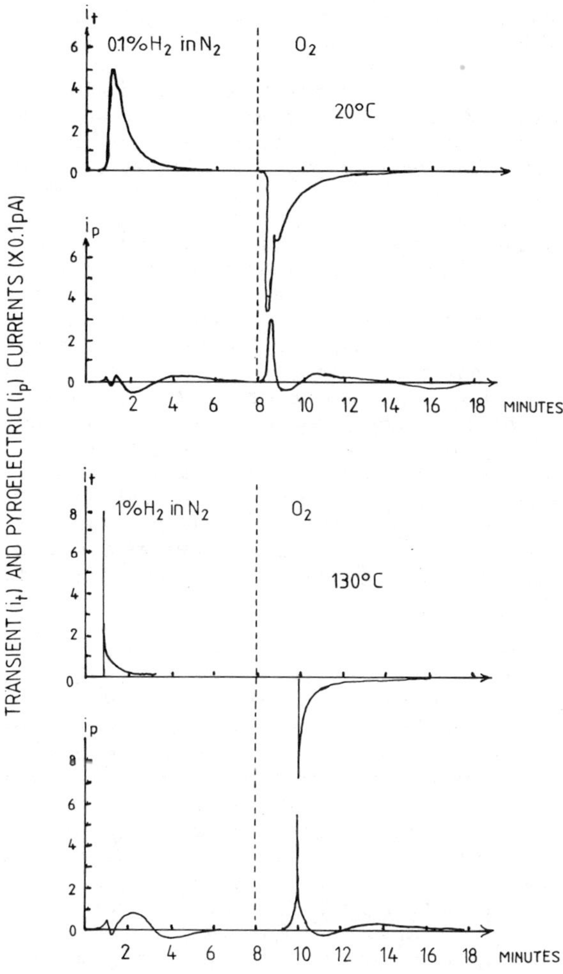

Figure 20. Simultaneous measurements of the transient current response of a Pd–MOSCAP (i_t) and the pyroelectric current from the pyroelectric enthalpiemeter (i_p) illustrated in Figure 18. Upper two curves are at room temperature and with 0.1% H_2 in N_2 and the lower two are for 1% H_2 in N_2.

in real time on a silicon chip no more than 2x5mm in area. The packaging of the integrated silicon based sensors structures appears to be under some degree of control and improvements are likely as the engineering effort in this area increases. The key problem appears to be tailoring the properties of the chemically sensitive layers to the specific problem at hand. We have seen that the interaction of water molecules with SiO_2 profoundly alters the electronic characteristics of this material. It is reasonable to expect that similar modifications are likely to occur in other materials when the highly polar water molecule diffuses into thin deposited layers, be they organic or inorganic. The same problem will undoubtedly occur with any chemically sensitive coating when it is immersed in aqueous solutions whether they are associated with integrated silicon based sensors, FOCS or thermal sensors. Clearly this will be an important research topic for these devices. There is another issue that was pointed out in Section 3. The diffusion of water into the chemically sensitive layer can also alter the chemical affinity of the layer for different ions. In the case of integrated silicon based sensors, this can lead to degradation of the selectivity and performance of the sensor even though the electronics remains secure. The problems of leakage and cross talk in chemfets are a major drawback which has limited their use to relatively short times. In this respect, the integrated shielded electrochemical sensor is more suitable and offers the additional advantage that multi-sensor integration can be conducted more easily. These leakage problems are not relevant for the FOCS and thermal sensors but the interaction between the solution and the chemically sensitive materials on ther sensors will need futher investigation.

The major advantage of the FOCS over the integrated silicon based sensors and the thermal class of sensors is that the energy source and the information processing involves light transmission over a chemically inert medium, the optical fiber. There is no need to transmit an electrical signal over a metallic conductor. Combined with the exquisite selectivity that spectroscopy can afford in the medium and near infrared, the FOCS offers an attractive alternative to the integrated silicon based sensors in many situations. The restriction on this device is the very advantage it offers. Unlike the integrated silicon based sensors that can make use of the well established microelectronic information processing technology, the FOCS must employ the considerably more expensive and delicate spectroscopy measuring system. In the long run, the overall system cost will be a major determinant of the use of a given chemical sensor element. The fiber optic and integrated optic industry is just now undergoing the major cost reductions associated with economies of scale as optical communications becomes widespread. Simply put,it is too early to tell which system is likely to dominate beyond the next decade. However, during the coming decade, it will probably be the integrated silicon based sensors because of its lower overall system cost.

The pyroelectric element is likely to be a specialty device, though it is far too soon to tell. Its primary advantage is that it can permit time integration of the species of interest and as a result can be used to detect extremely small concentrations. The most that can be said at the present is that more research is clearly needed for this structure.

<u>Literature Cited</u>

1. Advanced Study Institute on "Chemically Sensitive Electronic Devices", J. N. Zemel and P. Bergveld, Eds. (Elsevier Sequoia, Lausanne, Switzerland, 1981). Also published as Sensors and Actuators, Volume 1, (1981).
2. International Symposium on Solid State Transducers, Materials Research Society, Boston, November, 1981; also published as Volume 3, Sensors and Actuators.
3. 2nd International Conference on Solid State Transducers, Delft, Netherlands, May, 1983; also published as Volume 4, Sensors and Actuators.
4. International Conference on Chemical Sensors, Futuoka, Japan, September, 1983.
5. The Japanese Institute of Electrical Engineers has been sponsoring an annual conference on sensors since 1982.
6. The IEEE has begun a biannual conference series on solid state transducers which began in 1984.
7. I. Lundstrom, Sensors and Actuators, **1**, 403, (1981)
8. I. Lundstrom and D. Soderberg, Sensors and Actuators, **2**, 105 (1981)
9. L. G. Petersson, H. M. Dannetun and I. Lundstrom, Phys. Rev. Lett. **52**, 1806, (1984)
10. S. R. Hofstein, IEEE Trans. Elec. Dev., **ED-13**, 222, (1966), ibid., **ED-14**, 749, (1967)
11. S. I. Raider, L. V. Gregor and R. Flitsch, J. Electrochem. Soc. **120**, 425, (1973)
12 G. L. Holmberg, J. Electrochem. Soc., **117**, 677, (1970)
13. I. Lundstrom, Private communication
14. C. Nylander, M. Armgarth and C. M. Svensson, "Insulating Films on Semiconductors", M. Schulz and G. F. Pensl, Eds., (1981) p. 195
15. C. Nylander, M. Armgarth and C. M. Svensson, J. Appl. Phys., **56**, 1177, (1984)
16. R. H. Doremus, J. Non-Cryst. Sol., **19**, 137 (1975)
17. E. H. Nicollian and J. R. Brews, "MOS Physics and Technology", John Wiley, (New York, 1982)
18. D. E. Yates, S. Levine and T. W. Healy, J. Chem. Soc. Faraday Trans. I, **70**, 1807 (1974)
19. D. E. Bousse, N. F. de Rooij and P. Bergveld, IEEE Trans. Elec. Dev. **ED-30**, 1263, (1983)
20. N. M. Johnson, D. K. Biegelsen and M. D. Moyer, J. Vac. Sci. Tech. **19**, 390, (1981)
21. N. M. Johnson, D. K. Biegelsen, M. D. Moyer, S. T. Chang, E. H. Poindexter and P. J. Caplan, J. Appl. Phys., **43**, 563, (1983)
22. Y. Nishi, Jap. J. Appl. Phys., **10**, 52, (1971)
23. P. J. Caplan, E. H. Poindexter, B. E. Deal and R. R. Razouk, J. Appl. Phys. **50**, 5847, (1979)
24. E. H. Poindexter, G. J. Geraldi, M. E. Rueckel and P. J. Caplan, J. Appl. Phys., **56**, 2844, (1984)
25. K. L. Brower, P. M. Lenahan and P. V. Dressendorfer, Appl. Phys. Lett., **41**, 251 (1982)
26. C. C. Wen and J. N. Zemel, Thin Solid Films, **69**, 275, (1980)
27. J. R. Monkowski, M. D. Monkowski, I. S. T. Tsong and J. Stach, "Silicon Processing, ASTM STP 804", D. C. Gupta, Ed., American Society for Testing and Materials, (1983) p. 245
28. B. E. Deal, E. L. MacKenna and P. L. Castro, J. Electrochem. Soc. **116**, 997, (1969)

29. T. Fare, PhD Dissertation, Electrical Engineering, University of
    Penna., Philadelphia PA, (1985)
30. T. Fare, J. N. Zemel, I. Lundstrom and A. Feygenson, Submitted to
    Phys. Rev. Letters; ibid. 3rd Int. Conf. on Sol. St. Transducers,
    Philadelphia PA, (1985), Technical Digest
31. T. Fare and J. N. Zemel, In Preparation
32. B. Keramati and J. N. Zemel, J. Appl. Phys. **53**, 1091, (1982)
33. B. Keramati and J. N. Zemel, J. Appl. Phys. **53**, 1100, (1982)
34. M. C. Steele, J. W. Hile and B. A. MacIver, J. Appl. Phys. **47**,
    2537, (1976)
35. P. F. Ruths, S. Ashok, S. J. Fonash and J. M. Ruths, IEEE Trans.
    El. Dev. **ED-28**, 1003, (1981)
36. T. Poteat and B. Lalevic, Trans. IEEE on El. Dev., **ED-29**, 123,
    (1982)
37. P. Bergveld, IEEE Trans. Biomedical Eng., **BME-17**, 70, (1970)
38. T. Matsuo and K. D. Wise, IEEE Trans. Biomedical Eng., **BME-21**,
    2785 (1974)
39. D. Moss, J. Janata and C. C. Johnson, Anal. Chem., **47**, 2238,
    (1975)
40. Available from the the Kuraray Co., Lt., Japan.
41. C. C. Wen, T. C. Chen and J. N. Zemel, IEEE Trans. El. Dev., **ED-
    26**, 1945, (1979)
42. G. C. Chern and J. N. Zemel, IEEE Trans. El. Dev., **ED-29**, 115,
    (1982)
43. J. Van der Spiegel, I. Lauks, P. Chan and D. Babic, Sensors and
    Actuators, **4**, 291, (1983)
44. I. Lauks, J. Van der Spiegel, W. Sansen and M. Steyaert, 3rd Int.
    Conf. on Sol. St. Transducers, Philadelphia PA, (1985), Technical
    Digest
45. T. Katsube, I. Lauks and J. N. Zemel, Sensors and Actuators, 2,
    399, (1982)
46. T. Katsube, I. Lauks, J. Van der Spiegel and J. N. Zemel, Jap. J.
    Appl. Phys., **22**, 469, (1983)
47. I. Lauks, M. F. Yuen and T, Dietz, Sensors and Actuators, **4**, 375,
    (1983)
48. T. Matsuo and M. Esashi, Sensors and Actuators, **1**, 77, (1981)
49. I. Lauks, Proc. 9th Ann. Meet. of the Fed. of An. Chem. and
    Spect. Societies, Philadelphia PA, (1982)
50. D. Harame, J. Shott, J. Plummer and J. Meindl, IEDM, Washington
    DC, (1981), Technical Digest, p. 467
51. R. P. Buck and D. E. Hackleman, Anal. Chem., **49**, 315, (1977)
52. Y. Ohta, S. Shoji, M. Esashi and T. Matsuo, Sensors and
    Actuators, **2**, 387, (1982)
53. I.Lauks, Proc. Int. Soc. for Opt. Eng., SPIE, **387**, (1983)I.
54. J. Van der Spiegel and I. Lauks, Submitted for publication, IEEE
    Trans. Biomedical. Eng.(1985)
55. M. Itkis, S. Kugelmass and N. Blecherman, Senior Design Project,
    University of Pennsylvania, (1985)
56. S. Pace, Sensors and Actuators, **1**, 475, (1981)
57. S. L. Yankell, C. Ram and I. R. Lauks, Caries Research, **17**, 439,
    (1983)
58. N. J. Ho, L. Kratochil, G. F. Blackburn and J. Janata, Sensors
    and Actuators, **4**, 413, (1983)
59. T. Dietz, J. N. Zemel, I. Lauks and T. Carroll, Thin Solid Films,
    119, 439, (1984)
60. N. J. Harrick, U. S. Patent 3,491,366 (20 Jan 1970)

61. J. I. Peterson, S. R. Goldstein and R. V. Fitzgerald, Anal. Chem., **52**, 864 (1980)

62. J. I. Peterson and G. G. Vurek, Science, **224**, 123, (1984)

63. J. I. Peterson, R. V. Fitzgerald and D. K. Buckhold, Anal. Chem., **56**, 16A, (1984)

64. T. Hirshfeld, F. Deaton, F. Milanovich and S. Klainer, Opt. Eng. **22**, 27, (1983)

65. "Novel Optical Fiber Techniques for Medical Applications", A. Katzir, Ed. Proc. SPIE, **494**, (1984)

66. M. El Sherif, J. N. Zemel and S. Babin, In preparation

67. S. Babin, Masters Thesis, University of Pennsylvania, (1984)

68. M. El Sherif and J. N. Zemel, 3rd Int. Conf. Sol. St. Transducers, Philadelphia PA, (1985), Technical Digest

69. D. W. Lubbers and N. Opitz, Sensors and Actuators, **4**, 641, (1983)

70. H. W. Kroneis and H. J. Marsoner, Sensors and Actuators, **4**, 587, (1983)

71. F. P. Milanovich, T. Hirshfeld, T. B. Wang, S. M. Klainer and D. Walt, in "Novel Optical Fiber Techniques for Medical Applications", A. Katzir, Ed., Proc. SPIE, **494**, 18, (1984)

72. J. S. Shultz, S. Mansouri and I. J. Goldstein, Diabetes Care **5**, 245, (1982)

73. C. Wolfe, MSE Thesis, University of Pennsylvania, (1978)

74. J. N. Zemel, B. Keramati, C. W. Spivak and A. D'Amico, Sensors and Actuators, **1**, 427, (1981)

75. Lyndon B. Johnson Space Center, "Fabricating Grating Couplers on Optical Fibers", NASA Technical Briefs, **7**, MSC-20286, (1983)

76. J. F. Giuliani, P. P. Bey, H. Wohltjen, and D. Ballantine, 3rd Int. Conf. on Sol. St. Transducers, Philadelphia PA, (1985), Technical Digest

77. T. M. Freeman and W. R. Seitz, Anal. Chem., **53**, 98, (1981)

78. W. R. Seitz, Anal. Chem., **56**, 16A, (1984)

79. The principles of thermal measurements are discussed in all standard introductory texts in general chemistry and physical chemistry, e.g. F. Brescia, S. Mehlman, C. Pellegrini and S. Stambler, "Chemistry: A Modern Introduction", 2nd Ed., W. B. Saunders and Co., (Philadelphia PA, 1978)

80. J. Barthel, " Thermometric Titrations", John Wiley and Sons, (New York NY, 1975)

81. E. H. Putley, Topics in Appl. Phys., **19**, 71, (1977)

82. E. H. Putley, Semiconductors and Semimetals, [R. K. Willardson and A. C. Beer, Editors, Academic Press, NY, (1970)], **5**, p. 259

83. S. T. Liu and D. Long, Proc. IEEE, **66**, 14, (1978)

84. H. Rahnamai and J. N. Zemel, Sensors and Actuators, **2**, 1, (1981)

85. J. R. Frederick, J. N. Zemel and N. F. Goldfine, J. Appl. Phys. **57**, 4936, (1985)

86. P. Hesketh, J. N. Zemel and B. Gebhart, J. Appl. Phys., **57**, 4944, (1985)

87. A particularly valuable reference book, though dated, is S. B. Lang "Sourcebook of Pyroelectricity", Gordon and Breach, (New York NY, 1974)

88. J. C. Young, PhD Dissertation, Electrical Engineering, University of Penna., Philadelphia PA, In preparation

89. J. N. Zemel and A. D'Amico, J. Appl. Phys., **56, aaa**, (1985)

90. A. D'Amico, G. Fortunato, Wu GeiHua and J. N. Zemel, 3rd Int. Conf. on Sol. St. Transducers, Philadelphia PA (1985), Technical Digest

91. N. W. Ashcroft and N. D. Mermin, "Solid State Physics", Holt, Rinehart and Winston, (New York NY, 1976)
92. W. P. Mason "Crystal Physics of Interaction Processes", Academic Press, (New York NY, 1966)
93. A. J. Lovinger, Science, **220**, 1115, (1983)
94. W. Cochran, Phys. Rev. Letters, **3**, 412, (1959)
95. A. Glass, Phys. Rev. **172**, 564, (1968)
96. H. S. Carslaw and J. C. Jaeger, "Conduction of Heat in Solids", 2nd Ed. Oxford University Press, (Oxford, 1959)
97. M. Esashi and T. Matsuo, J. Jap Soc. Appl. Phys. **44**, 339, (1978)
98. H. Abe, M. Esashi and T. Matsuo, IEEE Trans. Elect. Dev. **ED-26**, 1939, (1979)
99. P. W. Cheung, W. H. Ko, D. J. Fung and S. H. Wong, "Theory, Design and Biomedical Applications of Solid State Chemical Sensors", P. Cheung et al., p. 91, (1978),
100. M. Esashi and T. Matsuo, IEEE Trans. Biomed. Eng., **BME-25**, 184, (1975)
101. J. S. Schenck, J. Colloid and Interface Sci., **61**, 569, (1977)
102. P. Bergveld, IEEE Trans. Biomed. Eng. **BME-19**, 342, (1972)
103. M. A. Afromowitz and S. S. Yee, J. Bioeng. **1**, 55, (1977)
104. M. A. Afromowitz and S. S. Yee, U.S. Patent 4,133,735, (1979)
105. J. H. Wang and E.Copeland, Proc. Nat. Acad. Sciences, USA, **70**, 1909, (1973)
106. D. Erne, K. V. Schenker, D. Ammann, E. Pretsch and W. Simon, Chimia, **35**, 178, (1981)
107. P. Schulthess, Y. Shijo, H. V. Pham, E. Pretsch, D. Ammann and W. Simon, Analytica Chemica Acta, **131**, 111, (1981)

RECEIVED February 26, 1986

Recent Advances in Gas Sensors in Japan

Tetsuro Seiyama and Noboru Yamazoe

Department of Materials Science and Technology, Graduate School of Engineering Sciences, Kyushu University 39, Kasuga, Fukuoka 816, Japan

The development of new gas sensors has shown great advances in Japan during the past decade. The first type of gas sensor has centered around the development of semiconductor gas sensors using SnO_2 elements, the commercial production of which has been growing steadily as detectors or alarms for leakage of domestic fuel gases. Many efforts have also been directed to the development of new gas sensors from semiconductor materials. The second type has evolved from the development of solid electrolyte gas sensors. In addition to oxygen sensors based on stabilized zirconia which are produced on a massive scale for control of car engines and steel making processes, solid electrolyte sensors for these applications are under investigation. The third type involves humidity sensors which have been become indispensable parts of several electrical appliances. Recent advances in the development of several gas sensors, i.e., lean-burn sensors, CO sensors, FET humidity sensors and solid electrolyte H_2 sensors are described.

Research and development of gas sensors in Japan started with semiconductor gas sensors. In 1962, Seiyama (<u>1</u>, <u>2</u>) and shortly after Taguchi (<u>3</u>) proposed independently gas sensors using semiconductive metal oxides, which aimed at the detection of inflammable or reducing gases such as LP (liquified propane) gas and town gas. Taguchi and his colleagues in Figaro Engineering Inc. expended utmost efforts to put semiconductor gas sensors into practical use and succeeded in commercial production of them in 1968. Since then, many researchers in Japan have been stimulated to investigations of improved or new types of semiconductor gas sensors as well as basic understandings of their sensing mechanisms. Semiconductor gas sensor devices and their applications have been inceasing year after year as needs for them expand in various fields as seen from the number of patents shown in Figure 1. The success of semiconductor gas sensors has encouraged the development of other types of gas sensors such as the oxygen sensor, humidity sensor, catalytic combustion type sensor, and so on. These sensors have marked great advances largely through the effective use in the automobile, steel, and electronics industries.

The present state of gas sensors in Japan

The production of gas sensors. The production records of various types of gas sensors for past five years in Japan are listed in Table I except for the oxygen and humidity sensors. The sensors produced in the largest quantity are of the semiconductive type, followed by the catalytic combustion and thermistor types. These sensors have been mostly applied to domestic uses such as gas leakage alarms or gas control systems for LP gas and town gas which are extensively used for cooking and heating in Japanese houses. This is why these sensors are manufactured on a large scale. Other electrochemical sensors have been developed mainly to monitor other gases.

Table I. Production records of various types of gas sensors[1]

(thousand units)

Fiscal year / Sensor	1979	1980	1981	1982	1983
Semiconductor gas sensor	1,220	3,140	4,330	4,170	5,430
Catalytic combustion type sensor	171	186	202	223	240
Galvanic cell type sensor	29	31	34	36	38
Electrolytic sensor	9	9	10	11	12
Solid electrolyte sensor	26	28	31	74	82
Thermistor sensor	39	45	51	60	55
Ion selective electrode sensor	14	11	10	9	8
Others	2	2	7	8	8

1) except oxygen sensors which are used in automobile and steel industries.

Metal oxide semiconductor gas sensors. The production of semi-conductor gas sensors has grown into a large industry, producing more than five million pieces per year in 1983. Besides SnO_2, γ-Fe_2O_3 and α-Fe_2O_3 have been put into practical use as sensor materials; γ-Fe_2O_3 sensors for LP gas alarms and α-Fe_2O_3 sensors for town gas alarms were commercialized in 1978 and 1981, respectively, by Matsushita Electronic Components Co., Ltd. However, the "FIGARO" sensor (a type of SnO_2 sensor) still has an overwhelming market share over Fe_2O_3 sensors and the catalytic combustion type sensors, as shown in Figure 2.

The application of semiconductor gas sensors (mainly the SnO_2 sensor) is ever expanding into various fields, domestic and indus-trial, as shown in Figure 3. Examples are the CO selective sensor and combustion monitor sensor, which are applicable to micro wave ovens or ventilation fans, kerosene or gas stoves, hot water supply systems, and so on.

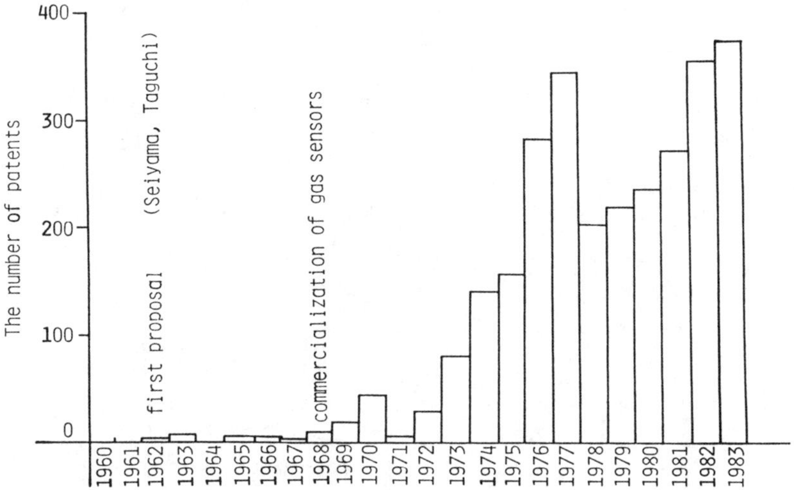

Figure 1. Change in annual number of patents concerning semiconductor gas sensors.

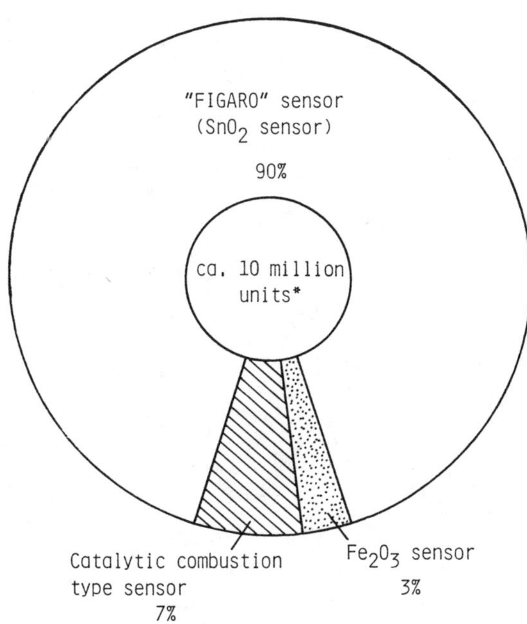

Figure 2. Market share of sensors equipped in household gas leakage alarms for both LP and town gases.

* the number of gas leakage alarms installed up to now.

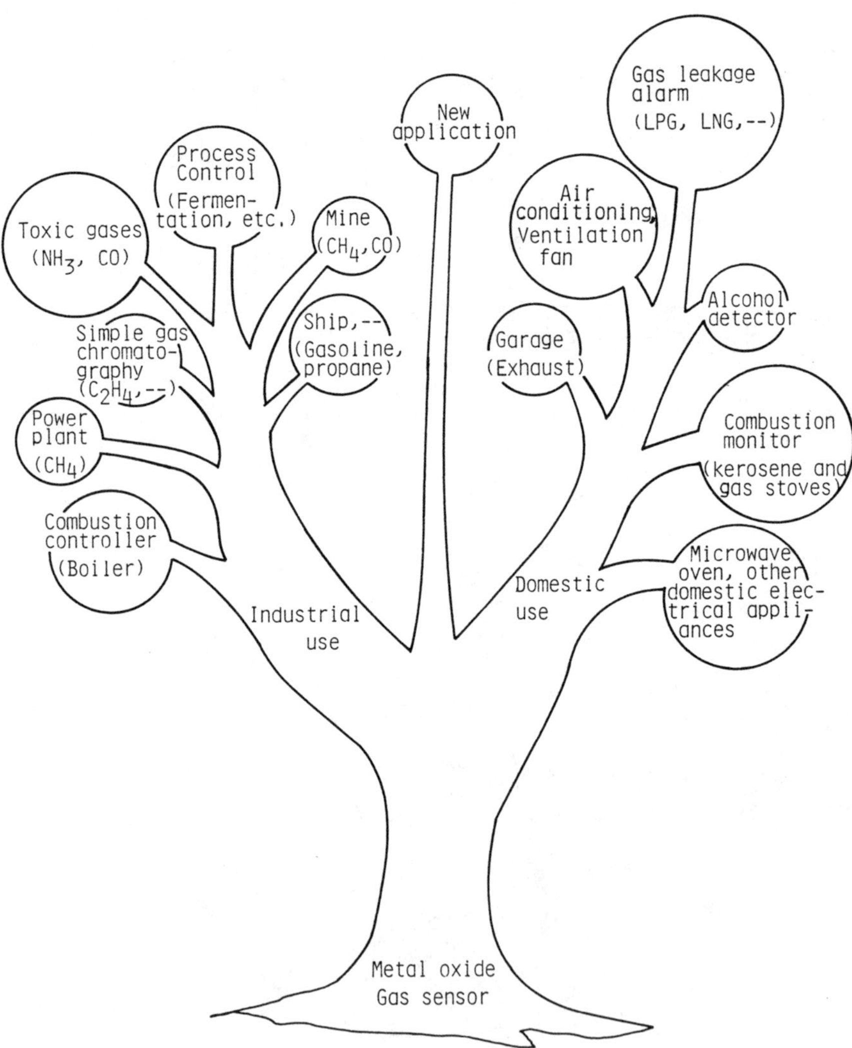

Figure 3. Application of semiconductor gas sensors.

<u>Oxygen sensors.</u>   The production of oxygen sensors using stabilized zirconia has been increasing rapidly since 1977, and as a whole about one million pieces were produced in Japan in 1980, as shown in Figure 4.   About 96 % of oxygen sensors are used for automobile exhaust control.   They are primarily supplied by two companies, Japan Electronic Control Systems and Nippon Denso Co., Ltd., as shown in Figure 5, though the share of the former company includes oxygen sensors imported from U.S.A.   More recent statistics have not been disclosed, but it is said that Nippon denso Co., Ltd. produced 2.4 million oxygen sensors in 1983.   In other applications, about two hundred thousand oxygen sensors are consumed annually for steel making process control in the Japanese steel industries.   The production of oxygen sensors is increasing mainly for application in car engine systems.

<u>Humidity sensor.</u>   There exists a strong demand for humidity sensors especially in Japan because of humid weather in summer.   The use of humidity sensors has been widely spreading in industrial fields as well as domestic.   The production record is listed in Table Ⅱ. Humidity sensors consist of two major types, i.e., one which measures relative humidity and the other which detects dew condensation.   The sensors for relative humidity made of ceramic or organic polymers are produced in amounts exceeding one million.   They are applied to controling systems in air conditioning, micro wave ovens, and so on (<u>4</u>).   Dew condensation sensors are usually made of hydrophilic or swelling organic polymers in which carbon particles are dispersed (<u>5</u>).   This type of sensor utilizes a drastic increase in resistance at the point of dew condensation, since water absorption swells the polymers to counteract ohmic contact between carbon particles.   This type sensor is now widely applied to humidity controling systems of video tape recorders or car windows.

Table Ⅱ.   Production Record of Humidity Sensors in Japan (thousand unit)

Fiscal year	Relative humidity		Dew condensation
	Ceramic	Organic polymer	
1983	750	560	8,500
1984	900[1)	---	12,000[1)

1) estimated

Examples of recently developed sensors

<u>Lean-burn sensor.</u>   Conventional oxygen sensors utilizing calcia- or yttria-stabilized zirconia have been based on a concentration cell (potentiometric) method.   When the sensor is located in car engine exhaust, the output emf undergoes a steep change at the stoichiometric air/fuel ratio as shown in Figure 6, and enables us to optimize the operating conditions of the three way catalysts.   On the other hand, recent investigations (<u>6</u>, <u>7</u>) for fuel economy have focused attention on the combustion under excess air (or lean fuel) conditions, which is also advantageous for depressing NO_x formation as shown in Figure 7.   For control of lean-burn conditions, however, the conventional oxygen sensor is not sensitive enough so that a lean-burn sensor based on an amperometric method is of current interest.

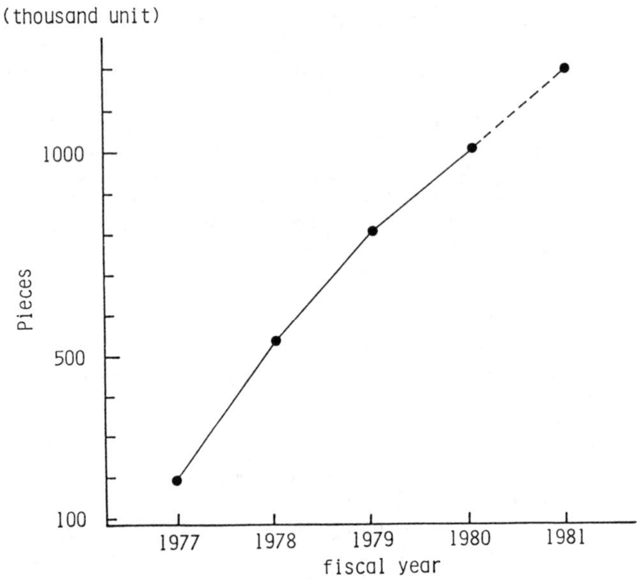

Figure 4.   Production record of oxygen sensors in Japan.

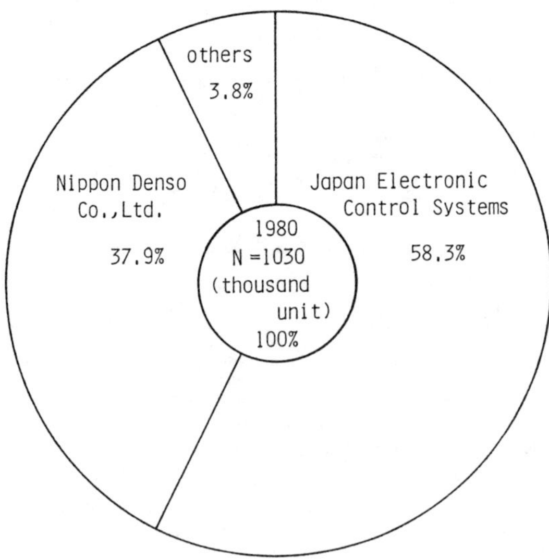

Figure 5.   Market share of oxygen sensors in Japan.

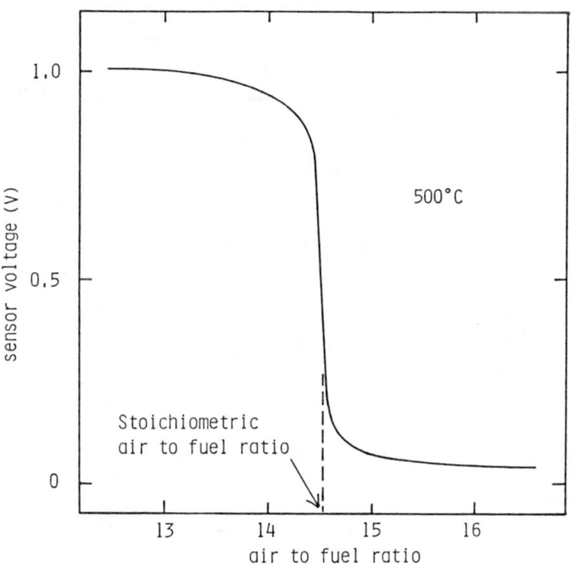

Figure 6. Sensor voltage versus air-to-fuel ratio for galvanic cell-type oxygen sensor.

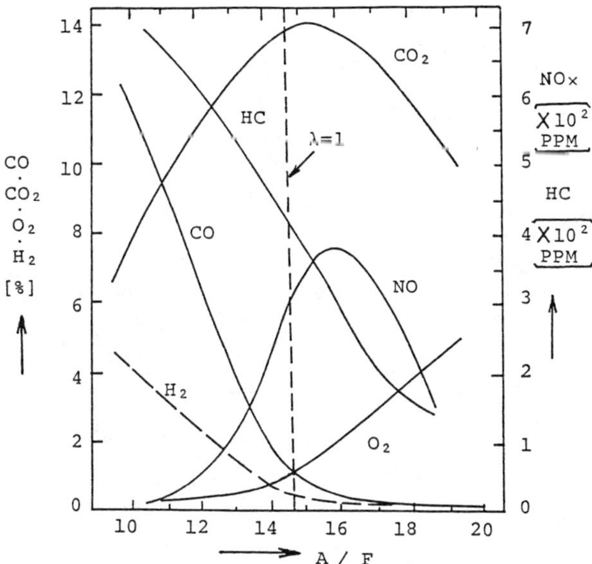

Figure 7. Gas concentrations in engine exhaust at varying air-to-fuel ratio (A/F). Reproduced with permission from Ref. 6. Copyright 1984 Japan Association of Chemical Sensors.

This very sort of lean-burn sensor has been developed success-fully by a Japanese automobile company, Toyota Motor Co. (<u>6</u>, <u>7</u>). This sensor utilizes electrochemical oxygen pumping, that is, electrochemical transfer of oxygen from cathode to anode under an applied voltage, as depicted schematically in Figure 8 . If an oxygen-containing gas is fed on the cathode through a small pin hole and the applied voltage is large enough, the rate of oxygen pumping is determined by the diffusion of gasous oxygen through the pin hole, and the external current (output) becomes proportional to the oxygen concentration outside the pin hole, as shown in Figure 9. The I-V characteristics as well as the basic mechanism of this amperometric oxygen sensor are quite similar to those of polarography. Hence, the sensor may be classified as a limiting current type. The diffusion limited current phenomenon also appears with a simple modification of the usual zirconia sensor. That is, a coating of the cathode with a porous ceramic layer, as carried out in the sensor for prac-tical use (see Figure 10), is enough to make an effective diffusion layer of oxygen. The output current of the practical sensor is al-most proportional to excess air ratio, λ, in the lean-burn region as shown in Figure 11, giving rise to responses much superior to the potentiometric sensor. The actual use of this type sensor has been begun in Toyota Motor Company. One remaining problem is its rela-tively high operating temperature of about 700 °C.

<u>Carbon monoxide sensor.</u>   Carbon monoxide is a toxic air pollutant originating from incomplete combustion of fuels in burners or engines. Despite a strong demand for a very reliable carbon monoxide sensor, the only available sensor was based on an electrochemical type until recently. In this type sensor, selectivity and sensitivi-ty for CO can often be enhanced by selecting elecrode materials and electrode potential appropriately, but several disadvantages are encountered, such as, short life, difficult maintenance and a rather expensive price.

In this circumstance, several investigations have been made to use SnO_2 elements for carbon monoxide detection. For instance, a SnO_2 element doped with a small amount of ThO_2 has been reported to show high sensitivity to carbon monoxide at room temperature. However, usually it is difficult to attain stable sensitivity and selectivity under these operation conditions. Recently carbon mon-oxide sensors have been commercialized by Figaro Engineering Inc. and Yazaki Meter Co., Ltd. independently. The sensor utilizes sintered SnO_2 elements which are essentially the same as those for LP gas detection but are operated in a different mode which combines low temperature CO sensing and periodic heat cleaning. As is well known, the gas sensitivity of semiconductor elements are influenced by operation temperature as well as by noble metal doping. The sensitivity-temperature characteristics for the TGS 203 element is shown in Figure 12 (<u>8</u>). Generally speaking, the sensitivity to CO, which is low at high temperature, increases with lower temperature, and the response becomes higher than other gases such as ethanol and hydrogen below 100 °C. At lower temperatures, on the other hand, the semiconductor surface tends to suffer more seriously from con-tamination due to adsorption of gases such as water vapor, resulting in poor stability and reproducibility in the sensor response. In

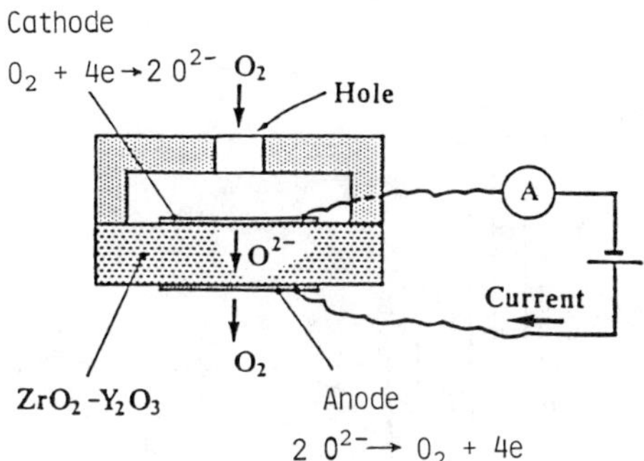

Figure 8.   Amperometric oxygen sensor utilizing electrochemical pumping. Reproduced with permission from Ref. 7. Copyright 1985 Institute of Electrical Engineering of Japan.

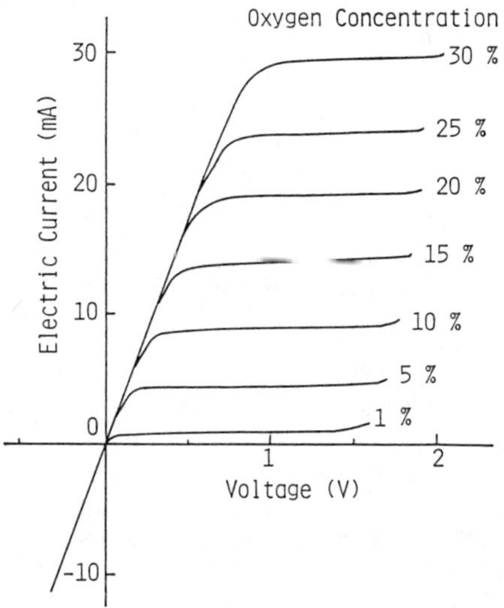

Figure 9.   I-V characteristics of amperometric oxygen sensor (700 $^{\circ}$C). Reproduced with permission from Ref. 6. Copyright 1984 Japan Association of Chemical Sensors.

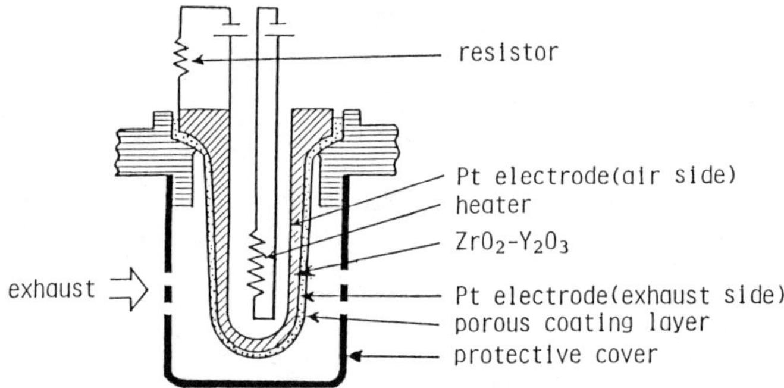

Figure 10. Practical amperometric oxygen sensor with porous ceramic coating. Reproduced with permission from Ref. 6. Copyright 1984 Japan Association of Chemical Sensors.

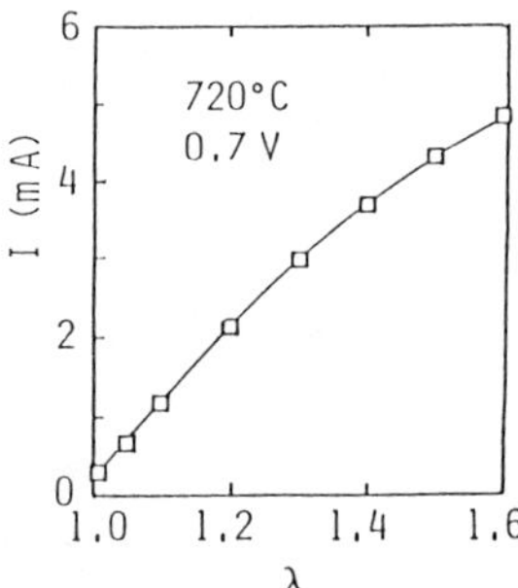

Figure 11. I –λ characteristics of amperometric oxygen sensor. λ: excess air ratio. Reproduced with permission from Ref. 7. Copyright 1985 Institute of Electrical Engineering of Japan.

order to overcome this disadvantage, the commercial sensor adapts intermittent heating (60 seconds) of the element up to ca 300 °C to keep the surface clean, before gas sensing at low temperatures for a period of 90 seconds. The sensor performance thus observed is shown in Figure 13. The electric resistance of the element to 100 ppm CO is similar to the response from 2000 ppm H_2 or 1000 ppm ethanol. This means that the sensor is free of false alarms due to H_2 and ethanol in ordinary use. An additional circuit device for temperature compensation can diminish the temperature effect in the temperature range 0 – 50 °C. An activated charcoal filter is also added to avoid the interfering effect of NO_x gas.

<u>FET type humidity sensor.</u> Although sensors based on a field–effect transistor (FET) appear to hold promise as a small and low–cost intelligent sensor, relatively few people have been engaged in the research on FET type sensors in Japan. In this respect, it is remarkable that a FET type humidity sensor was developed recently by Hijikigawa of Sharp Corp (<u>9</u>). The sensor is also worth notice as a new type of humidity sensor, which utilizes changes in electric capacitance of humidity sensitive membrane interposed between double gate electrodes.

A schematic cross–section of the FET humidity sensor integrated with a temperature sensing diode, which is about 1.5 mm square in size, is shown in Figure 14. The host FET device is an n–channel MIS–FET with a meandering–gate structure. The gate insulator is composed of double layers of silicon dioxide and silicon nitride films, the latter covering the whole surface to protect the FET characteristics in the highly humid environment. The humidity sensitive polymer membrane of one micron in thickness is stacked between the lower–gate electrode and upper–gate electrode (porous gold electrode). The equivalent circuit of this humidity sensor is shown in Figure 15. A D.C. voltage V_0, and a small A.C. voltage $\widetilde{\Delta v_0}$ is applied to the upper gate electrode. The upper- and lower–gates are electrically connected with a sufficiently large fixed resistance R_B. It can be shown that under adequate conditions, the output voltage, V_{out}, depends on the capacitance of the membrane, C_s, as follows.

$$V_{out} = \widetilde{\Delta v_0} \cdot R_L \cdot g_m / (1 + C_i/C_s) \tag{1}$$

Where R_L is the load resistor connected with the drain electrode, g_m the transconductance of the FET , and C_i the capacitance of the gate insulator. It has been shown experimentally that V_{out} is linearly dependent upon relative humidity in the almost whole humidity range.

<u>Solid–Electrolyte Hydrogen Sensor.</u> Most of solid gas sensors so far developed need high temperature operation because of limited ionic conductivities when the electrolyte is near room temperature. If solid electrolytes with sufficiently large ionic conductivities are available, unique gas sensors operative near room temperature can be fabricated. An example is the following proton conductor hydrogen sensor proposed by our group (<u>10</u>, <u>11</u>).

The sensor construction is shown in Figure 16. As a proton conductor, Nafion (DU PONT, Nafion 117 H–type) membrane or disks of several inorganic ion exchangers were used. With sample gas and reference air passing over the respective electrodes, the electric

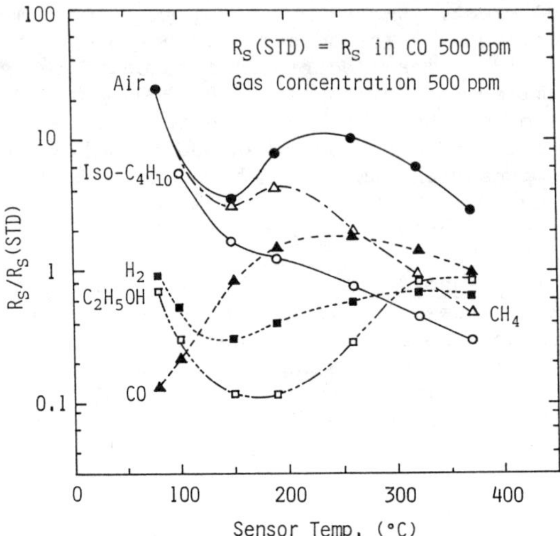

Figure 12. Temperature dependence of the response of TGS 203 element to sample gases.

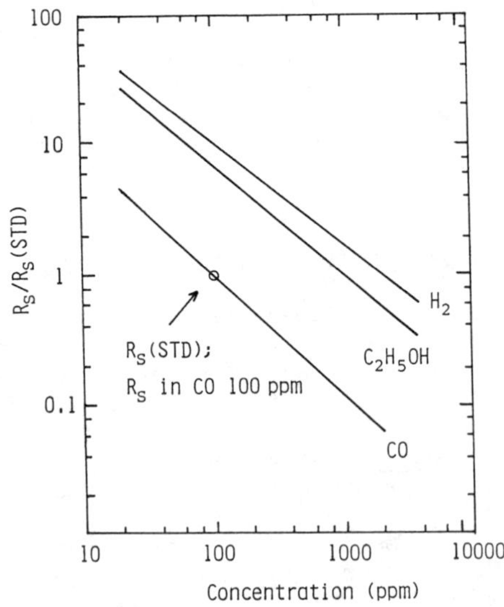

Figure 13. Gas concentration dependence of the response character-istics of TGS 203 element.

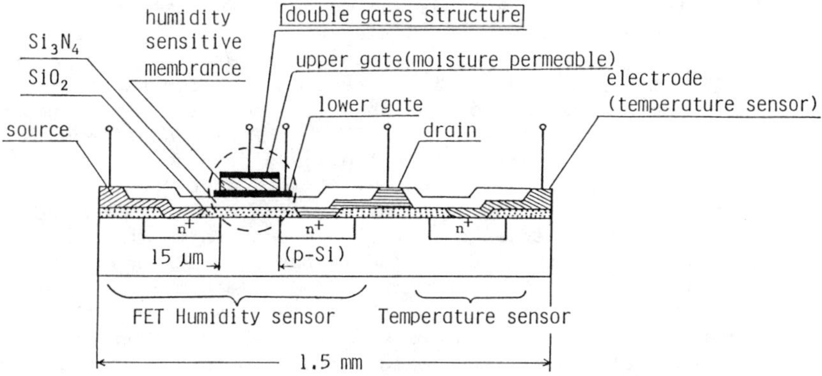

Figure 14. Structure of FET humidity sensor. Reproduced with permission from Ref. 9. Copyright 1985 Institute of Electrical Engineering of Japan.

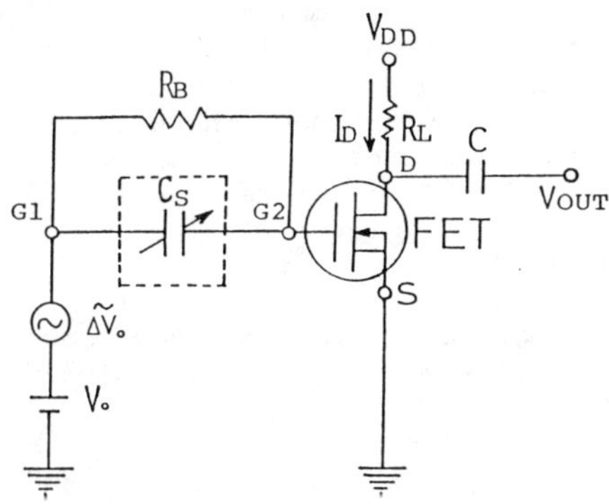

Figure 15. Equivalent circuit of FET humidity sensor. Reproduced with permission from Ref. 9. Copyright 1985 Institute of Electrical Engineering of Japan.

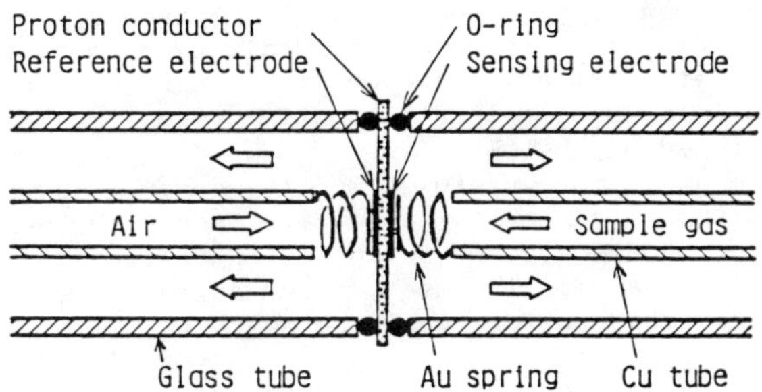

Figure 16. Structure of proton conductor sensor. Reproduced with permission from Ref. 10. Copyright 1983 Kodansha Ltd. (Tokyo).

potential of the sensing electrode was measured against the counter
electrode (reference) by means of an electrometer.   Figure 17 shows
the response curves to H_2 in air at room temperature of the sensor
using a Nafion membrane and Pt black electrodes.   The response time
to reach 2000 ppm H_2 was as short as about 10 seconds and the sensing
electrode potential reached about - 200 mV.   This sensor was also
sensitive to CO, but insensitive to methane and propane.

As shown in Figure 18, the potential is almost proportional to
the logarithm of H_2 concentration diluted in air.   When H_2 is
diluted in N_2, the observed potential corresponds to the electromotive force of a H_2-O_2 fuel cell, and in fact the EMF was as large
as about 1.0 V with a theoretical slope of 30 mV/decade, as shown in
the same figure.   It has been shown that in the case of H_2 diluted in
air, the following electrode reaction, i.e., electrochemical oxidation of hydrogen (2) and electrochemical reduction of oxygen (3), are
important.

$$H_2 \longrightarrow 2\ H^+ + 2\ e \qquad\qquad (\text{anodic}) \qquad\qquad\qquad (2)$$
$$1/2\ O_2 + 2\ H^+ + 2\ e \longrightarrow H_2O \qquad (\text{cathodic}) \qquad\qquad (3)$$

When the cell responds to the sample gas, both reactions should take
place on the sensing electrode to form a local cell, which results in
a mixed potential of the electrode.   This has been well confirmed by
the measurements of respective polarization curves.

Low temperature operation has been shown to give an advantageous
effect on the sensor structure.   The above sensor requires the
separation of the sample gas from the reference air.   We have found
that at room temperature, the silver electrode is active to oxygen,
but inert to low content of H_2 in air.   This indicates that, if one
uses the sensing Pt electrode together with the refernce Ag electrode, a reference air flow is not necessary, and therefore
simplification and miniaturization of sensor structure are possible
as shown in Figure 19.   The sensor element was fabricated by mixing
proton conductors such as zirconium hydrogen phosphate with Teflon
dispersion, followed by applying the resulting paste on the alumina
tube.   This modified sensor exhibited good performance with a long-
term stability as shown in Figure 20.   It is also possible to fabricate an amperometric type of proton conductor hydrogen sensors.

Future scope

Many problems and tasks, both scientific and technological, still
remain in the field of gas sensors.   For example, microsensors,
intelligent sensors, cord-less sensors, etc., seem to be very important as well as attractive.   However, these are not covered here,
and only trends in sensor application anticipated in the near future
in Japan are pointed out below.
1. For domestic use, more useful humidity sensors, CO sensors and
combustion monitoring sensors will be developed with their increased
integration into safety systems for homes and offices.
2. Gas sensors for automotive use , such as the λ sensor, lean-
burn sensor and sensors for car-room air-conditioning will be
utilized more extensively.
3. In the industrial field, the appropriate and effective application
of sensors for O_2, SO_2, CO, NH_3 and so on will be advanced for

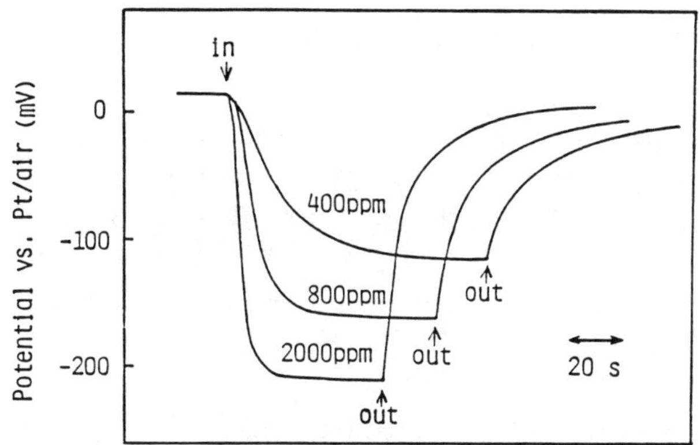

Figure 17.   Response curves to H_2 in air.   Reproduced with
permission from Ref. 10.   Copyright 1983 Kodansha Ltd. (Tokyo).

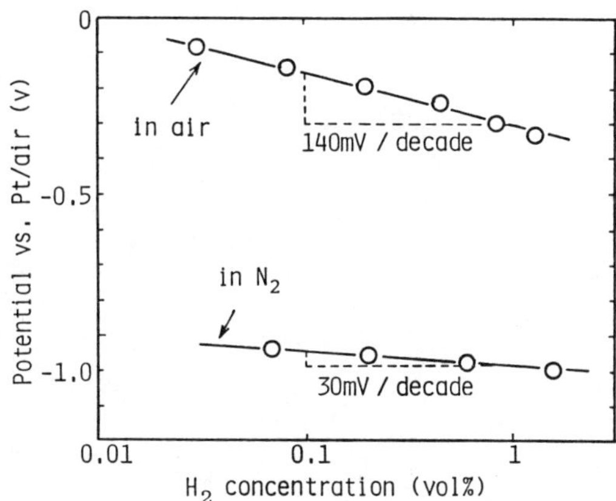

Figure 18.   Sensing electrode potential vs. H_2 concentration in
air or in N_2.   Reproduced with permission from Ref. 10.   Copyright
1983 Kodansha Ltd. (Tokyo).

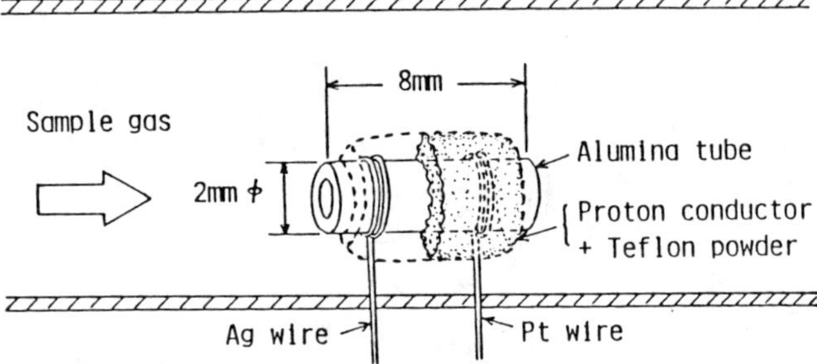

Figure 19.   Structure of the modified sensor with reference Ag
electrode.   Reproduced with permission from Ref. 10.   Copyright 1983
Kodansha Ltd. (Tokyo).

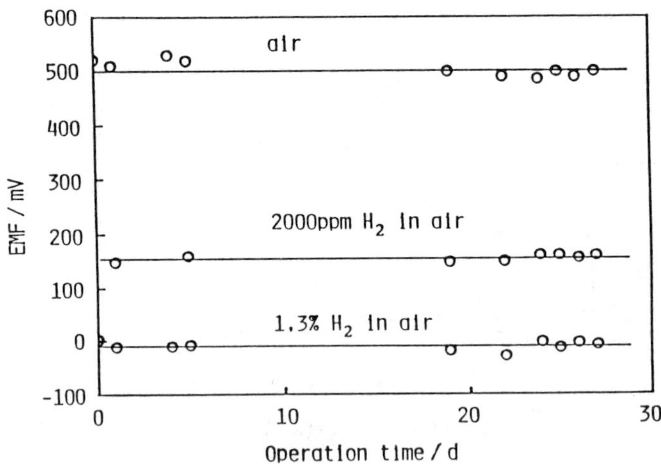

Figure 20.   Long-term stability of the modified sensors.   Reproduced
with permission from Ref. 10.   Copyright 1983 Kodansha Ltd. (Tokyo).

process   control   and security   monitoring.   For   instance,   sensors
detecting arsine gas and phosphine gas are needed in the electronics
industries.

Literature Cited

1.   Seiyama, T. ; Kato, A. ; Fujiishi, K. ; Nagatani, M. Anal.
     Chem., 1962. 34, 1502.
2.   Seiyama, T. ; Kagawa, S. Anal. Chem., 1966, 38, 1069.
3.   Taguchi, N. Japanese Patent 45-38200 (applied in 1962).
4.   Nitta, T. ; Terada, Z. ; Hayakawa, S. J. Am. Cerm. Soc., 1980,
     63, 295.
5.   Ishida, T. ; Kobayashi, T ; Kuwahara, K. ; Hatanaka, H.
     National Technical Reports, 1978, 24, 436.
6.   Igarashi, I. ; Takeuchi, T.   Preprint of 24th meeting of Japan
     Association of Chemical Sensors, 1984 ; p. 5.
7.   Saji, K. ; Takahashi, H. ; Kondo, H. ; Takeuchi, T. ;
     Igarashi, I. Proc. of the 4th Sensor Symposium, 1984, p.147.
8.   Chiba, A., personal communication.
9.   Hijikigawa, M. ; Furubayashi, H. ; Miyoshi, S. ; Inami, Y.
     Proc. of the 4th Sensor Symposium, 1984, p.135.
10. Miura, N. ; Kato, H. ; Yamazoe, N. ; Seiyama, T. Proc. of the
     Int. Mtg. on Chemical Sensors, Kodansha/Elsevier, 1983, p. 233.
11. Miura, N. ; Kato. H. ; Yamazoe, N. ; Seiyama, T. Chem. Lett.,
     1983, 1573.

RECEIVED October 31, 1985

GAS SENSORS

3

Tin Oxide Microsensors

Shih-Chia Chang and David B. Hicks

Electronics Department, General Motors Research Laboratories, Warren, MI 48090–9056

Tin oxide based semiconductor gas sensors were
fabricated on silicon wafers using conventional microelectronic processing technology. Thin film tin oxide
was delineated by a reactive ion etching technique
using $SiCl_4$ as the etch gas. The overall device size
is 450 μm x 350 μm with a sensing element feature size
of 50 μm x 50 μm. The sensor resistance changes
significantly when the sensor is exposed to alcohol
vapor or NO_x in ambient air. At sensor temperatures
ranging from 200 to 300°C, the values of R(air)/
R(200 ppm of alcohol vapor in air) are 50 and 140 for
tin oxide (SnO_x) and palladium-gold/SnO_x ($PdAu/SnO_x$)
sensors, respectively; while the values of R(50 ppm of
NO_x in air)/R(air) are 60 and 5 for SnO_x and $PdAu/SnO_x$
sensors, respectively. Possible sensing mechanisms
based on XPS, Kelvin probe, Hall and electrical
conductivity measurements are also discussed.

The resistivity of certain semiconductors such as tin oxide (SnO_x)
and zinc oxide (ZnO) can be strongly modulated by the presence of
certain gaseous species in the ambient. Several gas sensors have
been developed based on such material characteristics (1-5). The
principal advantages of semiconductor gas sensors are: (a) relative
simplicity of fabrication; (b) relative simplicity of operation; (c)
low cost (fabrication and maintenance). However, the major drawback
of these sensors is their low sensing selectivity among various
gases.

Sensor selectivity can be enhanced by devising a multi-sensor
scheme in conjunction with logic/control circuitry to form an integrated sensing system (6-7). In such a system the selectivity would
be achieved either by operating the individual sensing elements at
different temperatures, by altering or modifying the materials used

in the sensing elements, or by using a combination of both approaches. For example, in our recent tests on palladium-gold/tin oxide sensors, we found that the sensors responded quite differently to the presence of NO_x (50 ppm) and alcohol (C_2H_5OH, 200 ppm) in air when they were operated at different temperatures as indicated in Table I.

Table I. Resistance of PdAu/SnO$_x$ Sensor in Three Different
Ambients at Two Different Sensor Temperatures

	Gas Species in Air		
Thin Film PdAu/SnO$_x$ Sensor	NO_x	C_2H_5OH	NO_x + C_2H_5OH
Sensor resistance at 220°C	increase ↑	decrease ↓	increase ↑
Sensor resistance at 280°C	increase ↑	decrease ↓	decrease ↓

By simultaneously monitoring the resistance change of the PdAu/SnO$_x$ films at two different sensor temperatures, potentially we can differentiate three different ambient conditions: (a) NO_x present; (b) C_2H_5OH vapor present; (c) both NO_x and C_2H_5OH vapor present.

An integrated multi-sensor system requires precise control of sensor characteristics, and may require 10 or more sensors in close proximity. To achieve this, we have to rely on microelectronic processes in order to fabricate sensors with small and precisely controlled feature sizes on silicon.

In this paper, the application of microelectronic processing technology to the fabrication of SnO$_x$ and PdAu/SnO$_x$ microsensors on silicon wafers is described, sensor responses to various gases in air are presented, and the possible sensing mechanisms are briefly discussed.

Experimental

Device Fabrication. A four-mask process was used for the fabrication of tin oxide microsensors on silicon substrates. The major processing steps are shown in Figure 1:

1. A 1 μm thick SiO$_2$ layer was thermally grown on a silicon wafer.
    A 1 μm thick polycrystalline silicon (polysilicon) layer was then deposited by chemical vapor deposition (CVD). Phosphorus doping of polysilicon was done by ion implantation with a dosage of 10^{16} cm^{-2} and a voltage of 200 keV. The polysilicon sheet resistance of 50 Ω/□ was obtained after post-implant activation (Figure 1a).

2. Polysilicon (used as a sensor heater) was delineated by etching in a SF_6 plasma. A 1 µm layer of CVD SiO_2 was deposited, and a 100 nm tin oxide film was subsequently sputter-deposited (Figure 1b).

3. The tin oxide thin film was patterned by reactive ion etching (RIE) using either $SiCl_4$ or 7% H_2 in N_2 as the etch gas. The polysilicon contact holes were opened by wet-chemical etching in buffered hydrofluoric acid (BHF). A double-layer metallization (Cr ~50 nm plus Al ~1 µm) was done by electron beam evaporation to form the electrical interconnection (Figure 1c).

4. The metal interconnect feature was defined by wet chemical etching. A layer of PdAu, ~2.5 nm thick with a composition of Pd/Au = 4/6, was flash deposited on some of the samples to enhance the sensor sensitivity to C_2H_5OH (Figure 1d). Figure 2 is the SEM picture of a completed tin oxide microsensor with four metal bonding pads. The complete sensing device has a dimension of 450 µm x 350 µm with an active sensing element (either SnO_x or $PdAu/SnO_x$) feature size of 50 µm x 50 µm.

The sputter-deposition of the tin oxide thin film (~100 nm thick) was done under the following conditions:

a. target material: sintered tin oxide
b. RF power: 150 W
c. substrate-to-target distance: ~7.5 cm
d. argon pressure: ~1 Pa; oxygen pressure: ~0.2 Pa
e. sputter time: 10 min
f. substrate temperature: <150°C

The reactive ion etching system used for the delineation of tin oxide thin film is a diode RF sputtering system and was described in our earlier paper (8). For tin oxide etching, the etching parameters used are as follows:

a. etch mask: positive photoresist
b. etch gas: forming gas (4-10% H_2 in N_2), or $SiCl_4$ (preferably)
c. gas pressure: 3.3 Pa
d. RF power density: ~0.33 W/cm^2
e. cathode temperature: <150°C
f. cathode-to-anode distance: ~7.5 cm

The tin oxide etch rates obtained were ~18 nm/min and 4-8 nm/min for $SiCl_4$ plasma and forming gas plasma, respectively.

Since aluminum does not form an ohmic contact with tin oxide, a double layer metallization was used. By depositing a thin layer (~50 nm thick) of Cr underneath the Al, a good ohmic contact to the tin oxide thin film was formed.

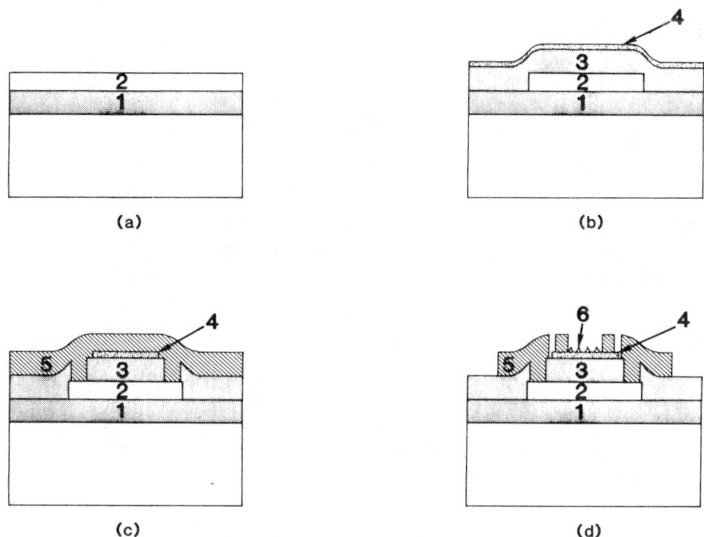

Figure 1. Processing steps for the fabrication of PdAu/SnO$_x$ microsensor (a) to (d). 1, SiO$_2$; 2, Polysi; 3, CVD SiO$_2$; 4, SnO$_x$; 5, Al/Cr; 6, Pd/Au islands.

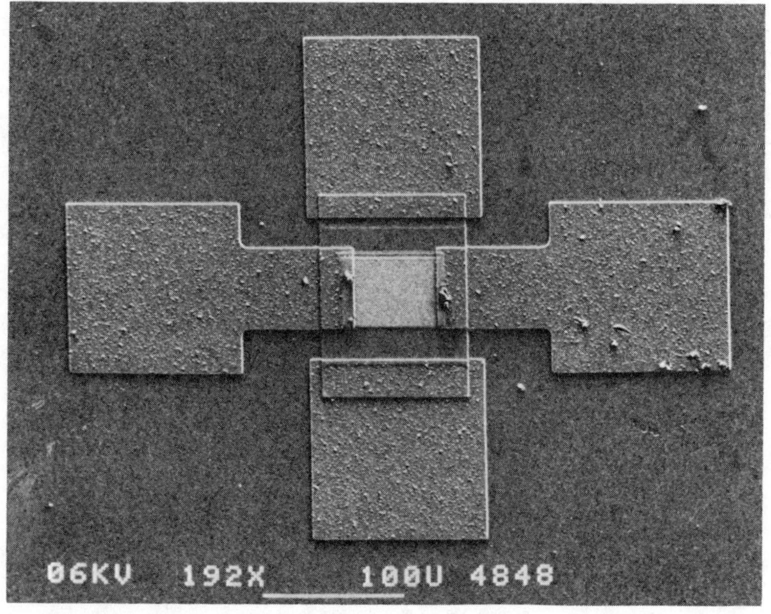

Figure 2. SEM picture of a finished tin oxide microsensor.

Device Characteristics and Discussion

A phosphorus-doped polysilicon layer was used as the sensor heater. Its temperature coefficient of resistivity was positive with a value of $\sim 6 \times 10^{-4} °C^{-1}$. The value of the heater resistance as a function of temperature was used to indicate the sensor temperature.

Sensor responses to 200 ppm of alcohol vapor in air are shown in Figure 3a (SnO_x at $\sim 250°C$) and Figure 3b ($PdAu/SnO_x$ at $\sim 250°C$). The sensor resistance decreases in the presence of C_2H_5OH vapor. The ratios of sensor resistance in air to that in 200 ppm of C_2H_5OH vapor in air are ~ 50 and ~ 140 for SnO_x and $PdAu/SnO_x$, respectively. While propylene (C_3H_6, 200 ppm) has little effect on the SnO_x sensor (both the sensor baseline and sensor response to 200 ppm of alcohol vapor are shifted slightly, as shown in Figure 3a), it affects the $PdAu/SnO_x$ sensor appreciably (Figure 3b). Evidently, by incorporating a thin, discontinuous film of $PdAu(\sim 2.5$ nm) on top of the SnO_x film (100 nm), the sensor sensitivities toward detecting C_2H_5OH vapor and C_3H_6 are greatly enhanced. The effect of carbon monoxide (CO) on the sensors is similar to that of C_3H_6 but to a much lesser extent. Methane (CH_4), a saturated hydrocarbon, with concentrations up to 1000 ppm has no detectable effect on the sensor baseline and sensor response to alcohol vapor. Figure 4a and 4b show sensor resistance versus concentration of alcohol vapor, C_3H_6 and CO in air for SnO_x and $PdAu/SnO_x$ sensors respectively.

Sensor responses to 50 ppm of NO_x in air are shown in Figure 5 for both SnO_x and $PdAu/SnO_x$ sensors. The sensor resistance increases in the presence of NO_x in air. The values of $R(NO_x)/R(air)$ are about 60 and 5 for SnO_x and $PdAu/SnO_x$ sensors, respectively. While a PdAu deposit enhances the sensor sensitivity to alcohol vapor and some other hydrocarbon gaseous species, it depresses the sensor sensitivity to NO_x.

The sensing mechanisms of the tin oxide based sensors have been discussed in many publications (9,10,11). The most widely accepted model for tin oxide based sensors operated at temperatures $<400°C$ is based on the modulation of the depletion layer width in the semiconductor (sensor) due to chemisorption as illustrated schematically in Figure 6. For C_2H_5OH and SnO_x (or $PdAu/SnO_x$) interaction, the possible reaction steps may be expressed by the following equations:

Step (1): $\frac{1}{2} O_2$ (gas) $+ 2e^- \rightarrow O^{-2}$ (ads)

Oxygen is chemisorbed on the SnO_x surface and at the inter-grain regions, capturing two electrons from the SnO_x conduction band. Depletion layers are formed in the surface as well as in the grain boundary regions, causing the carrier concentration and electron mobility (due to scattering) to decrease.

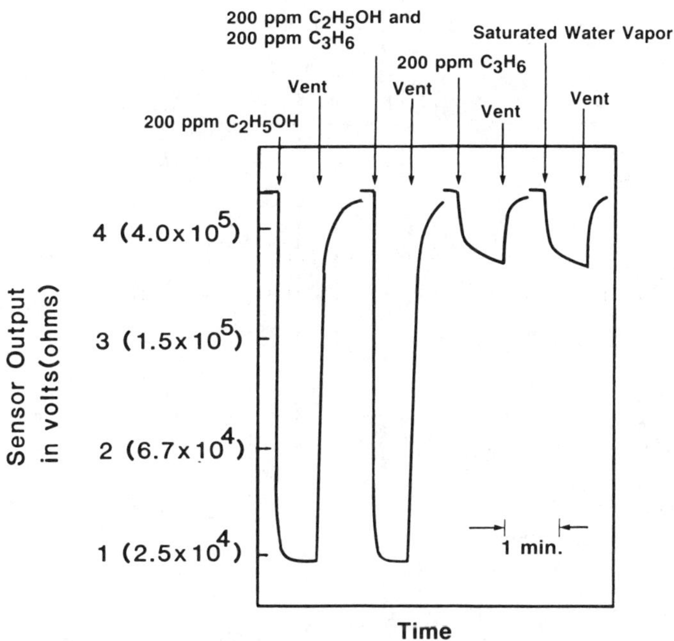

Figure 3a. Sensor responses to 200 ppm of alcohol vapor, 200 ppm of alcohol vapor plus 200 ppm of propylene, 200 ppm of propylene, and saturated water vapor in air for SnO_x sensor. Sensor temperature ~ 250 °C.

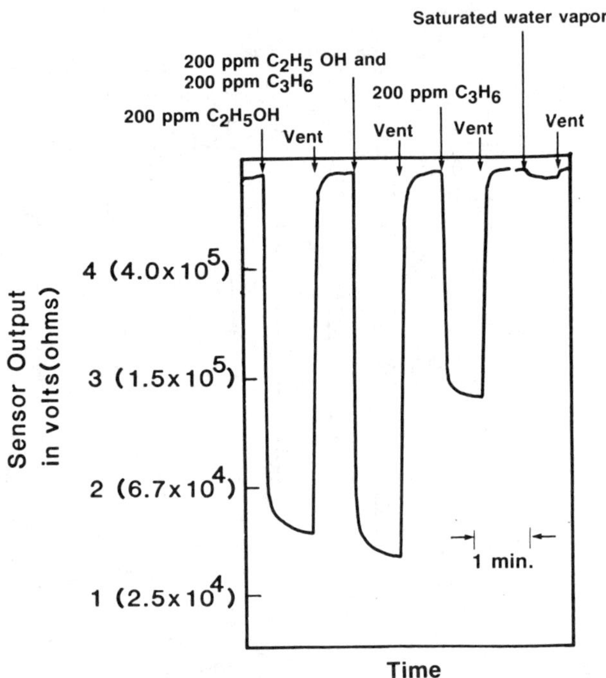

Figure 3b.   Sensor responses to 200 ppm of alcohol vapor, 200 ppm of alcohol vapor plus 200 ppm of propylene, 200 ppm of propylene, and saturated water vapor in air for PdAu/SnO$_x$ sensor. Sensor temperature   ~250 $^\circ$C.

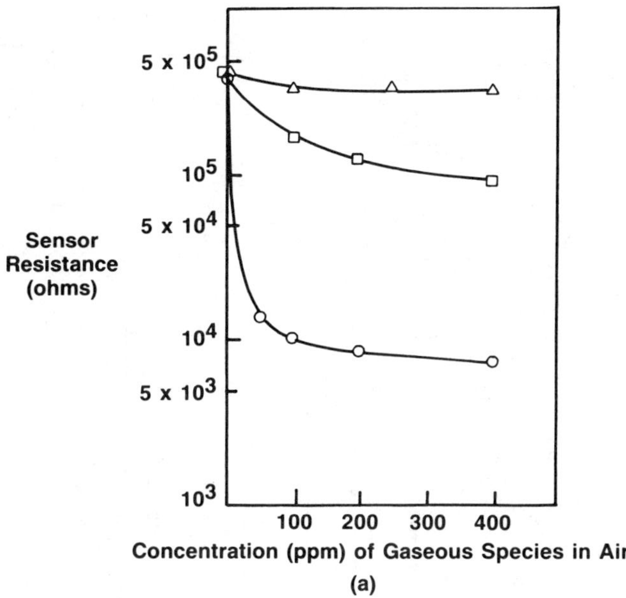

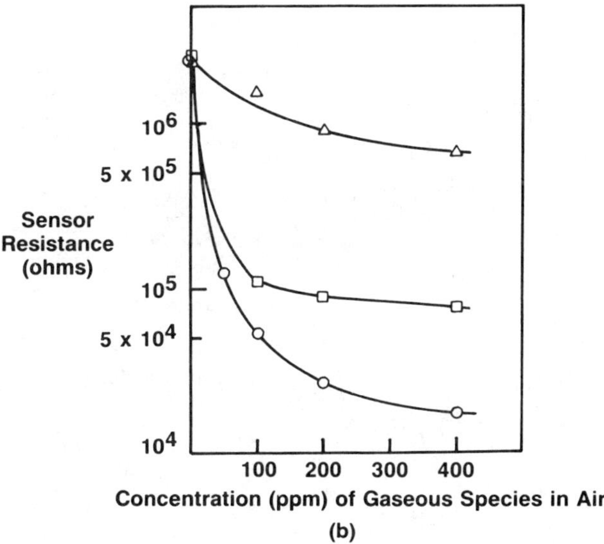

Figure 4.   Sensor resistance vs concentration (in ppm) of alcohol vapor (O), C_3H_6 (□) and CO (Δ):   (a) for SnO_x sensor, (b) for $PdAu/SnO_x$ sensor.   Sensor temperature ~250°C.

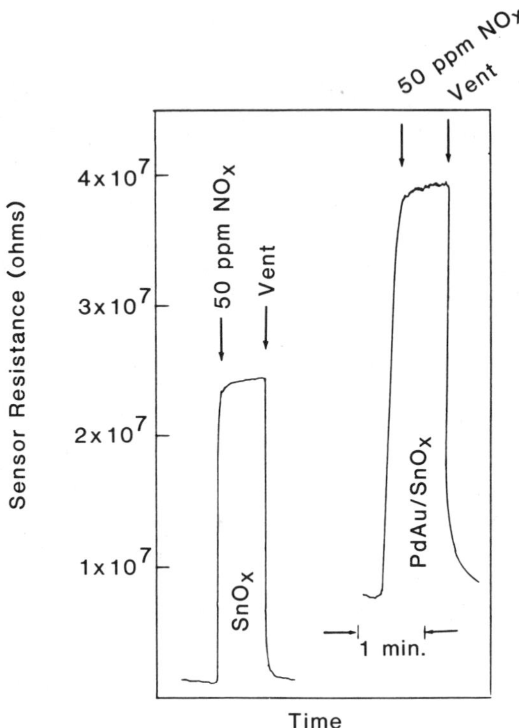

Figure 5.   Sensor responses to 50 ppm of NO_x in air for SnO_x and PdAu/SnO_x sensors.   Sensor temperature ~250°C.

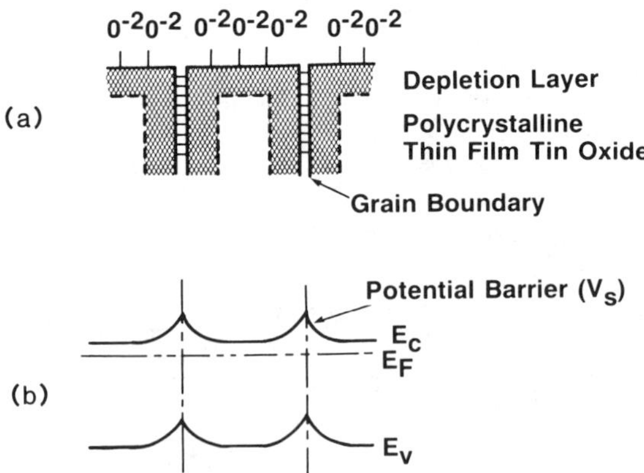

Figure 6.   (a) The formation of depletion layers in the surface and grain boundary regions due to oxygen chemisorption reduces the carrier concentration (n); (b) the formation of potential barriers at the grain boundaries reduces the carrier mobility (μ_H).   When alcohol vapor was introduced, both n and μ_H increased.

Step (2): C_2H_5OH (gas) → CH_3CHO + 2H*(ads)

C_2H_5OH is dissociatively chemisorbed on the SnO_x surface, producing active hydrogen H*.

Step (3):   2H* (ads) + O^{-2} (ads) → H_2O + $2e^-$

2H* reacts with O^{-2}, and the released electrons are injected back to the SnO_x conduction band, causing the depletion layer width to decrease. Hence, both the carrier concentration and carrier mobility increase.

Thus, the resistance of the SnO_x is higher in an ambient with higher oxygen concentration (or in an ambient containing strong oxidants, such as NO_x), (Step 1). The resistance of SnO_x is lowered by the presence of C_2H_5OH vapor, due to the reaction between H* and O^{-2}, which releases the localized electron back to the SnO_x conduction band (Step 3). The sensing mechanisms presented above are substantiated by the experimental results obtained from the Hall measurements (Table II). Table II gives the measured carrier concentration (n), and electron Hall mobility (μ_H) of a SnO_x sensor under different ambient conditions. As predicted, both electron concentration and electron mobility increase (or decrease) when alcohol vapor (or NO_x) is present in air. [In the Hall experiment, certain hydrocarbon gaseous species were released from the fiberglass/resin sample holder due to the sensor heating, causing the sensor conductivity in "air" to be unusually high as listed in the Table II].

The most likely effect of PdAu deposited on the $PdAu/SnO_x$ sensor surface is the promotion of the dissociative adsorption of C_2H_5OH (Step 2) due to the strong catalytic strength of Pd on hydrocarbon adsorbates. Hence, more active hydrogen species (H*) are created by Pd, and more localized electrons [O^{-2} (ads)] are released and injected back to the SnO_x conduction band (12,13).

Another possible effect of PdAu deposits on $PdAu/SnO_x$ sensors is through the formation of a Schottky barrier between PdAu and SnO_x, as in the case of the Pd/CdS hydrogen sensor. If such a barrier is formed, then a depletion layer is created inside the semiconductor tin oxide. Since the Pd work function can be reduced by hydrogen absorption through dipole or hydride formation (14,15), the width of the depletion layer in tin oxide may be reduced. The reduction of the depletion layer width causes the sample resistance to decrease. Such a possibility was checked and was ruled out, because a good ohmic contact was obtained between Pd (~50 nm thick) and SnO_x. It is also commonly known that gold forms an ohmic contact with tin oxide.

Recently, Yamazoe et al. observed an extremely high hydrogen sensitivity for $Ag-SnO_2$ sensors (16). They attributed this to the Fermi level of SnO_x being pinned at the redox potential of Ag^+/Ag° when the sensor was in air, and at the work function of Ag° (metallic) when the sensor was in air containing hydrogen. In our

Table II. Experimental Results Obtained from the Hall Measurements.

Sensor at 167°C

Ambient	Air	1000 ppm C_2H_5 OH in Air	2000 ppm C_2H_5 OH in Air	50 ppm NO_x in Air	100 ppm NO_x in Air
n (cm^{-3})	$3.6 \cdot 10^{18}$	$7.2 \cdot 10^{18}$	$8.15 \cdot 10^{18}$	$3.54 \cdot 10^{18}$	$2.5 \cdot 10^{18}$
u_H (cm^2/V.s)	3.14	4.9	5.5	2	1.54

Sensor at 230°C

Ambient	Air	1000 ppm C_2H_5 OH in Air	2000 ppm C_2H_5 OH In Air	50 ppm NO_x in Air	100 ppm NO_x in Air
n (cm^{-3})	$1 \cdot 10^{19}$	$2 \cdot 10^{19}$	$2.23 \cdot 10^{19}$	$2.4 \cdot 10^{18}$	$1.25 \cdot 10^{18}$
u_H (cm^2/V.s)	5.43	7	7.2	3	2.8

earlier study of $PdAu/SnO_x$ sensors (<u>12</u>), the observed chemical state of Pd (XPS data) was Pd° state (metallic) after the sensor was heat treated in air at 350°C. Hence, the electronic effect suggested by Yamazoe et al. for the $Ag-SnO_2$ sensor may not be applied to the $PdAu/SnO_x$ sensor discussed here.

Summary

We have demonstrated that tin oxide-based microsensors can be fabricated on silicon wafers using microelectronic processing tech-nologies. Alcohol vapor with a concentration of 50 ppm in air (which is about 1/10 of the alcohol concentration in the breath of a driver legally regarded as intoxicated) can be easily detected by the fabri-cated SnO_x and $PdAu/SnO_x$ sensors. The PdAu deposit on tin oxide thin films enhances alcohol vapor as well as propylene sensitivity appre-ciably. Such enhancement is mainly attributed to the strong cata-lytic effect of Pd on alcohol vapor and propylene. Although the SnO_x sensor showed reasonably high selectivity to alcohol vapor in the presence of water vapor, C_3H_6, CO, or CH_4, further improvement in alcohol selectivity is needed in order to make it a practicable and reliable alcohol detector. This is especially true with regard to the interference gases of hydrocarbons and CO which are the major constituents in cigarette smoke. One of the approaches to attain high selectivity and reliability is by devising a multi-sensor scheme in conjunction with logic/control circuitry to form an integrated sensing system. To achieve this, there are two prerequisites: A) a detailed (atomic scale) understanding of sensing mechanisms, B) an effective thermal-isolation scheme to minimize heat transfer to the surrounding environment. The benefit of such a scheme is two-fold: It reduces sensor power consumption and, at the same time, protects adjacent logic/control circuitry from extreme temperatures generated by the sensor heater.

Both sensors also showed high sensitivity to NO_x (NO + NO_2) content in air. NO_x with a concentration of 10 ppm in air is readily detected by both SnO_x and $PdAu/SnO_x$ sensors. While a PdAu deposit on tin oxide enhances alcohol vapor and propylene sensitivity, it depresses the sensitivity to NO_x.

Acknowledgments

We would like to thank Cheryl Puzio for the preparation of the mask set used in this work, John Biafora for the ion implantation doping of polysilicon, and W. Lange of the Analytical Chemistry Department for the SEM work.

Literature Cited

1. Seiyama, T.; Kato, A.; Fujiishi, K.; Nagatani, M. <u>Anal. Chem.</u>
    1962, 34, 1502.

2. Nitta, M.; Kanefusa, S.; Haradome, M. J. Electrochem. Soc.
   1978, 125, 1676.
3. Chang, S. C. IEEE Trans. Electron Devices 1979, ED-26, 1875.
4. Yamamoto, N.; Tonomura, S.; Matsuoka, T.; Tsubomura, H. Jap. J.
   Appl. Phys. 1981, 20, 721.
5. Heiland, G. Sensors and Actuators 1982, 2, 343.
6. Ko, W. H.; Fung, C. D.; Yu, D.; Yu, Y. H. Proc. of the Intl.
   Mtg. on Chemical Sensors, Seiyama, T.; Fueki, K.; Shiokawa, J.;
   Suzuki, S., Eds.; 1983, p. 496.
7. Clifford, P. K. Proc. of the Intl. Mtg. on Chemical Sensors,
   Seiyama, T.; Fueki, K.; Shiokawa, J.; Suzuki, S., Eds.; 1983,
   p. 153.
8. Chang, S. C. GM Research Publication GMR-4727, 1984.
9. Seiyama, T.; Yamazoe, N.; Futada, H. Denki Kagaku 1971, 21, 53.
10. Windischmann, H.; Mark, P. J. Electrochem. Soc. 1979, 126, 627.
11. Chang, S. C. Proc. of the Intl. Mtg. on Chemical Sensors,
    Seiyama, T.; Fueki, K.; Shiokawa, J.; Suzuki, S., Eds.; 1983,
    p. 78.
12. Chang, S. C. J. Vac. Sci. Technol. 1983, A1, 296.
13. Seiyama, T.; Futada, H.; Eva, F.; Yamazoe, N. Denki Kagaku
    1972, 40, 244.
14. Steele, M. C.; Hile, J. W.; MacIver, B. A. J. Appl. Phys. 1976,
    49, 2537.
15. Lundstrom, I. Sensors and Actuators 1981, 1, 403.
16. Yamazoe, N.; Kurokawa, Y.; Seiyama, T. Proc. of the Intl. Mtg.
    on Chemical Sensors, Seiyama, T.; Fueki, K.; Shiokawa, J.;
    Suzuki, S., Eds.; 1983, p. 35.

RECEIVED December 12, 1985

4

Oxygen Desorption and Conductivity Change of Palladium-Doped Tin(IV) Oxide Gas Sensor

Makoto Egashira, Masayo Nakashima, and Shohachi Kawasumi

Department of Materials Science and Engineering, Faculty of Engineering, Nagasaki University, 1-14 Bunkyo-machi, Nagasaki-shi, Japan 852

In relation to gas detection mechanism of the $Pd-SnO_2$ sensor, the adsorbed state of oxygen was studied by temperature programmed desorption (TPD) technique. Broadly, the TPD spectrum of oxygen comprised a small shoulder at 400 – 600 °C, a sharp peak at 600 – 700 °C, and an increase above 700 °C, though the spectrum was dependent upon various factors. The three species were assigned as oxygen adsorbates O^- or O^{2-} on SnO_2, oxygens as PdO, and a part of lattice oxygens of SnO_2, respectively. Conductivity of the sensor considerably increased by desorption of the adsorbates on SnO_2, but did not change so much by desorption of the two other species, indicating that consumption of the oxygen adsorbates on SnO_2 by inflammable gases was essential in the gas detection.

Electric conductivity of n-type semiconductive metal oxides such as tin(IV), zinc and iron(III) oxides is highly sensitive to the gaseous environment. The conductivity of them is low in normal air, but markedly increases when exposed to reducing gases. Seiyama et al. (1, 2) and Taguchi (3) first applied this phenomenon to detection of inflammable gases. Since then, many sensors have been widely used to monitor leakage of town gas or LPG. Most of them are based on the sintered polycrystalline tin oxide, and small amounts of noble metals like palladium and platinum are doped to enhance the gas sensitivity. However, mechanism of the gas detection and promotive functions of the noble metals are still obscure. Furthermore, there remain some important practical problems, including long-term stability, undesirable increases in conductivity and sensitivity especially under high temperature and high humidity, and improvement of gas selectivity. To answer these problems, it would be necessary to know about the nature of surface oxygens because it is believed that the oxygens play important roles in the gas detection (4, 5).

In the present work, factors influencing the adsorbed state of oxygen on the $Pd-SnO_2$ sensor have been investigated by the TPD technique. Relations between the TPD results and the conductivity changes of the sensor have also been examined.

Experimental

The sample used was 0.2 wt% palladium-doped tin oxide. It was prepared by impregnating a tin oxide hydrate gel, which was obtained by
hydrolysis of a solution of stannic chloride with aqueous ammonia,
with a palladium chloride solution, followed by calcination at 800°C
for 5 hr in air. Undoped tin oxide was also prepared. The sample
powder of about 50 mg was pressed into a disk-type sensor element of
4 mmϕ in diameter together with a couple of iridium-palladium alloy
coils as electrodes, as shown in Figure 1a. In the case of the
undoped sample, gold wire was employed as electrodes to avoid con-
tamination of the sample with iridium or palladium. The gold wire
was confirmed not to exert any influence on the adsorbent properties
of tin oxide. Electric conductance of the sensor was measured as
an output voltage across a load resistor connected with it in series,
as shown in Figure 1b. The circuit voltage was DC 2 V.

The apparatus for TPD experiments was essentially the same as
described elsewhere ($\underline{6}$, $\underline{7}$), except that the sample tube was designed
so that the conductivity change of the sample could be simultaneously
measured during the TPD run. More than ten sensor elements were used
at a time as the TPD sample, and one of them was attached on the
platinum lead terminals of the sample tube with gold paste. The
sample was first pretreated at 800°C in a stream of helium, then
cooled from a desired temperature to room temperature in a stream of
10% oxygen plus 90% helium to adsorb oxygen. If necessary, water
was adsorbed after or prior to the oxygen adsorption. Succeedingly,
the TPD run was carried out up to 800°C at a heating rate 10°C/min
in flowing helium by monitoring desorbed gases with two thermal
conductivity detectors of same sensitivity. A cold trap was placed
between the two detectors to obtain individual spectra of oxygen and
water. It was confirmed by gas chromatography and mass spectroscopy
that the desorbed species were oxygen and water alone.

Oxygen Adsorbates

Oxygen can be adsorbed in various forms on the surface of Pd-doped
SnO_2. Yamazoe et al. ($\underline{8}$) observed four kinds of oxygen species on
pure SnO_2, which were desorbed around 80°C, 150°C, 560°C, and above
600°C in the TPD experiments, and assigned the four species to O_2
(α_1), O_2^- (α_2), O^- or O^{2-} (β), and a part of lattice oxygen (γ),
respectively, based on electric conductivity and ESR measurements.
For Pd-SnO_2, an additional desorption peak due to thermal decomposition of PdO was reported at about 650°C ($\underline{9}$). Among the species, α_1
and α_2 may be insignificant in the gas detection, because they are
formed by adsorption only at lower temperatures than 200°C, and
because the practical sensors are operated usually at 300–400°C.
It may be said, therefore, that we have only to consider three other
species; β, γ, and oxygens as PdO.

First, how the population of these species on the Pd-SnO_2 surface varied with adsorption temperature was examined between 400
and 800°C in detail. Figure 2 shows a TPD spectrum of oxygen, which
was preadsorbed by cooling from 400°C to room temperature in the
presence of oxygen after pretreatment at 800°C in helium. For comparison, a result from the undoped oxide is also given, where oxygen
was preadsorbed from 800°C. The ordinate is the detector response

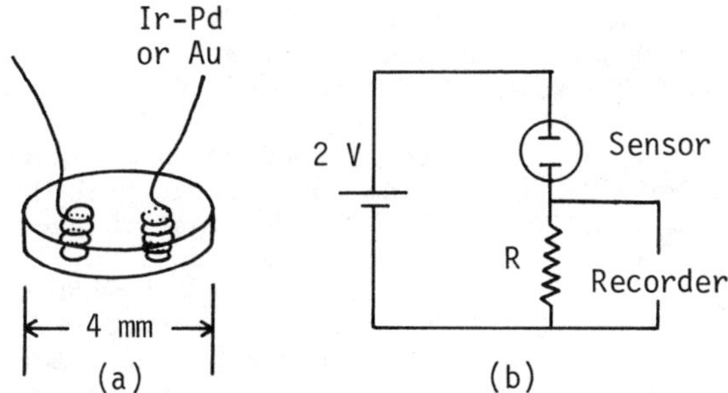

Figure 1. A sensor element and a circuit for conductivity measurement.

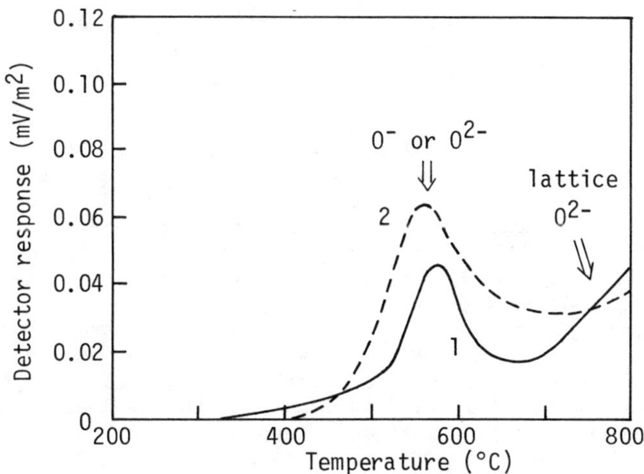

Figure 2. Comparison of the TPD spectra of oxygen. (1) Pd–SnO$_2$, O$_2$ ads. by 400 °C→R. T., (2) SnO$_2$, O$_2$ ads. by 800 °C→R. T.

mV per unit surface area of the sample. It is seen that the spectrum of Pd–SnO$_2$ is composed of a peak having its maximum at 580 °C and an increase above 700 °C. The spectrum well resembles that of the undoped oxide. In accordance with the report by Yamazoe et al. (8), the lower temperature peak may be assigned to adsorbed oxygen species O$^-$ or O^{2-} (β), and the increase above 700 °C to lattice oxygen (γ), though the desorption temperatures are somewhat higher in the present results. Thus, when oxygen was preadsorbed on Pd–SnO$_2$ by cooling from a temperature of 300 – 400 °C to room temperature, only oxygen species bonded to SnO$_2$ were observed; in other words, palladium did not exert any significant effect on the oxygen adsorptive property under these conditions.

However, if oxygen was preadsorbed from a higher temperature, for example, from 550 °C, an additional large peak was observed as shown by spectrum 2 of Figure 3. The response in the lower temperature region also increased, compared with spectrum 1 which was the same one as Figure 2. This large peak is considered to arise from thermal decomposition of PdO (9) which was formed during the adsorption procedure. In contrast, when oxygen was preadsorbed by the 800 °C to room temperature cooling, the lattice oxygen species markedly increased, whereas the two lower temprerature species considerably decreased. The reason for the appearance of two maxima in the region of PdO decomposition is not yet clear. Anyway, it is obvious that the desorption spectrum of oxygen is strongly dependent upon its adsorption temperature. Broadly, the low temperature adsorption gives rise to the appearance of β- and γ-oxygens, while oxygens as PdO are formed only by the high temperature adsorption.

<u>Variation of the TPD Spectrum of Oxygen with Repetition of Heating and Cooling</u>

Commercial gas sensors based on Pd–SnO$_2$ are usually kept at 300 – 400 °C in practical use in normal air. When inflammable gases are introduced, however, the temperature goes up to 500 – 600 °C, due to the increased Joule heating resulting from the conductance increase (a circuit voltage 100 V is usually applied). In this connection, it is interesting how the oxygen adsorptive property varies with the heating and cooling history.

Figure 4 shows the variation of the TPD spectrum when the TPD run was repeated. That is, the cooling was the oxygen adsorption procedure from 800 °C to room temperature in a stream of 10 % oxygen, and the heating the TPD run from room temperature to 800 °C in helium. The 1st spectrum gave a large and sharp peak at 650 °C, as well as a weak shoulder at 400 – 600 °C and a relatively large increase above 700 °C. When the TPD runs were repeated, however, the sharp peak at 650 °C slightly decreased, and instead another sharp peak appeared at 675 °C. As a result, the amount of oxygen molecules desorbed at 600 – 710 °C increased from 4.10 μmol/g of the 1st run to 5.56 μmol/g of the 7th run. In this sample, palladium atoms of 18.8 μmol/g are contained. If the all atoms exist as PdO, then its decomposition should release oxygen molecules of 9.4 μmol/g. Thus, it is seen that approximately a half of palladium atoms have combined with oxygen atoms. On the other hand, the amount of the 400 – 600 °C species slightly increased from 2.34 μmol/g of the 1st run to 2.52 μmol/g of the 7th run, while that of the lattice oxygens above 700 °C was almost constant.

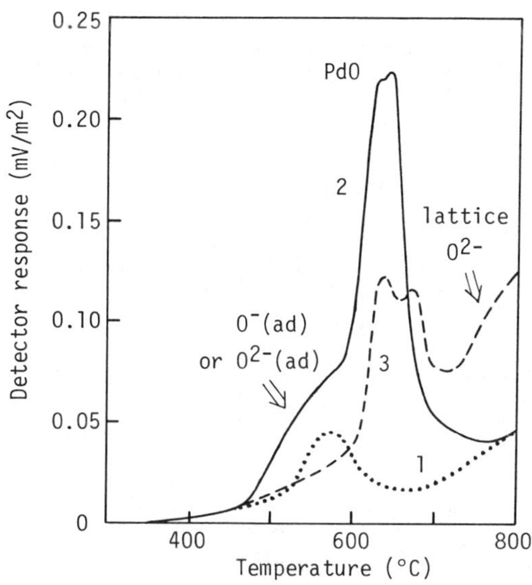

Figure 3. Variation of the TPD spectra of oxygen on Pd–SnO$_2$ with the adsorption temperature. (1) 400 °C→R. T., (2) 550 °C→R.T., (3) 800 °C→R. T.

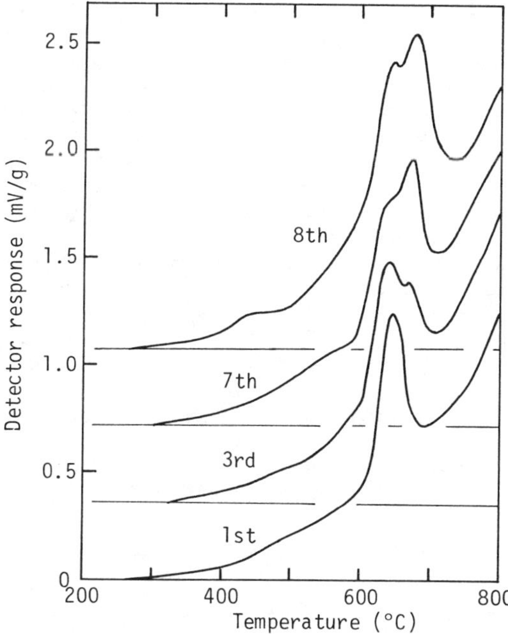

Figure 4. Change of the TPD spectrum of oxygen on Pd–SnO$_2$ with repetition of the TPD run. O$_2$ ads. by 800 °C→R. T. 1st – 7th runs: without H$_2$O(ad), 8th run: with H$_2$O(ad).

Although the reason for these variations is not clear, it is apparent that the oxygen adsorptive property of the Pd-SnO$_2$ sensor is somewhat changed by the heating and cooling cycle.

Influence of Water Adsorption

It is known that the nature of oxygen adsorbates on SnO$_2$ and on Ag-SnO$_2$ are considerably modified by coexistence of water adsorbates (7, 10, 11). Since the practical sensors are usually exposed to an atmosphere containing moisture under their operating conditions, it would be necessary to know how water influences the adsorbed state of oxygen on the present sample.

The 8th spectrum of Figure 4 is the result when water was adsorbed at room temperature on the oxygen-preadsorbed surface. From comparison with the 7th spectrum, it is noted that water has induced the appearance of a shoulder-like peak at about 450 °C. The peak was very weak, but reproducible. This is considered to arise from the rearrangement of oxygen adsorbates which desorb at higher temperatures if water is absent. A similar result has been observed on a commercial sensor Figaro TGS #109 (11). And also, the result was analogous to the water-induced rearrangement on Ag-SnO$_2$ (10, 11), in a sense that the oxygens were transformed into more weakly bonded species. Based on the analogy, it may be proposed that the rearrangement is due to a change of the adsorption mode of oxygen from the bridge-type coordination on Sn^{4+} and Pd^{2+} ions at the boundary of SnO$_2$ and PdO particles to the monodentate coordination on Sn^{4+} or Pd^{2+} through hydroxylation of either one of the two cations.

The above result was obtained on the sample on which oxygen was preadsorbed by the 800 °C to room temperature cooling. When water was adsorbed on the surface on which oxygen was preadsorbed by the 400 °C to room temperature cooling, however, very different variation of the TPD spectrum was observed as shown in Figure 5. Spectrum 1 is the normal TPD result in the absence of water adsorbates. The spectrum changed into spectrum 2 when water was adsorbed at room temperature. In contrast with the 8th run of Figure 4, it is seen that the oxygens are transformed into more strongly bonded species. This is essentially the same result as on the undoped oxide (7), where the rearrangement has been supposed due to the change of O^{2-} adsorbates into surface hydroxyls by a reaction of them with water molecules, because the desorption of oxygen was accompanied by concurrent desorption of water in a ratio H$_2$O/O$_2 \simeq 2$.

Conversely, if water was first preadsorbed at 400 °C and then oxygen was adsorbed by the 400 °C to room temperature cooling, the TPD spectrum of oxygen markedly increased as shown by spectrum 3 in Figure 5. The result undoubtedly indicates that water has increased somehow the adsorption sites for oxygen. The mechanism is not clear at present, but this influence of water is quite important when discussing the nature of oxygen adsorbates.

Relation between Oxygen Desorption and Conductivity Change

As described above, oxygen adsorption on Pd-SnO$_2$ occurred broadly in three different forms, O$^-$ or O^{2-} on SnO$_2$ (β), lattice oxygens (γ), and oxygens as PdO, though the population of them was dependent upon some factors. Since the gas-sensing property of Pd-SnO$_2$ is considered intimately related with the existence of oxygen adsorbates, it is very intriguing how the oxygens influence the conductivity of the sensor.

First, correlations were examined between the total amount of the oxygens, which desorbed up to 800 °C, and the conductivity (or resistivity). Figure 6 shows variation of the resistance at 250 °C in an atmosphere of 10 % oxygen plus 90 % helium, as well as of the amount of the oxygens, when the TPD run was repeated. As mentioned before, the amount of the oxygens, which was calculated by integrating the TPD spectrum of Figure 4 up to 800 °C, gradually increased with repetition of the TPD run. At the same time, the resistance increased. Thus, the more the oxygen adsorbates, the higher the resistance. This may be simply because donor electrons in the SnO_2 crystal decreased as a result of electron transfer to the adsorbed oxygen atoms.

A similar correlation was obtained also for the experiments of Figure 3, where oxygen adsorption temperature was varied. The sensor resistance at 300 °C is shown as a function of the adsorption temperature in Figure 7. The resistance increased with increasing adsorption temperature, and then decreased via a maximum. The variation corresponds roughly well with that of the total amount of surface oxygens.

Conductivity change due to the rearrangement of oxygen adsorbates by water was also measured during the experiments of Figure 5. However, it was not able to distinguish the change due to the rearrangement from the contribution of water adsorbates themselves.

In Figure 7, two maximum points of the sensor resistance and the amount of the oxygens do not agree with each other. This probably indicates that individual oxygen species has a different magnitude of contribution to the sensor conductivity. Taking this indication into account, the following experiments were performed: that is, a part of oxygen adsorbates on $Pd-SnO_2$, which were pre-adsorbed by the 550 °C to room temperature cooling, were desorbed little by little from the lower temperature species by heat-treating at a given temperature prior to the TPD run. The variation of the TPD spectrum of oxygen is shown in Figure 8. Thus varying the species or the amount of preadsorbed oxygens, electric resistance of the sensor was measured at a constant temperature 200 °C in 10 % oxygen to distinguish the contribution of each species.

Figure 9 shows the results as a function of the amount of the oxygens remained on the surface. The uppermost plot represents the initial oxygenated surface. When only a very small amount of the oxygens was desorbed from the initial state, the resistance drastically decreased about one order of magnitude. This region corresponds to the desorption of β-oxygens (O^- or O^{2-}) on SnO_2. In the succeeding region of PdO decomposition, on the other hand, the resistance did not change so much, presenting a striking contrast to the first region. If oxygen was further desorbed, that is if γ-oxygens (a part of lattice oxygens) were desorbed, the resistance considerably decreased again, but to a less extent than the initial decrease.

Similar results were obtained on the $Pd-SnO_2$ sample on which oxygen was preadsorbed by the 800 °C to room temperature cooling, and also on the undoped SnO_2. On the latter sample, the region of PdO decomposition was not observed as a natural result, but again the decrease of the resistance was more striking in the region of the adsorbed oxygens than the lattice oxygens, as shown in Figure 10.

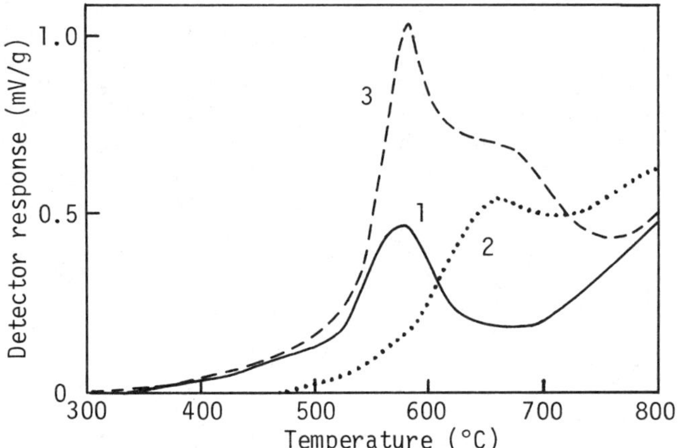

Figure 5.   Influence of water adsorption on the TPD spectrum of oxygen on Pd–SnO$_2$.   O$_2$ ads. by 400 °C → R. T.   (1) without H$_2$O(ad), (2) with H$_2$O(ad), (3) O$_2$ ads. after H$_2$O ads.

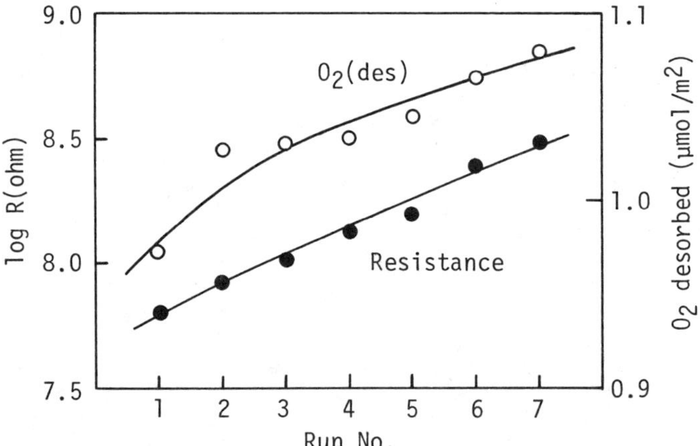

Figure 6.   Variations of resistance of a Pd–SnO$_2$ sensor at 250 °C in 10 % O$_2$ and of the amount of oxygen desorbed with repetition of the TPD run.

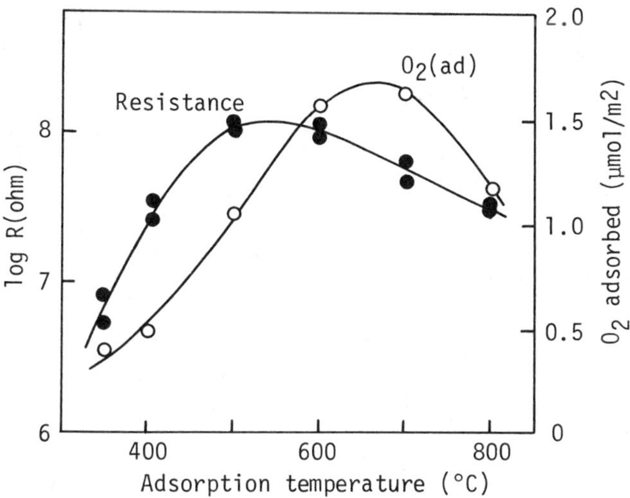

Figure 7.   Resistance of a Pd–SnO$_2$ sensor at 300 °C in 10 % O$_2$ and the amount of oxygen adsorbed as a function of adsorption temperature.

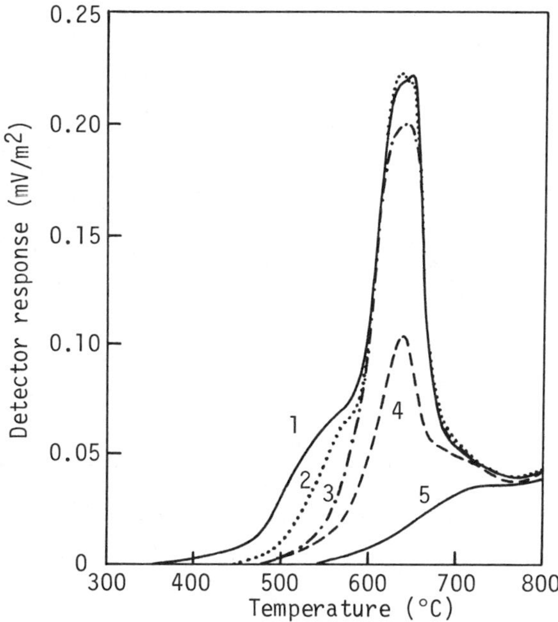

Figure 8.   Variation of the TPD spectrum of oxygen which was preadsorbed on Pd–SnO$_2$ by 550 °C → R. T. and then heat-treated at a given temperature before the TPD run.   (1) R. T., (2) 500 °C, (3) 550 °C, (4) 600 °C, (5) 650 °C.

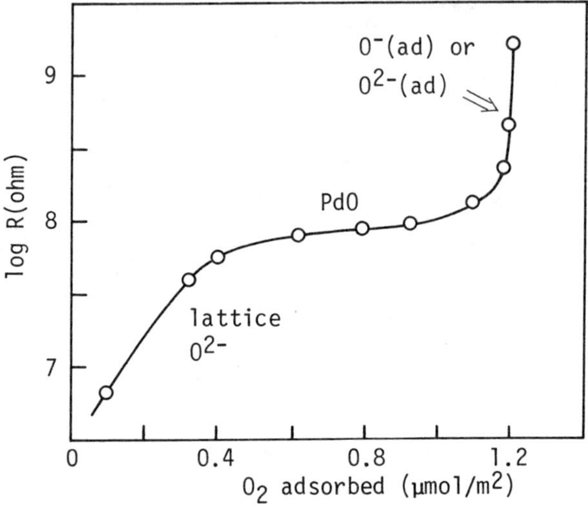

Figure 9.   Relation between resistance of a Pd–SnO$_2$ sensor at 200
°C and the amount of oxygen adsorbed.   Preadsorption of O$_2$ was by
550 °C → R. T.

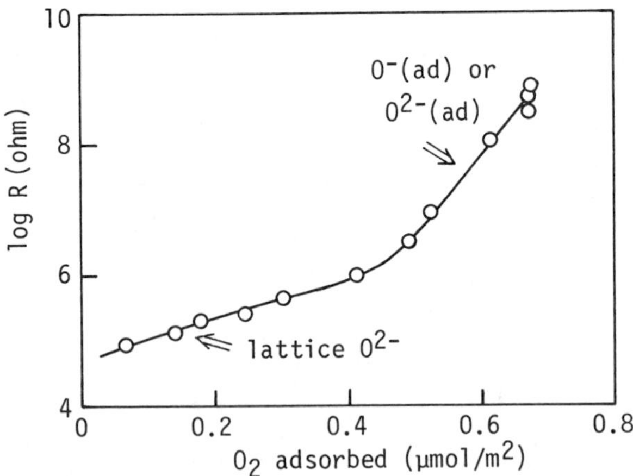

Figure 10.   Relation between resistance of an SnO$_2$ sensor at 200
°C and the amount of oxygen adsorbed.   Preadsorption of O$_2$ was by
800 °C → R. T.

It is generally accepted that surface electronic properties play an important role in gas detection with sintered n-type oxide sensors (4, 5). In normal air, a positively charged space layer is built up near the surface as a result of electron transfer from the oxide to chemisorbed oxygens. This causes a high potential barrier to electronic conduction through grain boundaries. Irrespective of the exact mechanism of the influence of oxygen adsorbates on the electric conduction, it is obvious from the above mentioned results that the β-oxygens on SnO_2 are most responsible for the buildup of the potential barrier, and that the oxygens bound to Pd do not exert any direct contribution to the barrier. Thus, it may be essential to gas detection that the negatively charged oxygen adsorbates are consumed by inflammable gases. As a result, the barrier decreases due to the increased carrier density in the surface layer, so that the conductivity markedly increases. Pd or PdO may play only a promotive function in the oxygen adsorption (spillover effect) and/or in the reaction between the oxygens and the gases (catalytic effect) (4, 12).

Conclusion

From TPD experiments combined with conductivity measurement, the following points have been revealed:
1. Oxygen adsorbates O^- or O^{2-} on SnO_2 play an important role in the inflammable gas detection.
2. Oxygens bound to Pd are not significant in the conductivity changes.
3. The adsorbed state of oxygen is strongly dependent upon various factors; adsorption temperature, heating and cooling history, and coexistence of water adsorbates.

Since the oxygen adsorbates are thus most important in the gas detection, it may be said that we have to focus our attention on the oxygen adsorptive properties of the sensor materials, in discussing mechanism of gas detection or in solving the practical problems mentioned in introduction.

Acknowledgments

This work was supported by a Grant-in-Aid for Scientific Research from the Ministry of Education of Japan and also by Figaro Engineering Inc.

Literature Cited

1. Seiyama, T.; Kato, A.; Fujiishi, K.; Nagatani, M. Anal. Chem. 1962, 34, 102.
2. Seiyama, T.; Kagawa, S. Anal. Chem. 1966, 38, 1069.
3. Taguchi, N. Japanese Patent 45-38200, 1962.
4. Yamazoe, N.; Kurokawa, Y.; Seiyama, T. Proc. Intern. Mtg. Chem. Sensors, 1983, p. 35.
5. Ihokura, K. New Mater. & New Process. 1981, 1, 43.
6. Egashira, M.; Kawasumi, S.; Kagawa, S.; Seiyama, T. Bull. Chem. Soc. Japan 1978, 51, 3144.
7. Egashira, M.; Nakashima, M.; Kawasumi, S.; Seiyama, T. J. Phys. Chem. 1981, 85, 4125.

8. Yamazoe, N.; Fuchigami, J.; Kishikawa, M.; Seiyama, T. <u>Surf. Sci.</u> 1979, 86, 335.
9. Wada, K.; Yamazoe, N.; Seiyama, T. <u>Nippon Kagaku Kaishi</u> 1980, 1597.
10. Egashira, M.; Nakashima, M.; Kawasumi, S. <u>J. Chem. Soc. Chem. Comm.</u> 1981, 1047.
11. Egashira, M.; Nakashima, M.; Kawasumi, S. <u>Proc. Intern. Mtg. Chem. Sensors</u>, 1983, p. 41.
12. Yamazoe, N.; Kurokawa, Y.; Seiyama, T. <u>Sensors & Actuators</u> 1983, 4, 283.

RECEIVED October 31, 1985

Oxygen Sensor Using Perovskite-Type Oxides
Measurements of Electrical Characteristics

Yasuhiro Shimizu, Yoshiki Fukuyama, Hiromichi Arai, and Tetsuro Seiyama

Department of Materials Science and Technology, Graduate School of Engineering Sciences, Kyushu University 39, Kasuga, Fukuoka 816, Japan

The perovskite-type oxides (ABO_3) were studied from the viewpoints of application for oxygen sensors. Among the perovskite-type oxides examined, $SrTiO_3$ showed a high sensitivity to oxygen in the temperature range from 550 to 800 °C in the "lean-burn" region (10^2 Pa < P_{O_2} < 10^4 Pa). The conductivity of $SrTiO_3$ was found to be proportional to $P_{O_2}^{1/4}$ at P_{O_2} > 10^2 Pa and showed a p-type semiconductive nature. As for $SrSnO_3$, no significant changes in the conductivity were observed in the "lean-burn" region, since the change of the conduction mechanism from p-type to n-type occurred in this region. The resistivity characteristics of the specimens were also investigated in the exhaust gas of propane-oxygen combustion. $SrSnO_3$ was promising for a combustion monitoring sensor, judging from the magnitude of the decrease in the resistivity at $\lambda = 1$ with decreasing λ and good reproducibility of resistivity characteristics (λ ; oxygen excess ratio). On the other hand, the magnitude of the decrease in the resistivity of $SrTiO_3$ at $\lambda = 1$ was smaller than that of $SrSnO_3$, because the conduction mechanism of $SrTiO_3$ changed from p-type to n-type at $P_{O_2} = 10^{-2}$ Pa. The effects of B site partial substitution of $SrTiO_3$ on the oxygen partial pressure dependence of the conductivity were also investigated.

The combustion of hydrocarbon fuels is widely used to obtain useful energy in various industries. As for an engine or a furnace, it is necessary to control the air to fuel ratio from the viewpoints of fuel conservation and pollution control ([1]). Oxygen sensors are widely used to detect an oxygen partial pressure in the exhaust gas ([1-6]). Among oxygen sensors, some are useful to monitor and control the stoichiometric air to fuel ratio as required for pollution control ([1-6]), and others are capable of monitoring the "lean-burn" region as required for enhancement of an energy efficiency ([3]). These oxygen sensors are classified into two types, i.e., solid electrolyte type such as calcia-stabilized zirconia and yitria-stablized zirconia ($\underline{7}$-$\underline{10}$), and semiconductor type such as titania ($\underline{1}$, $\underline{5}$, $\underline{6}$) and mixed metal oxides of CoO and MgO ($\underline{3}$). In the past several years, there have been increasing interests in the semiconductive type oxygen sensor, because of small size, simple structure and slight cost. This type of sensor utilizes electrical conductivity changes due to oxygen adsorption or desorption. The conductivity is proportional to oxygen partial pressure as the following equation.

$$\sigma = \sigma_{\circ} P_{O_2}^{1/m} \tag{1}$$

The m value depends on the semiconductive nature of metal oxides in the surrounding oxygen partial pressure. For example, m is 4 or 6 for p-type semiconductors and -4 or -6 for n-type semiconductors ($\underline{11}$).

Modification of redox properties

In order to use semiconductive metal oxides as an oxygen sensor, both thermal stability at elevated temperatures and atmospheric stability under reductive environments are required for reproducibility and accuracy of the sensor.

It is well known that redox properties of oxides can be modified by the formation of mixed oxides ($\underline{12}$). Effect of mixed oxide formation on $\Delta G°$ of M-O dissociation is illustrated in Figure 1. The free energy changes of the following reactions, $\Delta G_A°$ and $\Delta G_R°$, are calculated from thermodynamic data ($\underline{13}$, $\underline{14}$).

$$AO_n \longrightarrow AO_{n-1} + 1/2\ O_2 \qquad\qquad \Delta G_A° \tag{2}$$
$$ABO_{n+m} \longrightarrow AO_{n-1} + BO_m + 1/2\ O_2 \qquad \Delta G_R° \tag{3}$$

where component oxide, AO_n, is more reducible than BO_m. Figure 1 shows that $G_A°$ of CoO, ZnO, and TiO_2 alone are usually smaller than $\Delta G_R°$ of mixed oxides derived from the component oxides. The dotted lines, a), b), c), and d) are $-\Delta G_R°$ for these reactions

$$CO + 1/2\ O_2 \longrightarrow CO_2 \tag{a}$$
$$H_2 + 1/2\ O_2 \longrightarrow H_2O \tag{b}$$
$$1/13\ C_4H_{10} + 1/2\ O_2 \longrightarrow 4/13\ CO_2 + 5/13\ H_2O \tag{c}$$
$$1/10\ C_3H_8 + 1/2\ O_2 \longrightarrow 3/10\ CO_2 + 2/5\ H_2O \tag{d}$$

which are some possible oxidation reactions that may occur in an engine or a furnace. Among these oxides, Nb_2O_3, TiO_2, and mixed oxides of CoO and MgO are now studied for a semiconductive oxygen sensor. CoO is the most reducible among them and it is not suitable for an oxygen sensor to detect the air to fuel ratio due to a low stability toward reduction with hydrogen or hydrocarbons. However, as shown in Figure 1, the stability will be improved by means of mixing with other oxides. In fact, a mixed oxide of CoO and MgO was reported as an oxygen sensor to improve a stability toward reduction in the "rich-burn" region where oxygen partial pressure is usually below 10^{-15} Pa ($\underline{3}$). Exactly speaking, the above calculation method might not be applicable to actual oxides, since we do not know what kind of composition will be produced under reductive environments. However, if we deal with this problem first, the results calculated

from this method will best meet our needs.   These considerations
prompted us to investigate the oxygen sensor that consisted of mixed
oxides. The perovskite-type oxides were chosen because of the following reasons : 1) The electrical properties of these oxides can be
modified easily by selecting an appropriate combination of the
cation constituents. 2) They are stable under reductive environments
at high temperatures ($\underline{15}$).

Experimental

Perovskite-type oxides were prepared by calcining the mixtures of
TiO_2, SnO_2, Al_2O_3 and alkaline earth carbonates in a desired proportion at 800 - 1000 °C for 2 - 5 h ($\underline{16}$).   The calcined powders
were again ground with a ball mill and then pressed into discs of 10
mm in diameter and 1 mm in thickness.   The disc was covered with
powder of the same composition and sintered at 1200 °C for 6 h in an
alumina crucible.   Some of the discs were crushed, and then the
powder was identified by X-ray diffraction with filtered Cu kα radiation.   The surface area of the specimen was measured by BET method
using the nitrogen adsorption isotherms.   The Pt paste was applied
on both sides of these discs and was fired at 900 °C for 10 min for
conductivity measurements, while the Pt paste was applied on the same
side for thermoelectromotive coefficient measurements.   Temperature
differences, ΔT, between the two ends of the discs were established
with a small heater near one end of the disc.   The semiconductive
nature of the specimens was determined by the thermoelectromotive
coefficient, i.e., negative value means n-type semiconductor and
positive one means p-type.
    Figure 2 shows the gas-flow diagram for measuring conductivity
characteristics of the specimens.   The specimen was mounted in a
quartz vessel located in an electrical furnace. The D.C. conductivity of the specimen was measured using a constant voltage supply and
an electronic picoammeter in the temperature range from 400 °C to
800 °C under oxygen partial pressure between 10^2 and 10^5 Pa. The
oxygen partial pressure between 10^2 and 10^5 Pa was established by a
continuous flow of a mixed gas of nitrogen and oxygen of a total
pressure of 10^5 Pa.   An oxygen pump using a calcia-stabilized
zirconia was adopted to obtain an oxygen partial pressure between
10^{-10} to 10^2 Pa.   The resistivity characteristics of the specimen
were also investigated at 700 °C in the exhaust gas of propane-oxygen
combustion. Propane and oxygen were mixed at an appropriate ratio,
and then a flow rate of 200 ml/min with a total pressure of 10^5 Pa
was established by balancing with nitrogen. The mixture gas was
burned at 400 °C over a Pt/Al_2O_3 catalyst and the exhaust gas was
introduced into the quartz vessel in which the specimen had been
installed.   Water vapor originating from propane-oxygen combustion
was trapped with dry-ice and ethanol mixture before reaching the test
chamber. A calcia-stabilized zirconia oxygen sensor was installed in
the test chamber to obtain an actual equilibrium oxygen partial pressure.   A stoichiometric point is defined as $\lambda = 1$.   At this point
there is just enough oxygen to convert all of propane to CO_2 and H_2O.

Results and discussion

In order to find a material well suited to application for a "lean-

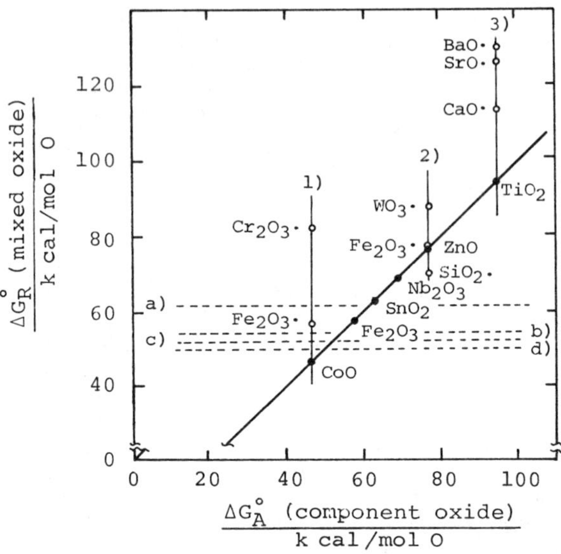

Figure 1.   Effect of mixed oxide formation on ΔG° of M-O
dissociation (298 K).   Lines 1)-3) show the mixed oxide
systems based on CoO, ZnO, and TiO_2, respectively.
a) - d) are - ΔG°_R for the following reactions.
    a) $CO + 1/2\ O_2 \rightarrow CO_2$
    b) $H_2 + 1/2\ O_2 \rightarrow H_2O$
    c) $1/13\ C_4H_{10} + 1/2\ O_2 \rightarrow 4/13\ CO_2 + 5/13\ H_2O$
    d) $1/10\ C_3H_8 + 1/2\ O_2 \rightarrow 3/10\ CO_2 + 2/5\ H_2O$

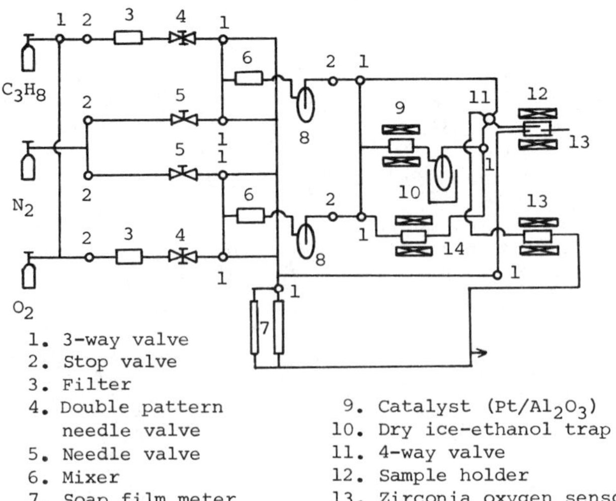

Figure 2.   Gas-flow diagram for measuring sensor characteristics.

burn" sensor,   the D.C. conductivity of the specimen was measured in the temperature range from 400 °C to 800 °C under oxygen partial pressure between 10^2 and 10^5 Pa.   Characteristics of specimens used for electrical measurements are listed in Table I (17).   Among the perovskite-type oxides examined, $SrTiO_3$ showed a high sensitivity to oxygen as shown in Figure 3.   Since the conductivity of $SrTiO_3$ was found to be proportional to $P_{O_2}^{1/4}$ above 10^2 Pa of oxygen partial pressure in the temperature range from 550 °C to 800 °C, the electrical conductivity of $SrTiO_3$ showed a p-type semiconductive nature under these conditions (18),   as also expected from the result of thermoelectromotive coefficient.   On the other hand,   as for $SrSnO_3$, no significant changes in the conductivity were observed in the oxygen partial pressure between 10^2 and 10^5 Pa as shown in Figure 4, while thermoelectromotive coefficient of this specimen indicated a p-type semiconductor at 2.1×10^4 Pa.   These results suggest that the change of the conduction mechanism of the specimen from p-type to n-type occurs in the oxygen partial pressure from 10^2 to 10^5 Pa. The change of the conduction mechanism was also observed for an undoped specimen of $BaTiO_3$ (19, 20),   shown in Figure 5.   The effect of A site partial substitution on the oxygen pressure dependence of the conductivity was examined for this specimen.   Plotting the logarithm of the conductivity against the logarithm of P_{O_2} as   shown in Figure 6,   a slope of approximately 1/5 was found in $Ba_{0.97}Na_{0.03}TiO_3$ specimen,   a 3 % sodium substituted compound derived from $BaTiO_3$. This result suggests that the increase of positive hole concentration by partially   substituting a univalent Na^+ for a divalent Ba^{2+} results in the enhancement of a p-type conduction mechanism.

The resistivity characteristics of the specimens were also investigated in the exhaust gas of propane-oxygen combustion. As shown in Figure 7,   the resistivity of all specimens decreased dramatically at the stoichiometric point of combustion with decreasing λ.   From these results,   we can classify these specimens into   the following three groups.   1) $SrTiO_3$, $CaTiO_3$, and $Ba_{0.97}Na_{0.03}TiO_3$; the resistivity increased with decreasing λ at first,   decreased abruptly at λ = 1, and then decreased with decreasing λ.   The change in the resistivity at λ = 1 was smaller than that of other specimens. 2) $SrSnO_3$, $BaTiO_3$, and $CaSnO_3$; the resistivity increased slightly with decreasing λ down to λ = 1, however, the magnitude of the change in the resistivity above λ = 1 was small compared with that of the first group, and then the resistivity decreased dramatically at λ = 1.   3) $BaSnO_3$ and $Sr_{0.9}La_{0.1}SnO_3$; the resistivity decreased slightly with decreasing λ and then decreased dramatically at λ = 1.   These resistivity characteristics depend on the semiconductive nature of the specimens.   As stated previously, $SrTiO_3$ showed the p-type conductivity in the oxygen partial pressure of 10^2 to 10^5 Pa and the change of the conduction mechanism into n-type occurred below 10^2 Pa. This phenomenon results in a slight decrease in the resistivity at λ = 1. On the other hand, this change for $SrSnO_3$ and $BaTiO_3$ was found in the oxygen partial pressure range from 10^2 to 10^5 Pa.   The enhancement of p-type conduction mechanism of $Ba_{0.97}Na_{0.03}TiO_3$ by the partial substitution of a univalent Na^+ for a divalent Ba^{2+} in the oxygen partial pressure region from 10^2 to 10^5 Pa resulted in a slight decrease in the resistivity at λ = 1 compared with an undoped $BaTiO_3$. In the case of $BaSnO_3$ and $Sr_{0.9}La_{0.1}SnO_3$, the n-type conduction mechanism was found to prevail even in a high oxygen partial

Table I.   Characteristics of sintered specimens used for electrical measurements

Specimen	Thermoelectromotive coefficient in air		Semiconductive type[a]	Relative dencity	Surface area	Conductivity[a]	m value[a]
	$d\theta/dT$ (mV·K^{-1})	T/K		%	m^2·g^{-1}	S·cm^{-1}	
$SrSnO_3$	0.41	983	p	69	2.7	9.1×10^{-8}[b]	—
$SrTiO_3$	0.47	833	p	70	1.0	1.8×10^{-4}[c]	4.2[c]
$Sr_{0.9}La_{0.1}SnO_3$	− 0.20	773	n	63	0.6	—	—
$BaSnO_3$	− 0.08	745	n	65	5.9	2.4×10^{-3}	− 5.1
$BaTiO_3$	0.50	953	p	57	0.4	4.6×10^{-6}	—
$Ba_{0.97}Na_{0.03}TiO_3$	0.32	743	p	63	0.9	8.0×10^{-6}	4.8
$CaSnO_3$	0.65	973	p	54	1.8	—	—
$CaTiO_3$	0.21	873	p	55	1.5	6.9×10^{-6}	4.4

a) At 973 K in air.   b) At 995 K.   c) At 873 K.

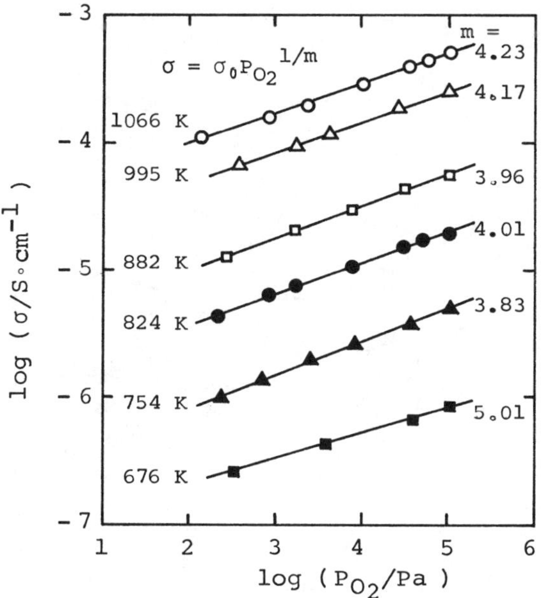

Figure 3. Electrical conductivity of $SrTiO_3$ as a function of oxygen partial pressure.
"Reproduced with permission from Ref. 17. Copyright 1985, 'The Chemical Society of Japan'."

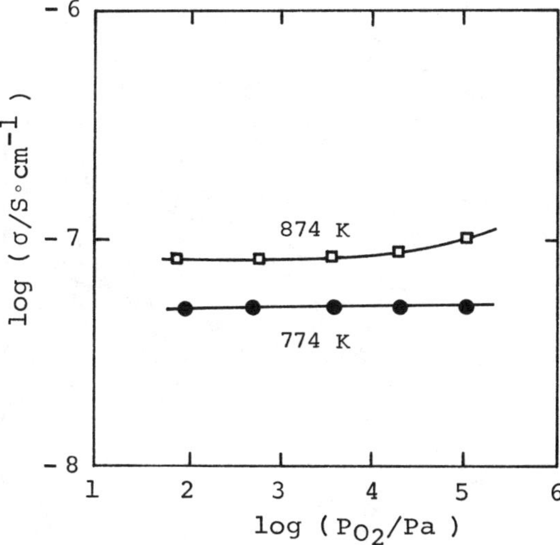

Figure 4. Electrical conductivity of $SrSnO_3$ as a function of oxygen partial pressure.
"Reproduced with permission from Ref. 17. Copyright 1985, 'The Chemical Society of Japan'."

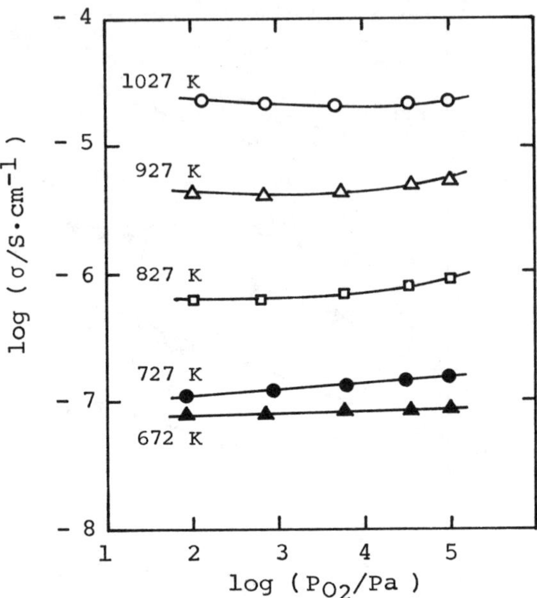

Figure 5.   Electrical conductivity of $BaTiO_3$ as a function of oxygen partial pressure.

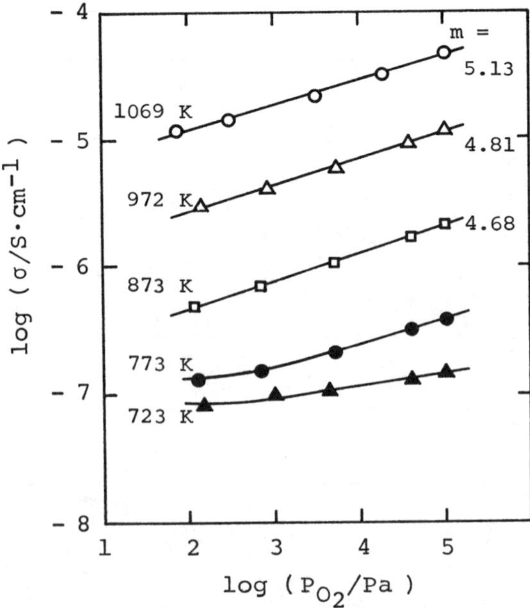

Figure 6.   Electrical conductivity of $Ba_{0.97}Na_{0.03}TiO_3$ as a function of oxygen partial pressure.

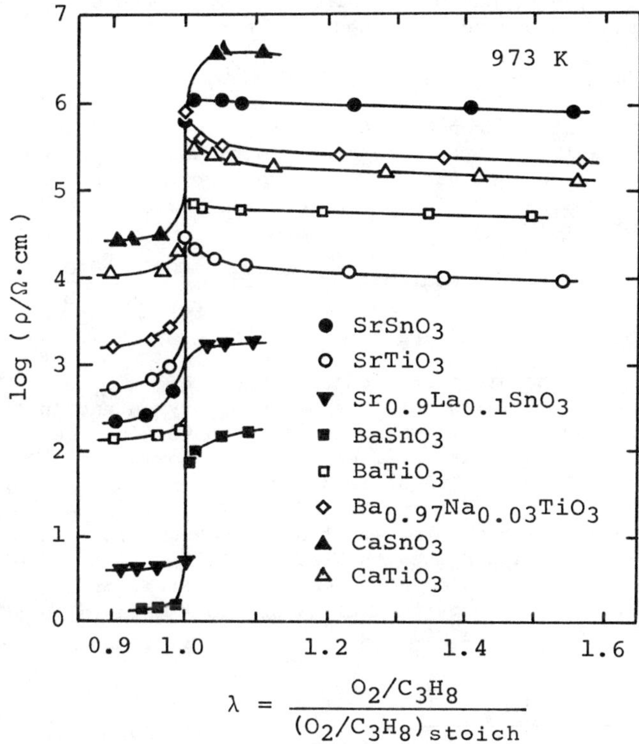

$$\lambda = \frac{O_2/C_3H_8}{(O_2/C_3H_8)_{stoich}}$$

Figure 7.   Dependence of the resistivity of perovskite-type oxides on the λ value.
"Reproduced with permission from Ref. 17.   Copyright 1985, 'The Chemical Society of Japan'."

pressure.   Therefore,   specimens whose conduction mechanism   change
from   p-type to n-type below 10^2 Pa can not be used for monitoring of
the stoichiometric ratio of air to fuel.   Among the   perovskite-type
oxides examined, $SrSnO_3$ was promising for a combustion monitoring
sensor, judging from the magnitude of the decrease in the resistivity
at $\lambda = 1$ and from excellent reproducibility of the resistivity   characteristics.

In order to clarify the resistivity characteristics of the
specimens, we obtained the relationship between an equilibrium oxygen
partial pressure and the oxygen excess ratio from both theoretical
calculations and measurements using the oxygen sensor. The complete
propane oxidation can be described by the following reaction.

$$C_3H_8 \; + \; 5\,O_2 \longrightarrow \; 3\,CO_2 \; + \; 4\,H_2O \tag{4}$$

If there is excess propane, CO will be one of residue products of
combustion, while if there is excess oxygen, free oxygen will be
present in the exhaust gas.   The oxygen partial pressure of the exhaust gas, P_{O_2}, at 700 °C was calculated using the following two
equations (13, 14), ignoring the presence of CO.

$$\Delta G^{\circ}_{973} \; = \; - \; 513.03 \text{ kcal/mol} = - \; RT \; \ln \frac{P_{CO_2}{}^3 \cdot P_{H_2O}{}^4}{P_{C_3H_8} \cdot P_{O_2}{}^5} \tag{5}$$

$$P_{O_2} = P_{O_2 i} - 5\,P_{C_3H_8 i}x \tag{6}$$

where ΔG°_{973} is the Gibbs free energy change of eq. (4), $P_{O_2 i}$ and
$P_{C_3H_8 i}$ are the initial partial pressure of oxygen and propane, respectively, and x is the fraction of propane consumed.   The free
energy change (ΔG°_{973}) was obtained by interpolation of values both at
900 K and at 1000 K.   A desired ratio of propane to oxygen was
established by the following procedure.   The flow rate of propane
was maintained at 7.5 ml/min, that of oxygen was changed to obtain a
desired value of λ, and then the total flow rate was maintained at
200 ml/min balancing with nitrogen.   Therefore, the oxygen excess
ratio, λ, is described by the following equation as a function of an
initial partial pressure of oxygen.

$$\lambda = \frac{P_{O_2 i}/P_{C_3H_8 i}}{(P_{O_2}/P_{C_3H_8})_{stoich}} = \frac{P_{O_2 i}}{5\,P_{C_3H_8 i}} = 5.33\,P_{O_2 i} \tag{7}$$

When λ is above unity and propane is oxidized completely,   the
equilibrium oxygen partial pressure can be given by the combination
of eqs. (6) and (7)

$$P_{O_2} = \; 5\,P_{C_3H_8 i}\,(\lambda - 1) \tag{8}$$

On the other hand, when λ is below unity, the initial oxygen partial
pressure can be written from eqs. (5) and (6) as a function of x.

$$P_{O_2 i} = 1.02 \times 10^{-24}\,\frac{x^{1.4}}{(1-x)^{0.2}} + 5\,P_{C_3H_8 i}x \tag{9}$$

And then the equilibrium oxygen partial pressure can be derived by
the combination of eqs. (6) and (9).

$$P_{O_2} = 1.02 \times 10^{-24} \frac{x^{1.4}}{(1-x)^{0.2}} \qquad (10)$$

Substituting some values for λ in eq. (7), P_{O_2i} can be obtained as a function of λ value. Using this P_{O_2i} and eq. (9), we can obtain the relationship between λ value and x value. In our experimental condition, x in eq. (9) can be substituted for λ approximately from the above calculations. Thus, when λ is below unity, eq. (10) can be rewritten by the following equation.

$$P_{O_2} = 1.02 \times 10^{-24} \frac{\lambda^{1.4}}{(1-\lambda)^{0.2}} \qquad (11)$$

As shown in Figure 8, the theoretical equilibruim oxygen partial pressure of the exhaust gas as a function of λ is given by eqs. (8) and (11) with the negative value of the logarithm of the oxygen partial pressure being shown on the left vertical axis. The observed oxygen partial pressure is also shown on the right vertical axis in the same figure. The difference between the theoretical and the observed oxygen partial pressures resulted from ignoring the presence of carbon monoxide in the "rich-burn region in this calculation. Although we did not have an actual ratio of carbon dioxide to carbon monoxide in the rich-burn region, $P_{O_2} = 10^{-15}$ Pa is estimated at 700 °C assuming that the ratio of carbon dioxide to carbon monoxide is about 10. In both cases, however, the oxygen partial pressures in the exhaust gas decreased dramatically from 10^3 to 10^{-15} Pa at $\lambda = 1$.

In order to confirm the relationship between the resistivity characteristics and the observed oxygen partial pressure, the electrical conductivities of three typical specimens, $SrTiO_3$, $SrSnO_3$ and $BaSnO_3$, were investigated at 700 °C in the oxygen partial pressure range from 10^{-10} to 10^5 Pa using the oxygen pump. The resistivity characteristics of these specimens can be explained clearly by the conductivity dependence of the observed oxygen partial pressure as shown in Figure 9. The change of the conduction mechanism from p-type to n-type for $SrTiO_3$ is found at 10^{-2} Pa. This phenomenon results in a slight decrease in the resistivity at $\lambda = 1$, as stated previously. However, this change for $SrSnO_3$ is found in the oxygen partial pressure range from 10^2 to 10^5 Pa. As for $BaSnO_3$, only the n-type conduction mechanism is observed in the conductivity dependence of the oxygen partial pressure.

In general, the electrical conductivity of semiconductive metal oxides can be illustrated as a function of oxygen partial pressure, as shown in Figure 10. The figure shows that equilibruim conductivity results of specimens are characterized by an oxygen-deficient, n-type region where the conductivity increases with decreasing P_{O_2}, and an oxygen excess, p-type region where the conductivity increases with increasing P_{O_2}, separated by a conductivity minimum at a P_{O_2}, which will be designated as $P_{O_2}^{\circ}$. The oxygen partial pressure, $P_{O_2}^{\circ}$ at which the change of the conduction mechanism is observed, depends on the variation of defect concentration of fully ionized atomic defects, electrons (n), and electron holes (p) in the specimen. The oxygen partial pressure $P_{O_2}^{\circ}$, the type of conduction mechanism both in the "lean-burn" and in the "rich-burn" and the magnitude of the decrease in the resistivity at $\lambda = 1$ were summarized in Table II. The materials whose conduction mechanism is unchangeable in the oxygen partial pressure below 10^2 Pa are useful for an oxygen

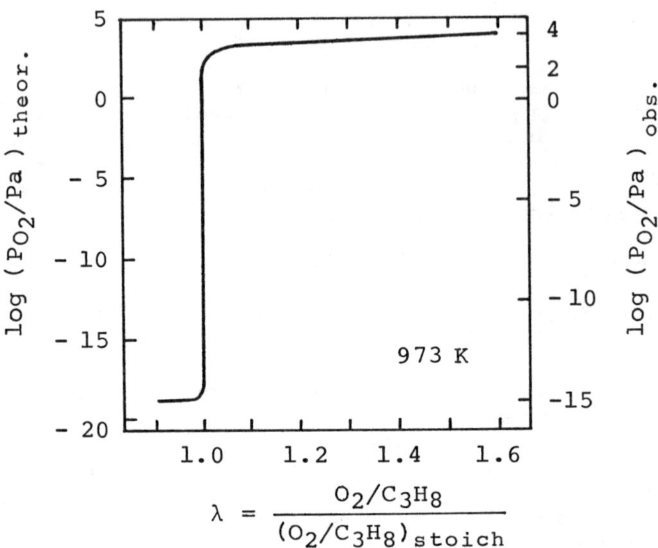

$$\lambda = \frac{O_2/C_3H_8}{(O_2/C_3H_8)_{stoich}}$$

Figure 8.   Relationship between equilibrium oxygen partial pressure and λ value.

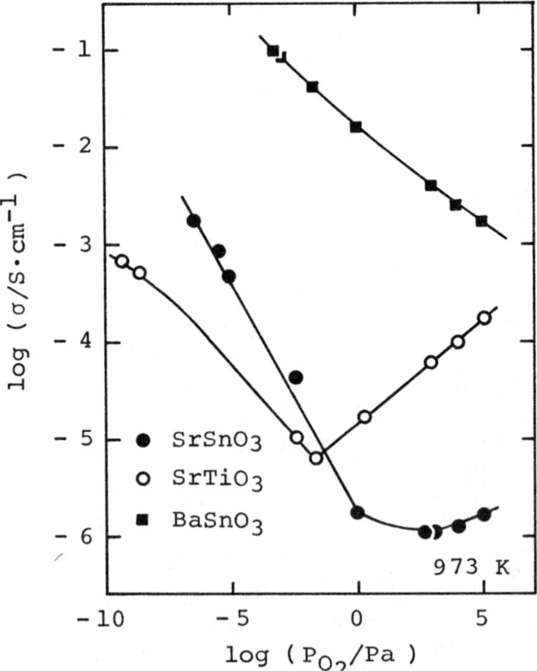

Figure 9.   Electrical conductivity of specimens as a function of oxygen partial pressure.

Table II.   The change of conduction mechanism of specimens

Specimen	Oxygen excess ratio $\lambda = 1$ $P_{O_2} \simeq 10^{-5} \sim 10^{-10}$ Pa	$n \rightleftarrows p$ transition $P_{O_2}°$/Pa	Atmosphere $P_{O_2} = 2 \times 10^4$ Pa	Magnitude of the decrease in the resistivity at $\lambda = 1$ $\log (\sigma_{\lambda>1}/\sigma_{\lambda<1})$
$SrSnO_3$	n	$10^2 \sim 10^4$	p	3.7
$SrTiO_3$	n	10^{-2}	p	1.3
$Sr_{0.9}La_{0.1}SnO_3$	n	—	n	2.6
$BaSnO_3$	n	—	n	2.1
$BaTiO_3$	n	$10^2 \sim 10^4$	p	2.6
$Ba_{0.97}Na_{0.03}TiO_3$	n	below 10^2	p	2.2
$CaSnO_3$	n	—	p	2.2
$CaTiO_3$	n	below 10^2	p	1.1

sensor to detect the stoichiometric ratio.   From this point of view, $BaSnO_3$ and $Sr_{0.9}La_{0.1}SnO_3$ are suitable for this sensor.   However, judging from the magnitude of the decrease in the resistivity at $\lambda = 1$, $SrSnO_3$ was the most suitable for a combustion monitoring sensor among these oxides.

The effects of B site partial substitution of $SrTiO_3$ on the oxygen partial pressure dependence of the conductivity were investigated for research materials which maintain the p-type conduction mechanism even at a extremely low oxygen partial pressure.   Lattice constant, relative density, and surface area of the specimens used for electrical measurements were listed in Table III.   The relative density was calculated using the apparent density data and the X-ray density. The lattice constant was calculated from the (312) peak of the perovskite-type structure obtained by X-ray diffraction measurements using Si as an internal standard.   In this calculation, the cubic system is assumed to maintain independent of the y value.   The lattice constant was slightly decreased by the partial substitution of a trivalent Al^{3+} for a tetravelent Ti^{4+} and was nearly constant above 0.01 of aluminum content y.   Surface area of the specimens increased and relative density decreased with an increase in the partial substitution of aluminum.   Only $SrAl_{0.01}Ti_{0.99}O_3$ was light yellowish rather than white, while all of other specimens were white.   Therefore, some segregation of the additional alumina above y = 0.01 might counteract any coarsening of particle.

Table III.   Characteristics of $SrAl_yTi_{1-y}O_3$ specimens

y	Lattice constant	Relative density[a]	Surface area
in $SrAl_yTi_{1-y}O_3$	nm	%	$m^2 \cdot g^{-1}$
0	0.3906	70	1.0
0.01	0.3905	66	1.2
0.1	0.3905	65	3.0
0.2	0.3905	64	5.2

a) Density(XRD) = 5.116 $g \cdot cm^{-3}$

The effects of the partial aluminum substitution on the resistivity dependence of λ were also examined at 700 °C.   The conduction mechanism of $SrTiO_3$ itself was a p-type in an atmospheric environment as stated previously,   therefore, the enhancement of positive hole concentration will be expected by the partial substitution of aluminum.   All specimens showed similar resistivity characteristics as shown in Figure 11.   With increasing the additional amount of alumina, the resistivity of specimens tended to increase with de-

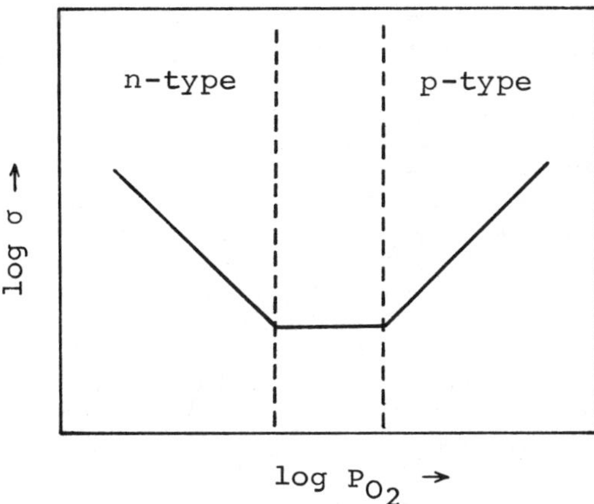

Figure 10. Dependence of the electrical conductivity of semiconductive metal oxides on oxygen partial pressure.

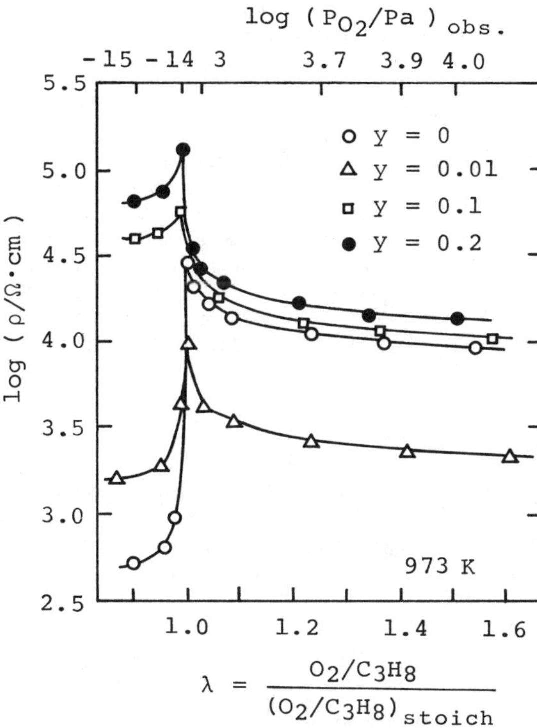

$$\lambda = \frac{O_2/C_3H_8}{(O_2/C_3H_8)_{stoich}}$$

Figure 11. Dependence of the resistivity of $SrAl_yTi_{1-y}O_3$ on the λ value.

creasing λ down to $\lambda = 1$ at first, increase dramatically at $\lambda = 1$ and then decrease slightly with decreasing λ. The resistivity of specimen, whose aluminum content, y, equaled 0.01, was about 1/5 smaller than that of the undoped specimen, while the resistivity of the specimen whose aluminum content was above 0.1 was the same as that of the undoped specimen above $\lambda = 1$. On the other hand, the resistivity of specimens below $\lambda = 1$ depended on the aluminum content, i.e., resistivity increased with increasing aluminum content. These experimental results were explained by the following idea. The partial substitution of aluminum enhances positive hole concentration, and an oxygen partial pressure at which the change of conduction mechanism from p-type to n-type was observed might shift to lower values. However, some segregation of the additional aluminum content above 0.01 in the grain boundary counteract any decrease in the resistivity.

Electrical characteristics of sepcimens can be also expressed using an electrical current measured in closed electric circuit in which the specimen, a D.C. voltage supply, and a picoammeter were directly connected. As shown in Figure 12, the magnitude of the current decreased with increasing aluminum content below $\lambda = 1$. The dotted line is calculated from an ideal specimen assuming that the conductivity is proportional to the quarter power of oxygen partial pressure even at an extremely low oxygen partial pressure of about 10^{-20} Pa. It is found that the characteristic of the ideal specimen is suitable for a lean-burn sensor, since the current varies greatly with the value of λ in the lean-burn region, while the magnitude of the current and the changes are almost ignored in the rich-burn region. Among these specimens, the current-λ characteristics of $SrAl_{0.2}Ti_{0.8}O_3$ agree well with the ideal one above $\lambda = 1$, and it is suitable for the "lean-burn" sensor because of the large change in the current above $\lambda = 1$ and the small change below $\lambda = 1$. However, stronger enhancement of the p-type conduction mechanism by further substituting aluminum will not be expected in this composition, because the formation of solid solution might be a limiting factor above $y = 0.01$. Therefore, further investigations should be directed to research on new materials which exhibit a p-type conduction mechanism even at an extremely low oxygen partial pressure in order to obtain an excellent lean-burn sensor.

Conclusion

In conclusion, among the perovskite-type oxides examined, $SrSnO_3$ was promising for a combustion monitoring sensor, judging from the magnitude of the decrease in the resistivity at $\lambda = 1$ and from excellent reproducibility of the resistivity characteristics. As for an oxygen sensor capable of detecting the "lean-burn" region, $SrAl_{0.2}Ti_{0.8}O_3$ was suitable because the small change in the current below $\lambda = 1$ compared with that of above $\lambda = 1$ and thermal stability under reductive environments. In order to develop excellent lean-burn sensors, however, further investigations are required for researching new materials which remain p-type conductors even at a extremely low value of oxygen pressure.

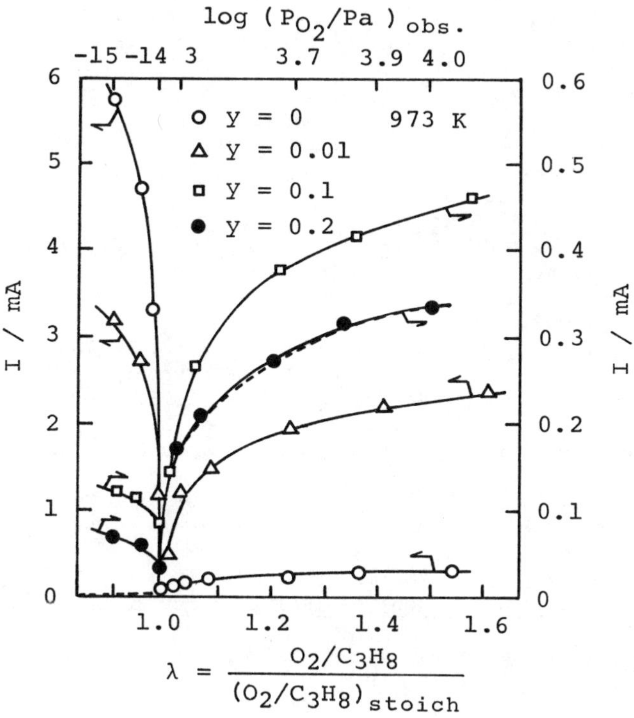

$$\lambda = \frac{O_2/C_3H_8}{(O_2/C_3H_8)_{stoich}}$$

Figure 12. Dependence of the current of $SrAl_yTi_{1-y}O_3$ on the λ value.
     Supplied voltage ; 2 V, Specimen thickness ; 1 mm, Electrode area ; 30 mm^2.

Literature Cited

1. Tien, T. Y. ; Stadler, H. L. ; Gibbons, E. F. ; Zacmanidis, P.
    J. Ceram. Bull. 1975, 54, 280.
2. Logothetis, E. M. ; Park, K. ; Meitzler, A. H. ; Laud, K. R.
    Appl. Phys. Lett. 1975, 26, 209.
3. Park, K. ; Logothetis, E. M.   J. Electrochem. Soc.   1977, 124,
    1443.
4. Murakami, N. ; Tanaka, K. ; Sasaki, K. ; Ihokura, K.   Proc. of
    the Int. Mtg. on Chemical Sensors, Kodansha/Elsevier 1983,
    p. 165.
5. Saji, K. ; Takahashi, H. ; Kondo, H. ; Takeuchi, T. ; Igarashi,
    I.   Proc. of the Int. Mtg. on Chemical Sensors, Kodansha/
    Elsevier, 1983, p. 171.
6. Logothetis, E. M. ; Kaiser, W. J. Sensors and Actuators, 1983,
    4, 333.
7. Obayashi, H. ; Okamoto, H. Solid State Ionics, 1981, 3/4, 631.
8. Schouler, E. J. L. ; Mesbahi, N. ; Vitter, G.   Solid State
    Ionics, 1983, 9/10, 989.
9. Dietz, H.   Solid State Ionics, 1982, 6, 175.
10. Haaland, D. M.   J. Electrochem. Soc., 1980, 127, 796.
11. Kröger, F. A. ; Vink, H. J. In "Solid State Physics" ; Seitz,
    F. ; Turnbull, D., Ed. ; Academic Press : Ney York, 1956 ;
    Vol. 3, p. 307.
12. Seiyama, T.   Oxidation Communication, 1982, 2, 239.
13. Barin, I. ; Knache, O. "Thermochemical Properties of Inorganic
    Substances", Springer-Verlag : New York, 1973.
14. Stull, D. R. ; Westrum, E. F. ; Jr. & Sinke, G. C.   "The
    Chemical Thermodynamics of Organic Compounds", John Wiley &
    Sons. Inc. : New York, 1967 ; Chap. 9.
15. Galasso, F. S. "Structure, Properties and Preparation of
    Perovskite-Type Compounds", Pregamon Press : Hungary, 1969 ;
    Chap. 4.
16. Arai, H. ; Ezaki, S. ; Shimizu, Y. ; Shippo, O. ; Seiyama, T.
    Proc. of the Int. Mtg. on Chemical Sensors, Kodansha/Elsevier,
    1983, p. 393.
17. Shimizu, Y. ; Fukuyama, Y. ; Narikiyo, T. ; Arai, H. ;
    Seiyama, T. Chem. Lett., 1985, 377.
18. Chan, N. -H. ; Sharma, R. K. ; Smyth, D. M.   J. Electrochem.
    Soc., 1981, 128, 1762.
19. Chan, N. -H ; Sharma, R. K. ; Smyth, D. M. J. Am. Ceram. Soc.,
    1981, 64, 556.
20. Eror, N. G. ; Smyth, D. M. J. Soild State Chem., 1978, 24, 235.

RECEIVED October 31, 1985

Principles and Development of a Thick-Film Zirconium Oxide Oxygen Sensor

Shinji Kimura, Shigeo Ishitani, and Hiroshi Takao

Materials Research Laboratory, Central Engineering Laboratories, Nissan Motor Company, Ltd., 1 Natsushima-cho, Yokosuka 237, Japan

The newly-developed oxygen sensor consists of laminated, porous, thick film zirconia, reference and measurement electrode layers on an alumina substrate in which a thick film heater is embedded. Measurement of the oxygen concentration is accomplished by positioning the sensor entirely in the exhaust gas, and sending a continuous flow of DC current through the porous zirconia layer between two electrodes. Reference oxygen gas instead of air or other standard materials is then generated electrolytically at the reference electrode/ziroconia interface.

The sensor has voltage characteristics which are nearly identical to the usual crucible-type sensor. Detailed analysis of the steady-state voltage chatacteristics of both the thick film oxygen sensor and the crucible type oxygen sensor are shown.

Monitoring oxygen content in exhaust gas from an automotive internal combustion engine has been widely used as the basis for controlling the air-fuel ratio of the combustible mixture fed to the engine. An oxygen sensor is used to produce an electrical signal representing the oxygen content in the exhaust gas. (1,2)

The zirconia sensor operates primarily on the principle of a concentration cell. It consists of a non-porous solid electrolyte layer fabricated from zirconia stabilized with yttria or calcia and exhibits high oxygen ion mobility. This layer is sandwiched between two porous and electrically conductive electrodes.

In one of the most common sensors, the non-porous solid electrolyte layers takes the form of a crucible closed at one end so that air used as a reference gas can be introduced into the interior of the crucible while the outside of the crucible is exposed to the exhaust gas. A schematic drawing and E versus the air-fuel ratio curve for this sensor are shown in Figure 1. E is the electromotive force (EMF) between the two electrodes in accordance with the Nernst equation:

$$E = \frac{RT}{4F} \ln \frac{Po_2(A)}{Po_2(B)} \tag{1}$$

where R is the gas constant, T is the absolute temperature, F is the Faraday constant, and $Po_2(A)$ and $Po_2(B)$ are oxygen partial pressures in air and in exhaust gas, respectively.

The sensor voltage varies greatly at the stoichimetric point which is an air-fuel ratio of approximately 14.7 for an ordinary engine.

In an atmosphere with an air-fuel ratio smaller than 14.7, the CO gas concentration in the exhaust gas increases; Such an atmosphere is called a rich atmosphere. In an atmosphere with an air-fuel ratio larger than 14.7, the oxygen gas concentration in the exhaust gas increases; This is called a lean atmosphere.

The objectives of this study were twofold; (1) to develop a new type of oxygen sensor which makes the crucible-type sensor more compact and which does not use air as a reference gas, (2) to analyze output characteristics of both the newly-developed zirconia oxygen sensor and the crucible-type oxygen sensor.

Thick Film Zirconia Oxygen Sensor and Measurements

A cross-section schematic drawing of the newly-developed thick film oxygen sensor is shown in Figure 2. The platinum film heater is embedded in the alumina substrate. Electrical resistance of the heater is about 6 ohms at room temperature.

Arranged in layered fashion on the alumina substrate are the zirconia underlayer, the platinum reference electrode, the zirconia solid electrolyte stabilized with 5.1 mole % Y_2O_3, the platinum measurement electrode, and finally, the protective spinel ($A_2O_3 \cdot MgO$) layer. The zirconia layer is 4mm long, 4mm wide and 30μm thick. The element itself measures 5mm by 9mm and is 1.2mm thick.

A plane schematic drawing of the thick film oxygen sensor is shown in Figure 3. The protective layer is eliminated. A part of each platinum lead wire is embedded in the alumina substrate. The earth line of the heater and sensor is common. Figure 4 shows a production flow chart for the thick film oxygen sensor.

The heater, the underlayer, the zirconia solid electrolyte and the two electrodes are formed by screen printing and sintering. The sintering condition is at 1,480°C for 2HR in air. The temperature of sensor surface rises to 600°C with plasma spraying.

As a result, the zirconia solid electrolyte, two electrodes and protective spinel layer become porous.

Measurements of the oxygen concentration are made with the sensor positioned entirely in the exhaust gas from an ordinary engine. A direct current is applied between the two electrodes from a DC power source. (See Figure 2.)

Measurement conditions are as follows:

o Atmosphere – In the exhaust gas from an ordinary engine.
o Gas temperature – 1,000°K
o Direct current 0 – 20μA
o Air-Fuel Ratio (λ) 12 – 17

This oxygen sensor can function when the gas temperature is higher than 200°C.

Figures 5 and 6 present the experimental results obtained with the thick film oxygen sensor.

Figure 5 shows the relationship between DC sensor current and

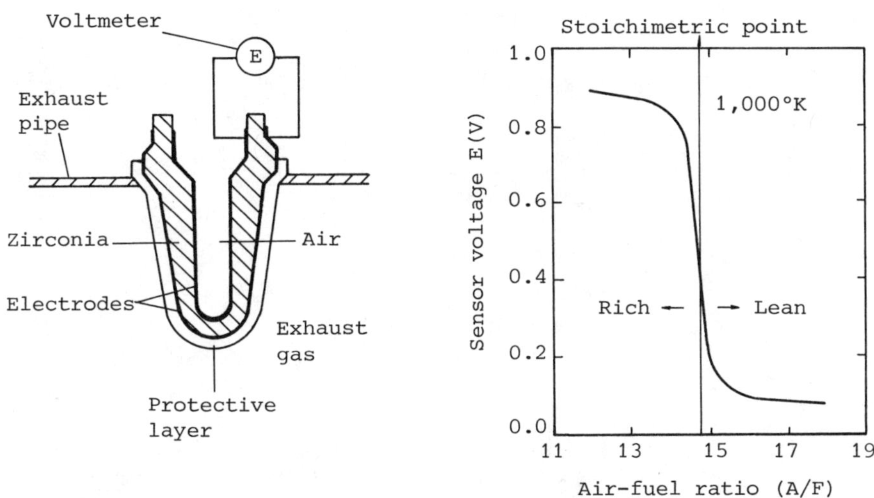

Figure 1.   Schematic drawing and voltage curve for crucible-type oxygen sensor

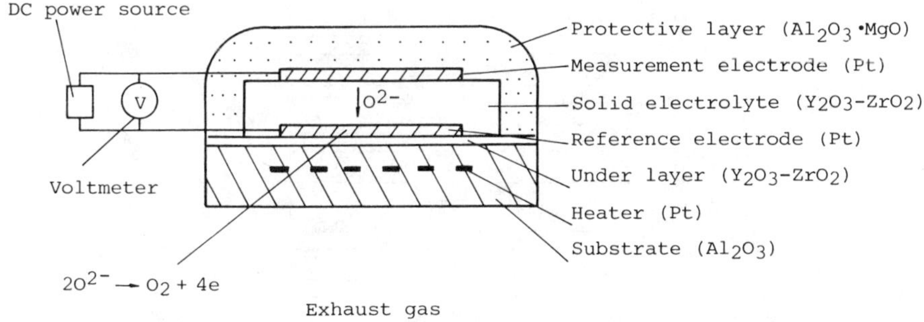

Figure 2.   Schematic drawing of thick film oxygen sensor

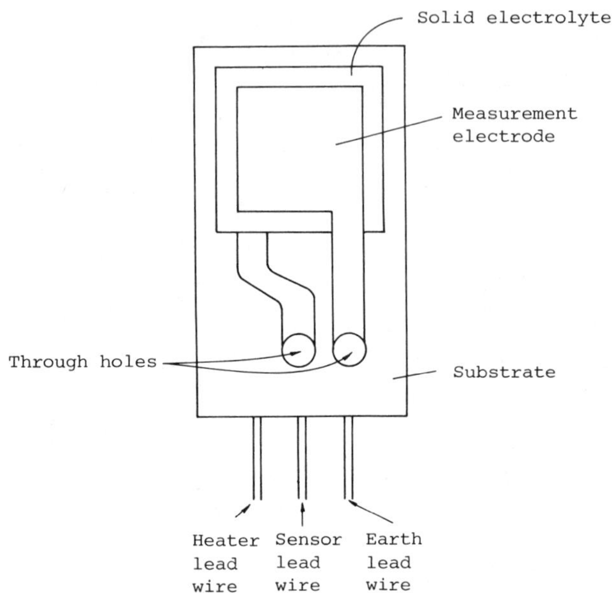

Figure 3.   Schematic drawing of thick film oxygen sensor

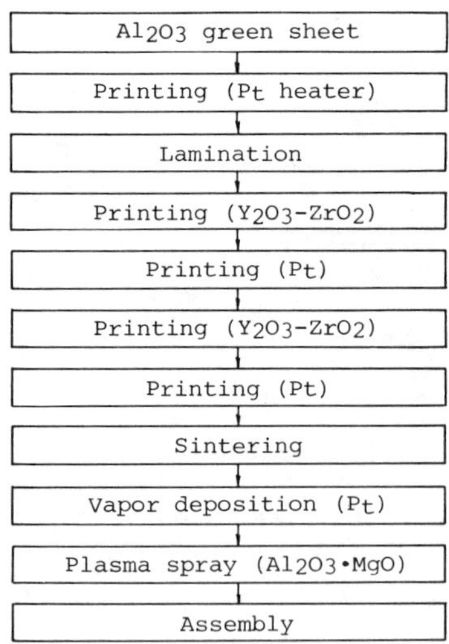

Figure 4.   Flow chart for preparation of thick film oxygen sensor

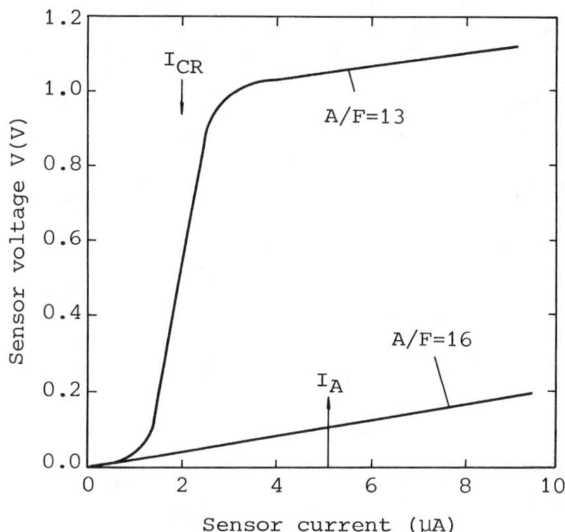

Figure 5. Experimental voltage curves for thick film oxygen sensor (Sensor current vs V curve)

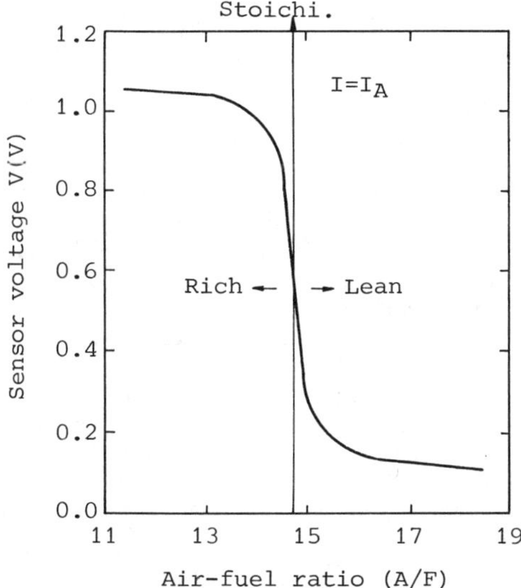

Figure 6. Experimental voltage curves for thick film oxygen sensor (A/F vs V curve)

sensor voltage at 1,000°K. With a lean atmosphere (A/F = 16), the sensor voltage rises steadily as applied current increases.

On the other hand, with a rich atmosphere (A/F = 13), the sensor voltage rises dramatically. The current which produces sudden change in sensor voltage is called I_{CR}. In this case, $I_{CR} = 2\mu A$. The output characteristics for the sensor versus the air-fuel ratio for a sensor current $I_A = 5\mu A$ larger than I_{CR} is shown in Figure 6. This voltage curve correlates well with the output characteristic of the crucible-type oxygen sensor. (See Figure 1.)

Zirconia Oxygen Sensor Model

Many types of oxygen sensor models have been proposed. ([3] - [8])
A schematic drawing of the zirconia oxygen sensor model used in this study is shown in Figure 7. The steady-state voltage characteristic of the thick film oxygen sensor can be explained analytically using this schematic drawing. In this analysis the following assumptions are made: (Note: symbols used in this paper are listed in the legend of symbols.)

① , Gases are CO, O_2, CO_2 and N_2. Total pressure on the measurement electrode and in the exhaust gas is 1 atm, and partial pressure of nitrogen gas is a constant 0.87 atm.

② , At the reference electrode and the measurement electrode, chemical equilibrium of the following reaction is maintained:

$$CO + 1/2\ O_2 \rightleftharpoons CO_2 \tag{2}$$
$$P_{CO} \cdot \sqrt{P_{O2}} \ / \ P_{CO2} = K \tag{3}$$

③ , The distributions of P_{CO}, P_{O2} and P_{CO2} in the porous solid electrolyte and porous protective layer are linear in the steady state. (See Figure 8.)

④ , O_2 gas is generated electrolytically at the interface between the reference electrode and the solid electrolyte layer. The mass of the O_2 gas is equal to the mass of O_2 gas which diffuses through the porous thick film zirconia, plus the mass of O_2 gas which reacts with CO gas at the reference electrode.

⑤ , At the reference electrode/solid electrolyte layer interface, the mass of the CO gas which diffuses from the porous thick film zirconia is equal to the mass of CO_2 gas which diffuses into the porous thick film zirconia.

⑥ , At the measurement electrode/solid electrolyte layer interface, the mass of O_2 gas which diffuses from the porous protective layer is equal to the mass of O_2 which changes to oxygen ion plus the mass of CO gas which diffuses from the porous protective layer.

⑦ , At the measurement electrode/solid electrolyte layer interface, the mass of CO gas which diffuses from the porous protective layer is equal to the mass of CO_2 gas which diffuses into the porous protective layer.

⑧ , Sensor voltage V can be expressed by the following equation

$$V = E + IR_0 \tag{4}$$
$$E = \frac{RT}{4F} \ln \frac{P_{O2}(I)}{P_{O2}(0)} \tag{5}$$

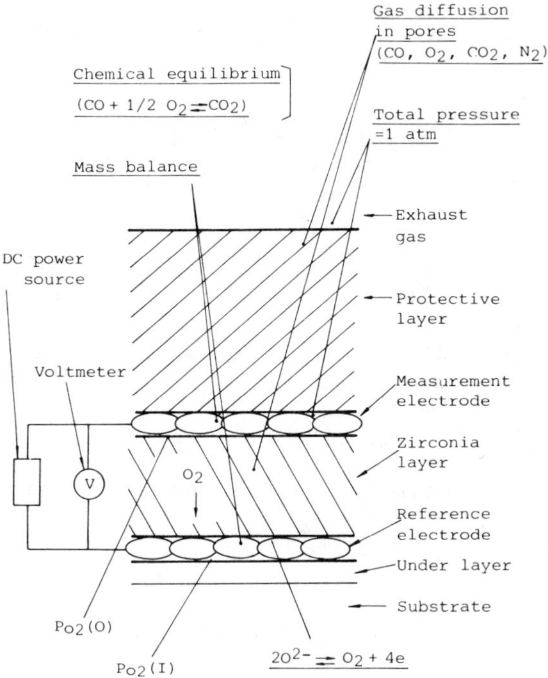

Figure 7. Oxygen sensor model

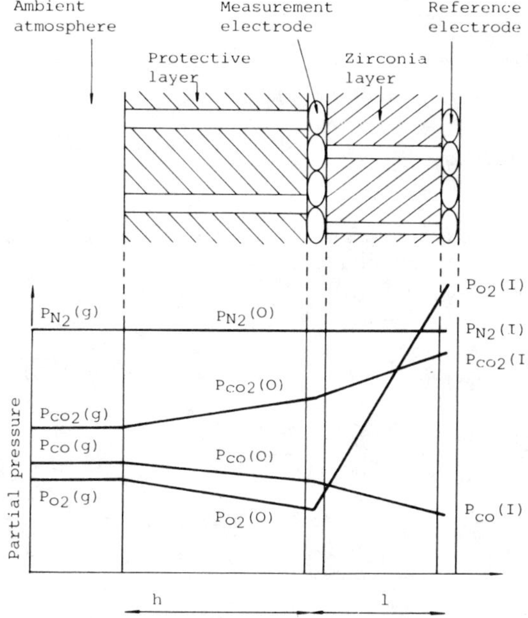

Figure 8. Partial pressure distribution model

From assumption ①

$$P_{O2}(0) + P_{co}(0) + P_{co2}(0) + P_{N2}(0) = 1 \tag{6}$$

From assumption ②

$$\frac{P_{co}(0) \cdot \sqrt{P_{O2}(0)}}{P_{co2}(0)} = K \tag{7}$$

$$\frac{P_{co}(I) \cdot \sqrt{P_{O2}(I)}}{P_{co2}(I)} = K \tag{8}$$

From assumption ③

$$P_{co}(I) = -b_1 l + P_{co}(0) \tag{9}$$
$$P_{O2}(I) = b_2 l + P_{O2}(0) \tag{10}$$
$$P_{co2}(I) = b_3 l + P_{co2}(0) \tag{11}$$

From assumption ④

$$\frac{IRT}{4FS} = b_2 D_{O2}(SE) + \frac{1}{2} b_1 D_{co}(SE) \tag{12}$$

Derivation of Equation 12 is shown in the appendix.

From assumption ⑤

$$b_1 \cdot D_{co}(SE) = b_3 D_{co2}(SE) \tag{13}$$

From assumption ①

$$P_{O2}(g) + P_{co}(g) + P_{co2}(g) + P_{N2}(g) = 1 \tag{14}$$

From assumption ③

$$P_{co}(0) = -a_1 h + P_{co}(g) \tag{15}$$
$$P_{O2}(0) = -a_2 h + P_{O2}(g) \tag{16}$$
$$P_{co2}(0) = a_3 h + P_{co2}(g) \tag{17}$$

From assumption ⑥

$$\frac{IRT}{4FS} + \frac{1}{2} a_1 D_{co}(PL) = a_2 D_{O2}(PL) \tag{18}$$

From assumption ⑦

$$a_1 D_{co}(PL) = a_3 D_{co2}(PL) \tag{19}$$

Theoretical oxygen partial pressures are given by the following simultaneous cubic equations:

$$X^3 + \frac{D_{co}(SE)}{D_{co2}(SE)} KX^2 + \left(\frac{D_{co}(SE)}{2D_{O2}(SE)} P_{co}(0) - Y^2 - \frac{RTl}{4FSD_{O2}(SE)} I \right) X$$

$$- \frac{D_{co}(SE)}{D_{co2}(SE)} K \left(\frac{D_{co2}(SE)}{2D_{O2}(SE)} P_{co2}(0) + Y^2 + \frac{RTl}{4FSD_{O2}(PL)} I \right) = 0 \tag{20}$$

$$Y^3 + \frac{D_{co}(PL)}{D_{co2}(PL)} KY^2 + \left(\frac{D_{co}(PL)}{2D_{O2}(PL)} P_{co}(g) - P_{O2}(g) + \frac{RTh}{4FSD_{O2}(PL)} I \right) Y$$

$$- \frac{D_{co}(PL)}{D_{co2}(PL)} K \left(\frac{D_{co2}(PL)}{2D_{O2}(PL)} P_{co2}(g) + P_{O2}(g) - \frac{RTh}{4FSD_{O2}(PL)} I \right) = 0 \tag{21}$$

where

$$X \equiv \sqrt{P_{O2}(I)} \tag{22}$$

$$Y \equiv \sqrt{P_{O2}(0)} \tag{23}$$

Equation 20 can be derived from Equations 6 – 13 and Equation 21 can be derived from Equations 14 – 19.
Thus, sensor voltage E is given:

$$E = \frac{RT}{4F} \ln \frac{P_{O2}(I)}{P_{O2}(0)} = \frac{RT}{4F} \ln \frac{X^2}{Y^2} = \frac{RT}{2F} \ln \frac{X}{Y} \tag{24}$$

Furthermore, internal pressure P_T at the reference electrode can be expressed by the following equation:
$$P_T = P_{O2}(I) + P_{co}(I) + P_{co2}(I) + P_{N2}(I)$$
$$= 1 + X^2 - Y^2 + 21 \left(\frac{1}{D_{co2}(SE)} - \frac{1}{D_{co}(SE)} \right) \left(\frac{IRT}{4FS} - \frac{D_{o2}(SE)}{1}(X^2 - Y^2) \right) \tag{25}$$

We can obtain the calculated values of the sensor voltage and the internal pressure P_T from Equations 20, 21, 24, and 25.

Calculated Results

Crucible-type oxygen sensor with catalytic electrode. In this case, the solid electrolyte is non-porous and the sensor current $I = 0$.
Air is used as a reference gas, $P_{O2}(I)$ is a constant $0.21\,\text{atm}(X^2)$.
Thus, only Equation 21 is considered.

Substituting 0 into I in Equation 21 yields
$$Y^3 + \frac{D_{co}(PL)}{D_{co2}(PL)} KY^2 + \left(\frac{D_{co}(PL)}{2D_{o2}(PL)} P_{co}(g) - P_{co2}(g) \right) Y$$
$$- \frac{D_{co}(PL)}{D_{co2}(PL)} K \left(P_{o2}(g) + \frac{D_{co2}(PL)}{2D_{o2}(PL)} P_{co2}(g) \right) = 0 \tag{26}$$

Considering that the value of K is extremely small ($K = 6.33 \times 10^{\overline{1}1}$ at $1{,}000°K$), solutions for Y in Equation 26 are divided into the following three cases: Regarding A as the coefficient of Y, i, e.
$$A \equiv \frac{D_{co}(PL)}{2D_{o2}(PL)} P_{co}(g) - P_{o2}(g) \tag{27}$$

I) $A < 0$ (lean atmosphere)
The terms of both Y^2 and the constant may be desregarded.
$$Y^2 = P_{o2}(0) = P_{o2}(g) - \frac{D_{co}(PL)}{2D_{o2}(PL)} P_{co}(g) \tag{28}$$
$$\doteqdot P_{o2}(g) \tag{29}$$

II) $A = 0$ (stoichiometric point)
The terms of Y^2 may be disregarded.
$$Y^2 = P_{o2}(0) = \left\{ \frac{D_{co}(PL)}{D_{co2}(PL)} K \left(P_{o2}(g) + \frac{D_{co2}(PL)}{2D_{o2}(PL)} P_{co2}(g) \right) \right\}^{2/3} \tag{30}$$

III) $A > 0$ (Rich atmosphere)
The terms of both Y^3 and Y^2 may be disregarded.
$$Y^2 = P_{o2}(0) = \left\{ \frac{D_{co}(PL)}{D_{co2}(PL)} K \left(P_{o2}(g) + \frac{D_{co2}(PL)}{2D_{o2}(PL)} P_{co2}(g) \right) \right\}^2 \Big/ \left\{ \frac{D_{co}(PL)}{2D_{o2}(PL)} P_{co}(g) - P_{o2}(g) \right\}^2$$
$$\doteqdot \left\{ P_{co2}(g) \cdot K / P_{co}(g) \right\}^2 \tag{31} \tag{32}$$

The calculated $P_{O2}(0)$ vs A/F curve and E vs A/F curve are shown in Figure 9. Total pressure at both the measurement and the reference electrode is 1 atm.

Crucible-type oxygen sensor with non-catalytic electrode. A non-catalytic electrode (e. g. Au) is thought to delay the reaction rate in the following reaction

$$CO + 1/2O_2 \rightleftharpoons CO_2$$

Ideally, the reaction to produce CO_2 cannot proceed. Therefore, the value of $P_{CO2}(g)$ decreases and the value of K ($\equiv P_{CO} \cdot \sqrt{P_{O2}}/P_{CO2}$) increases.
Dividing Equation 26 by K

$$\frac{Y^3}{K} + \frac{D_{CO}(PL)}{D_{CO2}(PL)} \, Y^2 + \left(\frac{D_{CO}(PL)}{2D_{O2}(PL)} \, P_{CO}(g) - P_{O2}(g) \right) \frac{Y}{K}$$

$$- \frac{D_{CO}(PL)}{D_{CO2}(PL)} \left(P_{O2}(g) + \frac{D_{CO2}(PL)}{2D_{O2}(PL)} \, P_{CO2}(g) \right) = 0 \tag{33}$$

The terms of both Y^3 and Y may be desregarded since the value of K is large.
Then

$$Y^2 = P_{O2}(0) = P_{O2}(g) + \frac{D_{CO2}(PL)}{2D_{O2}(PL)} \, P_{CO2}(g) \tag{34}$$

$$\fallingdotseq P_{O2}(g) \tag{35}$$

This result produces a continuous sensor voltage change at the stoichimetric point.

Thick film oxygen sensor with catalytic electrode (when I = 0).

Substituting I = 0 into Equation 20 yields

$$X^3 + \frac{D_{CO}(SE)}{D_{CO2}(SE)} \, KX^2 + \left(\frac{D_{CO}(SE)}{2D_{O2}(SE)} \, P_{CO}(0) - Y^2 \right) X$$

$$- \frac{D_{CO}(SE)}{D_{CO2}(SE)} \, K \left(Y^2 + \frac{D_{CO2}(SE)}{2D_{O2}(SE)} \, P_{CO2}(0) \right) = 0 \tag{36}$$

$$\therefore (X - Y) \left\{ (X + Y)\left(X + \frac{D_{CO}(SE)}{D_{CO2}(SE)} \, K \right) + \frac{D_{CO}(SE)}{2D_{O2}(SE)} \, P_{CO}(0) \right\} = 0 \tag{37}$$

Then

$$X = Y \tag{38}$$

$$E = \frac{RT}{4F} \, \ln \frac{P_{O2}(I)}{P_{O2}(0)} = \frac{RT}{2F} \, \ln \frac{X}{Y} = 0 \tag{39}$$

when I equals 0, the sensor voltage is always 0. The dependence of X and Y on the oxygen concentration in the exhaust gas is the same as that for the measurement electrode of the crucible-type oxygen

Table I. Calculation results for $P_{O_2}(I)$ and $P_{O_2}(0)$
(Rich atmosphere)

	$P_{O_2}(0)$ (Measurement electrode)	$P_{O_2}(I)$ (Reference electrode)
$0 \leq I < I_{CR}$		$\left\{ \dfrac{D_{CO}(SE)K\left(P_{O_2}(0) + \dfrac{D_{CO_2}(SE)}{2D_{O_2}(SE)}P_{CO_2}(0)\right.}{D_{CO_2}(SE)\left(\dfrac{D_{CO}(SE)}{2D_{O_2}(SE)}P_{CO}(0) - P_{O_2}(0)\right.} \; \left. + \dfrac{\dfrac{RT1I}{4FSD_{O_2}(SE)}}{-\dfrac{RT1I}{4FSD_{O_2}(SE)}} \right\}^{2}$ (41)
$I = I_{CR}$	$\left\{ \dfrac{D_{CO}(PL)K\left(P_{O_2}(g) + \dfrac{D_{CO_2}(PL)}{2D_{O_2}(PL)}P_{CO_2}(g)\right.}{D_{CO_2}(PL)\left(\dfrac{D_{CO}(PL)}{2D_{O_2}(PL)}P_{CO}(g) - P_{O_2}(g)\right.} \; \left. \dfrac{-\dfrac{RThI}{4FSD_{O_2}(PL)}}{+\dfrac{RThI}{4FSD_{O_2}(PL)}}\right)\right\}^{2}$ (40)	$\left\{ \dfrac{D_{CO}(SE)}{D_{CO_2}(SE)}K\left(P_{O_2}(0) + \dfrac{D_{CO_2}(SE)}{2D_{O_2}(SE)}P_{CO_2}(0)\right. \; \left. + \dfrac{RT1I}{4FSD_{O_2}(SE)}\right\}^{2/3}$ (42)
$I_{CR} < I \leq I_{LR}$		$P_{O_2}(0) - \dfrac{D_{CO}(SE)}{2D_{O_2}(SE)}P_{CO}(0) + \dfrac{RT1I}{4FSD_{O_2}(SE)}$ (43)
$I > I_{LR}$	Non-existent	Non-existent

Discussion

Sensor voltage characteristics of the crucible type oxygen sensor.
According to the oxygen sensor model used in this analysis, the
oxygen partial pressure $P_{O_2}(0)$ at the measurement electrode can be
expressed by Equations 28 – 32, 34, and 35. Calculated results for
the sensor voltage are shown in Figure 9.

D. S. Eddy calculated the sensor voltage characteristics using
a chemical reaction equilibrium model. (1) His results correlate
well with the results shown in Figure 9.

In the analysis of this work, $P_{O_2}(0)$ can be expressed using both
partical pressure and the diffusion coefficient of each gas. The
oxygen sensor with a catalytic electrode shows abrupt change in
sensor voltage at the stoichiometric point as shown in Figure 1. On
the other hand, the oxygen sensor with a non-catalytic electrode
shows continuous change of sensor voltage at that point, this conti-
nuity of sensor voltage can be explained with the K gap from an ideal
value in equilibrium condition.

Sensor voltage characteristics of the thick film oxygen sensor.
Experimental data show that sensor voltage characteristics of the
thick film oxygen sensor vary greatly with the value of the sensor
current.

Table II.   Calculation results for $P_{O_2}(I)$ and $P_{O_2}(0)$
(Lean atmosphere)

	$P_{O_2}(0)$ (Measurement electrode)	$P_{O_2}(I)$ (Reference electrode)
$0 \leqq I < I_{CL}$	$$P_{O_2}(g) - \frac{D_{CO}(PL)}{2D_{O2}(PL)}\, P_{CO}(g) - \frac{RThI}{4FSD_{O2}(PL)} \qquad (44)$$	
$I = I_{CL}$	$$\left\{ \frac{D_{CO}(PL)}{D_{CO2}(PL)} K\left(P_{O2}(g) + \frac{D_{CO2}(PL)}{2D_{O2}(PL)} P_{CO2}(g) \right. \right.\\ \left. \left. - \frac{RThI}{4FSD_{O2}(PL)} \right\}^{2/3} \qquad (45) \right.$$	$$P_{O2}(0) + \frac{D_{CO}(SE)}{2D_{O2}(SE)}\, P_{CO}(0) + \frac{RT1I}{4FSD_{O2}(SE)} \qquad (47)$$
$I_{CL} < I \leqq I_{LL}$	$$\left\{ \frac{D_{CO}(PL)K\left(P_{O2}(g) + \frac{D_{CO2}(PL)}{2D_{O2}(PL)} P_{CO2}(g)\right)}{D_{CO2}(PL)\left(\frac{D_{CO}(PL)}{2D_{O2}(PL)} P_{CO}(g) - P_{O2}(g) \right.}\right. \\ \left. \left. - \frac{RThI}{4FSD_{O2}(PL)} \right)^2 + \frac{RThI}{4FSD_{O2}(PL)} \right) \right\} \qquad (46)$$	
$I > I_{LL}$	Non-existent	Non-existent

At $I = 0$, in a steady state condition, the sensor voltage is a
constant zero and is independent of the oxygen partial pressure in
the exhaust gas. As shown in the calculated results (See Figure 10),
both $P_{O_2}(I)$ and $P_{O_2}(0)$ depend on the air-fuel ratio. But $P_{O_2}(I)$ is
always equal to $P_{O_2}(0)$ (See Equation 38); thus, E becomes zero (See
Equation 39). In this case, the internal total pressure P_T is equal
to 1 atm. Experimental data show on-off type sensor voltage characteristics in a rich atmosphere when a current is applied to the
sensor. (See Figures 5 and 6)

In the I - V curve (Figure 11), the sensor voltage varies greatly
at $I = I_{CR}$. I_{CR} means the mass of CO gas that diffuses through porous
zirconia (See Equation 49). Such amount is proportional to the CO
concentration when $I < I_{CR}$, the mass of oxygen that is generated at
the interface between zirconia/the reference electrode is smaller
than the mass of CO gas. And the oxygen is consumed in the reaction
($CO + 1/2\ O_2 \rightleftharpoons CO_2$). Thus, the oxygen partial pressure $P_{O_2}(I)$ at the
reference electrode is maintained at an extremely low level. (See
Equation 41). For this reason, P_T is maintained at 1 atm.

When $I > I_{CR}$, the mass of oxygen is larger than that of CO gas.
Thus, $P_{O_2}(I)$ is maintained at a high level and P_T rises steadily as

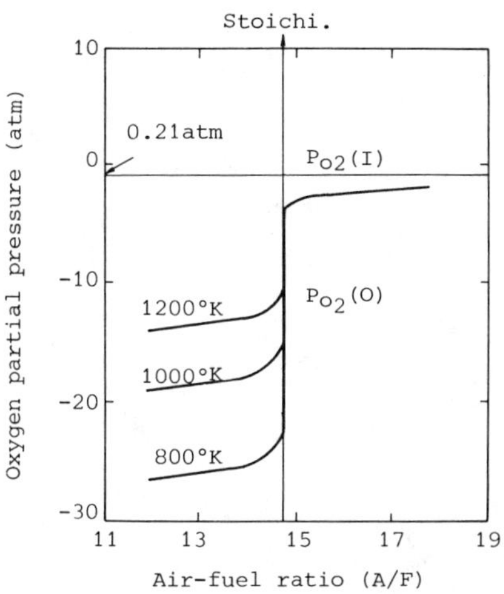

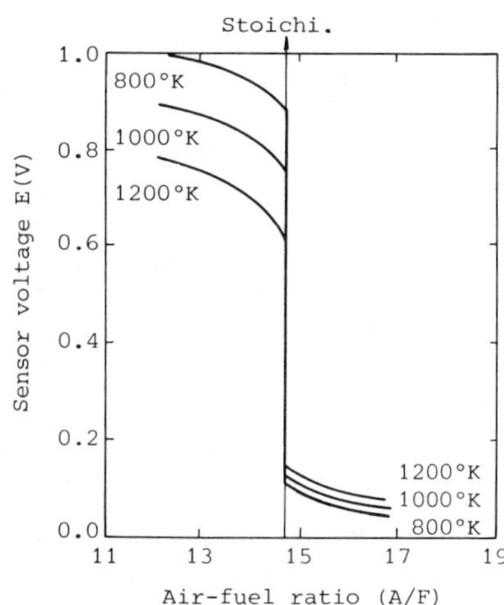

Figure 9. Calculated oxygen partial pressure and sensor voltage versus air-fuel ratio curves for crucible-type oxygen sensor

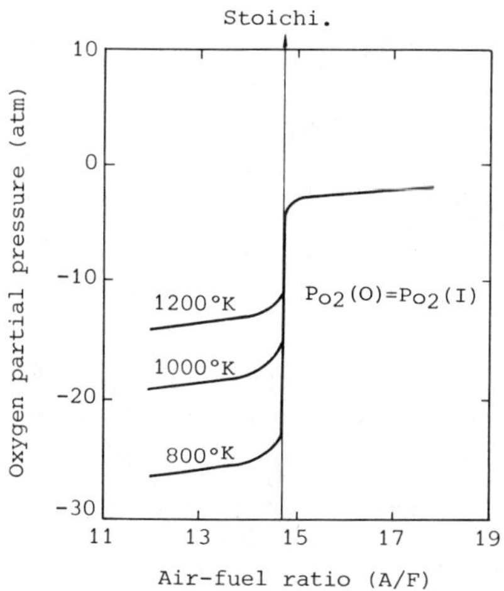

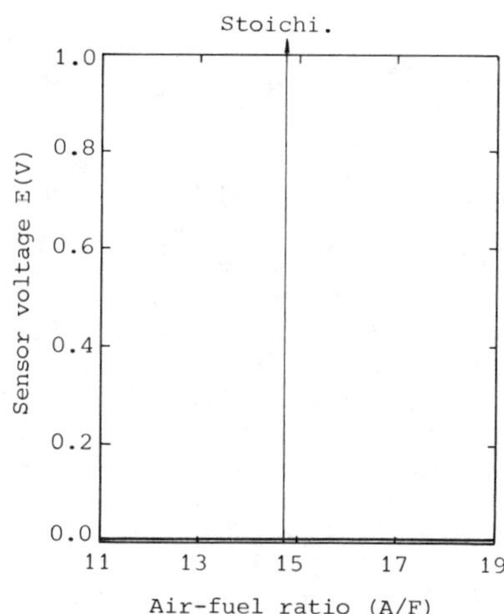

Figure 10. Calculated oxygen partial pressure and sensor voltage versus air-fuel ratio curves for thick film oxygen sensor (I = 0)

sensor with a catalytic electrode. Figure 10 shows $P_{O_2}(0)$, $P_{O_2}(I)$ vs
A/F curve and E vs A/F curve. Substituting $I = 0$ and $X = Y$, Equation
25 yields $P_T = 1$ (i. e. Total pressure at both the reference and the
measurement electrode are 1 atm).

Thick film oxygen sensor with catalytic electrode (when $I > 0$).

The calculated results are shown in Tables I and II. In Tables I
and II, I_{CR}, I_{LR}, I_{CL} and I_{LL} can be expressed by the following
equations:

$$I_{CR} \equiv \frac{4FSD_{O_2}(SE)}{RTl} \left(\frac{D_{CO}(SE)}{2D_{O_2}(SE)} P_{CO}(0) - P_{O_2}(0) \right) \tag{48}$$

$$\doteqdot \frac{2FSD_{CO}(SE)P_{CO}(0)}{RTl} \tag{49}$$

$$I_{LR} \equiv \frac{4FSD_{O_2}(PL)}{RTl} \left(\frac{D_{CO}(PL)}{2D_{O_2}(PL)} P_{CO}(g) + P_{O_2}(g) \right) \tag{50}$$

$$\doteqdot \frac{2FSD_{CO}(PL) P_{CO}(g)}{RTl} \tag{51}$$

$$I_{CL} \equiv \frac{4FSD_{O_2}(PL)}{RTh} \left(P_{O_2}(g) - \frac{D_{CO}(PL)}{2D_{O_2}(PL)} P_{CO}(g) \right) \tag{52}$$

$$\doteqdot \frac{4FSD_{O_2}(PL)P_{O_2}(g)}{RTh} \tag{53}$$

$$I_{LL} \equiv \frac{4FSD_{O_2}(PL)}{RTh} \left(P_{O_2}(g) + \frac{D_{CO_2}(PL)}{2D_{O_2}(PL)} P_{CO_2}(g) \right) \tag{54}$$

$$\doteqdot \frac{2FSD_{CO_2}(PL) P_{CO_2}(g)}{RTh} \tag{55}$$

Using the results in Tables I and II yields the calculated curves
in Figures 11 and 12.

The sensor current vs sensor voltage (V) curve and A/F vs sensor
voltage (V) curve are shown in Figure 11. The thick lines show the
calculated results and the thin lines show the experimental results.
Tendencies shown by the calculated results correlate well with the
experimental data.

The sensor current vs the sensor voltage (E) curve for the larger value of the sensor current is shown in Figure 12. In the case
of a lean atmosphere with the larger value of the sensor current, it
is seen that sensor voltage (E) changes greatly. This voltage
characteristic gives the A/F vs sensor voltage (E) characteristic
shown in Figure 13.

Internal total pressure P_T at the reference electrode can also
be calculated using the results in Tables I and II snf Equation 25.
Results of this calculation are shown in Figure 14.

When the sensor current I is small, P_T in the lean atmosphere is
larger than P_T in the rich atmosphere. As the sensor current I becomes larger, P_T becomes independent of the oxygen partial pressure
in the exhaust gas.

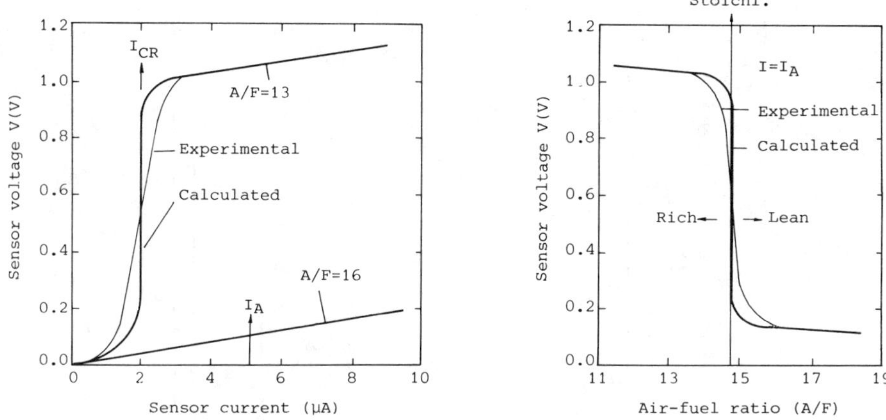

Figure 11.   Calculated and Experimental results for thick film oxygen sensor

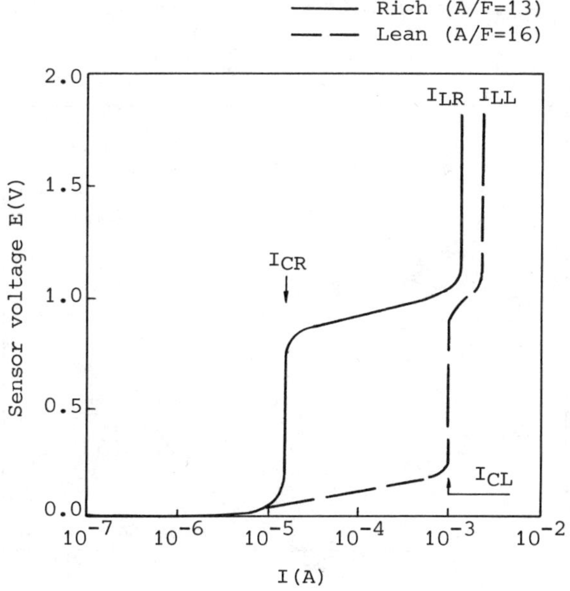

Figure 12.   Calculated E vs I curve

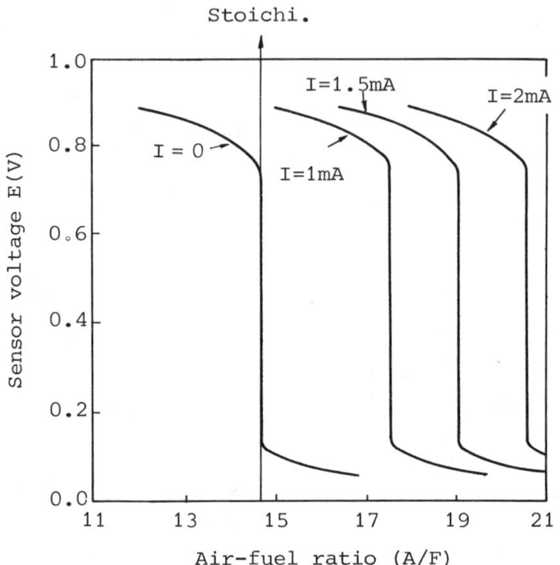

Figure 13.   Calculated E vs Air-fuel ratio curve

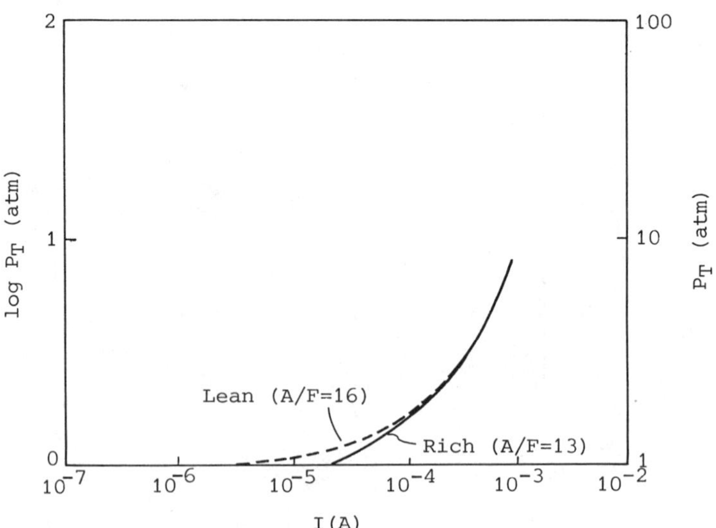

Figure 14.   Calculated P_T vs I curve

$I > I_{CR}$. $P_{O2}(I)$ is expressed by Equation 43. As I becomes larger, it
can generally be expressed by the following equation.

$$P_{O2}(I) = \frac{RTlI}{4FSD_{O2}(SE)}$$

(56)

Clearly, it becomes an almost constant oxygen partial pressure and is
independent of the oxygen partial pressure in the exhaust gas.
Therefore, it can be used as the reference value for the on-off type
stoichimetric point oxygen sensor.

I_{CR} depends on $D_{CO}(SE)$, T, S and l. As $D_{CO}(SE)$ becomes larger.
i. e., the zirconia layer becomes more porous, and I_{CR} becomes larger.
In this calculation, the value of $D_{CO}(SE)$ is assumed to be $10^{-7}cm^2/$
sec. The value of $D_{CO}(SE)$ obtained in experimental results and
calculated results closely matches.

As the sensor current I becomes larger, $P_{O2}(0)$ in a rich atmosphere becomes smaller. (See Equation 40). But when I is larger than
100μA, the effect of a current on $P_{O2}(0)$ must be considered.

When I is larger than I_{CR}, the cubic Equation 21 produces no
answers. I_{CR} means the mass of CO gas that diffuses through the
porous protective layer. As the protective spinel layer is more
porous than zirconia layer, $I_{LR} > I_{CR}$.

The depencence of $P_{O2}(I)$ and $P_{O2}(0)$ on sensor current I in a
lean atmosphere differs from that in a rich atmosphere.

In a rich atmosphere, $P_{O2}(I)$ changes abruptly. On the other
hand, in a lean atmosphere, $P_{O2}(0)$ changes greatly. This fact
suggests the possibility for a "lean oxygen sensor."

In the relationship between the sensor current and sensor
voltage (See Figure 12.), sensor voltage changes at I_{CL}. I_{CL} means
the mass of O_2 gas that diffuses through the porous protective layer
toward the measurement electrode. The change of sensor voltage at
I_{CL} shows a limiting current characteristics by oxygen gas diffusion.
The stoichimetric point in the A/F vs sensor voltage curve (See
Figure 13.) is shifted toward a lean atmosphere.

In the case of a different combustible gas atmosphere. Tests by
Takeuchi, et al produced the following esperimental phenomena: (9)
The zirconia oxygen sensor showed an on-off voltage character-
istic at $\lambda = 1$ in $N_2 - O_2 - 1\%$ CO gas.

$$\lambda = \frac{P_{O2}/P_{CO}}{(P_{O2}/P_{CO})_0}$$

(57)

$(P_{O2}/P_{CO})_0 = $ The stoichiometric point of P_{O2}/P_{CO}

(58)

On the other hand, the characteristic was seen at $\lambda = 4$ in N_2 gas.
In our model, this phenomena can be explained as follows: when N_2-
$O_2 - CO$ gas is used, A must be zero at the on-off point. (See Equation 27)

$$A = \frac{D_{CO}(PL)}{2D_{O2}(PL)} \; P_{CO}(g) - P_{O2}(g)$$

(27)

When $N_2 - O_2 - H_2$ gas is used, M must be zero at the on-off point.
M is defined by

$$M \equiv \frac{D_{co}(PL)}{2D_{o2}(PL)}\ P_{H2}(g) - P_{o2}(g)' \tag{59}$$

Thus
$$P_{o2}(g) = \frac{P_{co}(g)}{2D_{o2}(PL)}\ D_{co}(PL) \tag{60}$$

$$P_{o2}(g)' = \frac{P_{H2}(g)}{2D_{o2}(PL)}\ D_{H2}(PL) \tag{61}$$

$D_{H2}(PL)$ is considered larger than $D_{co}(PL)$. Assuming $D_{H2}(PL)/D_{co}(PL) = 4$, $P_{o2}(g)'/P_{o2}(g) = 4$ when $P_{co}(g) = P_{H2}(g) = 1\%$. Thus, λ differs from each gas atmosphere, i. e. the difference of diffusion coefficient in the protective layer of combustible gas makes λ different.

<u>Conclusions</u>

①　A more compact thick film zirconia oxygen sensor with a built-in heater has been developed. In this sensor, the reference oxygen gas is not air; the oxygen gas is generated electrolytically at the interface between the reference electrode and the porous zirconia electrolyte.
②　The voltage characteristics of the sensor are almost identical to those of the conventional crucible-type oxygen sensor.
③　Analysis of the steady-state voltage characteristics of the thick film oxygen sensor and conventional crucible-type oxygen sensor indicates agreement between the theoretical curves and the experimental curves.
④　In the case of a larger sensor current, the model for the thick film oxygen sensor used in this analysis showed favorable possibilities for a "lean oxygen sensor."

Appendix
Dervation of Equation 12
Equation 12 can be derived as follows:
The mass of oxygen which is converted to oxygen ions = $I/4FS$.
The mass of oxygen gas which diffuses through the porous zirconia layer = J

$$J = D\ \frac{\partial C}{\partial X} = D\ \frac{\partial}{\partial X}\ \left(\frac{P}{RT}\right) = \frac{D}{RT}\ \frac{\partial P}{\partial X} = \frac{D}{RT}\ b_2\ \left(\frac{\partial P}{\partial X} \equiv b_2\right)$$

From the assumption of chemical equilibrium in the following reaction

$$CO + 1/2\ O_2 \rightleftharpoons CO_2$$

the coefficient of the mass of CO gas is $1/2$.

We obtain　　$\dfrac{I}{4FS} = \dfrac{D_{o2}(SE)}{RT}\ b_2 + \dfrac{D_{co}(SE)}{2RT}\ b_1$

Therefore,　$\dfrac{IRT}{4FS} = D_{o2}(SE)\ b_2 + 1/2\ D_{co}(SE)b_1$

This equation is the same as equation 12.   Equation 18 can also be introduced in this way.

Legend of Symbols

A	Equation 27
a_1	Pressure coefficient of CO gas in protective layer
a_2	Pressure coefficient of O_2 gas in protective layer
a_3	Pressure coefficient of CO_2 gas in protective layer
A/F	Air Fuel ratio
$(A/F)_o$	Ideal air fuel ratio
b_1	Pressure coefficient of CO gas in solid electrolyte
b_2	Pressure coefficient of O_2 gas in solid electrolyte
b_3	Pressure coefficient of CO_2 gas in solid electrolyte
C	Gas concentration P/RT
D	Diffusion coefficient
$D_{CO}(PL)$	Diffusion coefficient of CO gas in protective layer
$D_{O_2}(PL)$	Diffusion coefficient of O_2 gas in protective layer
$D_{CO_2}(PL)$	Diffusion coefficient of CO_2 gas in protective layer
$D_{H_2}(PL)$	Diffusion coefficient of H_2 gas in protective layer
$D_{CO}(SE)$	Diffusion coefficient of CO gas in solid electrolyte
$D_{O_2}(SE)$	Diffusion coefficient of O_2 gas in solid electrolyte
$D_{CO_2}(SE)$	Diffusion coefficient of CO_2 gas in solid electrolyte
E	Equation 1
F	Faraday constant
h	Thickness of protective layer
I	Current
I_{CL}	Equation 52
I_{CR}	Equation 48
I_{LL}	Equation 54
I_{LR}	Equation 50
K	Equation 3
l	Thickness of solid electrolyte
M	Equation 59
P_{CO}	Partial pressure of CO gas
P_{O_2}	Partial pressure of O_2 gas
P_{CO_2}	Partial pressure of CO_2 gas
$P_{CO}(g)$	Partial pressure of CO gas in exhaust gas
$P_{O_2}(g)$	Partial pressure of O_2 gas in exhaust gas
$P_{CO_2}(g)$	Partial pressure of CO_2 gas in exhaust gas
$P_{N_2}(g)$	Partial pressure of N_2 gas in exhaust gas
$P_{H_2}(g)$	Partial pressure of H_2 gas in $N_2 - H_2 - O_2$ gas
$P_{O_2}(g)'$	Partial pressure of O_2 gas in $N_2 - H_2 - O_2$ gas
$P_{CO}(0)$	Partial pressure of CO gas at measurement electrode
$P_{O_2}(0)$	Partial pressure of O_2 gas at measurement electrode
$P_{CO_2}(0)$	Partial pressure of CO_2 gas at measurement electrode
$P_{N_2}(0)$	Partial pressure of N_2 gas at measurement electrode
$P_{CO}(I)$	Partial pressure of CO gas at reference electrode
$P_{O_2}(I)$	Partial pressure of O_2 gas at reference electrode
$P_{CO_2}(I)$	Partial pressure of CO_2 gas at reference electrode
$P_{N_2}(I)$	Partial pressure of N_2 gas at reference electrode
P_T	Total pressure at reference electrode
R	Gas constant
R_o	Resistance of sensor
S	Area of electrode
T	Absolute temperature
V	Equation 4

x   Diffusion length
X   Equation 22
Y   Equation 23
λ Equation 57

<u>Literature Cited</u>

1. Eddy, D. S., IEEE Transactions on Vehicular Tech., VT - 23, 1974, 125
2. Hamann, E., Manger, H. and Steinke, L., SAE paper 770401, 1977
3. Fleming, W. J., SAE paper 770400, 1977
4. Fleming, W. J., J. Electrochem. Soc., 1977, 124, 21
5. Fleming, W. J., SAE paper 800020, 1980
6. Wang, D. Y. and Nowick, A. S., J. Electrochem, Soc., 1979, 126, 1155
7. Verkerk, M. J. and Burggraaf, A. J., J. Electrochem. Soc., 1983, 130, 78
8. Mizusaki, J., Amano, K., Yamauchi, S. and Fueki, K., Proceedings of the International National Meeting on Chemical Sensor, Fukuoka, 1983, 279
9. Takeuchi, T., Saji, K. and Igarashi, I., abstract 74, p196, The Electrochemical Society Extended abstracts, Pittsburgh, Oct., 1978

RECEIVED December 12, 1985

A Solid Electrolyte for Sulfur Dioxide Detection
Sodium Sulfate Mixed with Rare Earth Sulfates and Silicon Dioxide

Nobuhito Imanaka, Gin-ya Adachi, and Jiro Shiokawa

Department of Applied Chemistry, Faculty of Engineering, Osaka University, Yamadaoka 2-1, Suita, Osaka 565, Japan

A new solid state chemical sensor for sulfur dioxide utilizing a sodium sulfate/rare earth sulfates/silicon dioxide electrolyte has been developed. The addition of rare earth sulfates and silicon dioxide to the sodium sulfate electrolyte was found to enhance the durability and electrical conductivity of the electrolyte. The electrolyte exhibits a Nernstian response in the range of SO_2 gas concentrations from 30 ppm to 1 %.

As is well-known, sulfur oxides and nitrogen oxides exhausted into air, which can result in acid rain, have caused serious deterioration of the environment. The potential need for regulation of SO_x and NO_x gases in combustion emissions is, nowadays, becoming an important research area.

For practical measurements, several techniques for SO_2 analysis have been widely adopted as follows:
   (1) The electrical conductivity measurement of an
       absorbed solution
   (2) Infrared absorbtion analysis
   (3) Ultra-violet spectrum photometric analysis
   (4) Flame photometry
   (5) Stationary potential electrolysis
However, the apparatus for these methods is expensive and complicated. Recently, a concentration cell method using a solid electrolyte has become of interest for SO_2 gas detection. A potential advantage of this technique is that monitering for SO_2 can be undertaken simply, selectively, and continuously with low cost.

As the electrolytes, alkali metal sulfates(M=Li, Na, and K)(1-11), β-Alumina(12), and NASICON(13, 14) have been examined. Alkali metal sulfates are cation conductors at elevated temperature($>700^{\circ}$C). However, they have several disadvantages. One is the phase transformation of the sulfates(15-18). By this transformation, cracks occur in the electrolyte body and result in the permeation of the ambient gases. The other disadvantage is their low electrical conductivity. Mono, di, or tri-valent cations(19-24) have been doped so as to enhance their conductivity. Furthermore, they become ductile at a temperature higher than approximately 800°C. β-Alumina is one of the other representative cation conductors. NASICON is one of the most widely used materials(25-32) that have been utilized as cation conductors. However, both of them are not commercially available at present. In addition, β-Alumina and NASICON materials are considerably more expensive than alkali metal sulfates.

In our investigation, sodium sulfate was selected as the electrolyte. Rare earth sulfates $Ln_2(SO_4)_3$(Ln=Y and Gd) were added in order to increase the electrical conductivity. Silicon dioxide was added so as to obtain the network structure which is effective for Na^+ cation conduction and to prevent the electrolyte from becoming too soft. A thinner electrolyte was possible to prepare by mixing in SiO_2. The suppression of the phase transformation(15, 16) from Na_2SO_4-I(a high temperature phase) to Na_2SO_4-III(a low temperature phase) was also achieved by mixing rare earth sulfates(Ln=Y and Gd) and silicon dioxide into sodium sulfate.

The application of the Na_2SO_4-$Ln_2(SO_4)_3$-SiO_2(Ln=Y and Gd) electrolyte samples as the solid electrolyte for an SO_2 gas detector was investigated. The EMF measurements were conducted by both the SO_2 gas concentration cell(33) and the solid reference electrode(34) methods. Several efforts have been concentrated on the development of the appropriate reference electrode. In our study, the sulfate-oxide solid reference electrode method was adopted.

<u>Experimental</u>

<u>Materials.</u> Sodium sulfate(purity: 99.99 %), silicon dioxide(purity: 99.9 %) were bought from Wako Pure Chemical Industries Ltd.. Yttrium (purity: 99.9 %) and gadlinium(purity: 99.99 %) oxides were purchased from Shiga Rare Metal Industries Ltd.. Rare earth sulfates(Ln=Y and Gd) were prepared by adding the concd. sulfuric acid into rare earth oxides(Ln=Y and Gd). Before weighing, sodium sulfate and silicon dioxide were dried. Rare earth sulfates were also heated for dehydration. Since $Ln_2(SO_4)_3$(Ln=Y and Gd) are considerably hygroscopic, the actual concentration of rare earth cation in the mixture was determined by the EDTA titration. Preheated materials were cooled in a desiccator, weighed, and mixed thoroughly in an agate mortar. The mixture was melted at 1473°K for 1 h and then quenched. The resulting material was reground, made into pellets under hydrostatic pressure (2.65×10^8 Pa), and then sintered at 1073°K for 1 h in air. Platinum sputtering on the center surface of the electrolyte was conducted using a Shimadzu's ion coater IC-50.

<u>Measurements.</u> X-ray and thermal analyses were carried out for the Na_2SO_4-SiO_2 and the Na_2SO_4-$Ln_2(SO_4)_3$-SiO_2 systems so as to investigate their phases with a Rigaku's Rotaflex diffractometer(Cu target) and a Rigaku's Thermoflex, respectively. Electrical conductivities were measured by means of the complex impedance method with a Hewlett Packard vector impedance meter 4800A. The apparatus for the electrical conductivity measurements are illustrated in Figure 1. The sample was fastened by spring loading the quartz rod.

Results and Discussion

Electrical conductivity, phases, and thermal properties.

Na_2SO_4-SiO_2 systems: The results of the phases and thermal properties are summarized in Table I.

Table I. The phases and thermal properties of Na_2SO_4-SiO_2

Sample #	Na_2SO_4 (mol%)	SiO_2 (mol%)	Phases	DTA peak $T/^OK$
1	90	10	Na_2SO_4-III+SiO_2	513
2	70	30	Na_2SO_4-III+SiO_2	513
3	50	50	Na_2SO_4-III+SiO_2	513

All samples show Na_2SO_4-III phase together with SiO_2. The typical phase transformation from I to III at 513^OK was observed from the measurement. The mixing of SiO_2 alone into sodium sulfate cannot suppress the I to III transition. The temperature dependences of the electrical conductivity for the Na_2SO_4-SiO_2 systems are shown in Figure 2. The addition of SiO_2 into Na_2SO_4 does not enhance the conductivity. The discontinuity in the $\log(\sigma T)$-1/T relation which results from the phase transformation between III to I, also appears in the Na_2SO_4-SiO_2 systems. The temperature at a break is nearly the same as the DTA result. Sodium sulfate mixed with SiO_2 does not seem to be an appropriate solid electrolyte because of its low electrical conductivity and the presence of a transformation.

Na_2SO_4-$Y_2(SO_4)_3$-SiO_2 systems: Phases and thermal properties are listed in Table II.

Table II. The phases and thermal properties of
Na_2SO_4-$Y_2(SO_4)_3$-SiO_2

Sample #	Na_2SO_4 (mol%)	$Y_2(SO_4)_3$ (mol%)	SiO_2 (mol%)	phases	DTA peak $T/^OK$
1	55.1	4.9	40.0	A'+(small SiO_2)	593
2	52.2	7.7	40.1	A+(small SiO_2)	———
3	50.1	9.9	40.0	A+(small SiO_2)	———
4	48.1	11.8	40.1	A+(small SiO_2)	———
5	45.1	14.8	40.1	A+(small SiO_2)	———

A and A': the phase which includes the peaks of Na_2SO_4-I-like phase. The simbol A' applied instead of A because some peaks of A phase split.

From the X-ray analyses, all five samples exhibit a new phase, A, and a small amount of SiO_2 phase. This A phase includes the phase similar to the Na_2SO_4-I phase(15, 16), which is superior in Na^+ cat-

ion conduction. From the DTA measurements, endo, and exo-thermal
peaks were observed at 593°K without any gravimetric deviation,
indicating that a phase transition exists. On the other hand, the
samples(# 2-5) exhibit no peak, that is, no phase transition occurs.
The samples from # 2 to 5 have a potential for the solid electrolyte because they hold Na_2SO_4-I-similar phase with no transition.
The $\log(\sigma T)$ vs. $1/T$ relation is presented in Figure 3. The electrical conductivity for the sample # 1 is relatively high at low temperature. However, the break in the curve exists approximately at
593°K, which coincides with the DTA results. The samples # 2-4
show almost the same results. The curves for their electrical conductivity are almost straight, indicating that no phase transition
appears. Their conductivities are about 20 times higher than that
of sodium sulfate at 873°K. In the case of 14.8 mol% $Y_2(SO_4)_3$, the
conductivity has decreased. The cation vacancies produced by the
$Y_2(SO_4)_3$ doping, come to form clusters and do not contribute to the
cation conduction. From DTA and the electrical conductivity results,
the samples # 2-4 are the candidates in the Na_2SO_4-$Y_2(SO_4)_3$-SiO_2
systems.

Na_2SO_4-$Gd_2(SO_4)_3$-SiO_2 systems: The summary of the X-ray and DTA analyses are tabulated in Table III.

Table III. The phases and thermal properties of
Na_2SO_4-$Gd_2(SO_4)_3$-SiO_2

Sample #	Na_2SO_4 (mol%)	$Gd_2(SO_4)_3$ (mol%)	SiO_2 (mol%)	phases	DTA peak $T/^{O}$K
1	55.2	4.7	40.1	α+(small SiO_2)	——
2	52.0	8.0	40.0	α+(small SiO_2)	——
3	51.0	8.2	40.8	α+(small SiO_2)	——
4	48.2	11.6	40.2	α+(small SiO_2)	——
5	43.8	17.2	39.0	α'+(small SiO_2)	——

α and α': the phase which includes the peaks of Na_2SO_4-I-like phase.
The peaks of phase α' are slightly deviated from those of phase α
toward low degree side except a peak at 31.6 degree.

All samples(# 1-5) show a new phase, α, in which some peaks of the A
phase disappear, besides a small amount of SiO_2 phase. DTA measurement indicates that no phase transformation exists. The $\log(\sigma T)$-$1/T$
results are shown in Figure 4. Almost straight relation and highest
conductivity(at temperature higher than 781°K) were obtained for
sample # 1-3. This means that a phase transition does not exist.
The conductivity of the sample # 4, decreases by the formation of
the cation vacancy clusters. In the case of 17.2 mol % $Gd_2(SO_4)_3$
mixed sample(# 5), the σT becomes higher at relatively lower temperature. However, the discontinuity exists at about 543°K and the conductivity becomes lower than that of the samples # 1-3. From these
results, the samples which maintain potential characteristics for
the electrolyte are # 1-3.

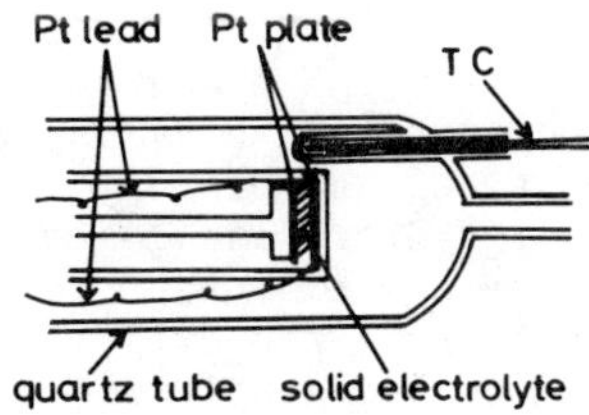

Figure 1. The apparatus for the electrical conductivity measurements.

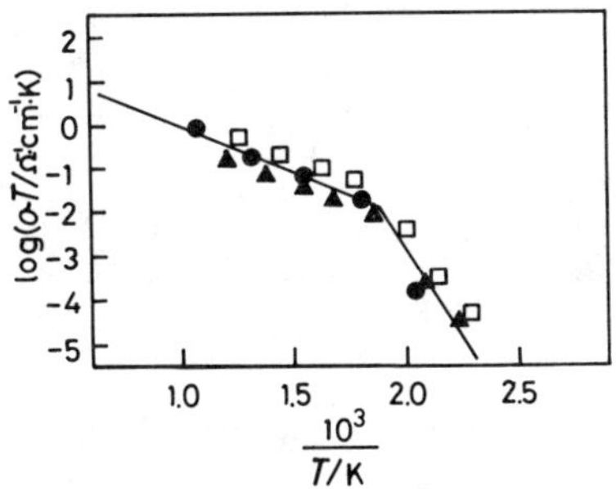

Figure 2. Temperature dependences of electrical conductivities for the Na_2SO_4-SiO_2.
● Na_2SO_4:SiO_2 = 90:10
□ Na_2SO_4:SiO_2 = 70:30
▲ Na_2SO_4:SiO_2 = 50:50
—— Na_2SO_4

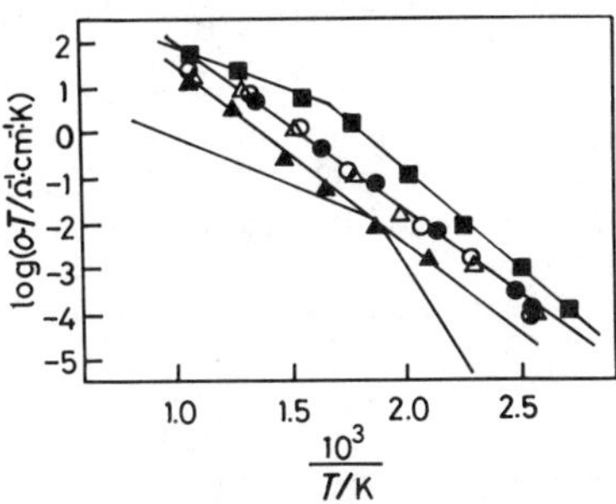

Figure 3. Temperature dependences of electrical conductivities for the Na_2SO_4-$Y_2(SO_4)_3$-SiO_2.
■ Na_2SO_4:$Y_2(SO_4)_3$:SiO_2 = 55.1: 4.9:40.0
○ Na_2SO_4:$Y_2(SO_4)_3$:SiO_2 = 52.2: 7.7:40.1
● Na_2SO_4:$Y_2(SO_4)_3$:SiO_2 = 50.1: 9.9:40.0
△ Na_2SO_4:$Y_2(SO_4)_3$:SiO_2 = 48.1:11.8:40.1
▲ Na_2SO_4:$Y_2(SO_4)_3$:SiO_2 = 45.1:14.8:40.1
—— Na_2SO_4

EMF measurements.

Na_2SO_4-$Y_2(SO_4)_3$-SiO_2 systems: EMF measurements were conducted with the apparatus illustrated in Figure 5 by using the SO_2 gas concentration cell method. Test SO_2 and O_2 gas mixtures were introduced from tube A depicted in Figure 5. The gas was led through the holes to reach the surface of the electrolyte. The Pt net was inserted in tube A in order to accelerate the oxidation from SO_2 to SO_3. Quartz wool(B) was placed between the tube(A) and the electrolyte so as to diffuse the test gas uniformly and to avoid the temperature gradient on the electrolyte. Pt net electrodes were applied so that the electrolyte could maintain good contact with the ambient gas. The solid electrolyte was fixed between the outer quartz tube and the quartz rod by spring loading. The ringed glass packing was applied to prevent the test gas from mixing with the reference gas. The test SO_2 gas concentration from 30 ppm to 1.11 % was generated using a Standard Gas Generator(SGGU-711SD) from Standard Technology Co.. The reference SO_2 gas content from 6 % to 23 % was controlled with the self-made flow meters. The EMF properties for the appropriate Na_2SO_4-$Y_2(SO_4)_3$-SiO_2(50.1:9.9:40.0) are presented in Figure 6. The measured EMF coincides with the calculated EMF(1) in the inlet SO_2 gas concentration from 1000 ppm(0.1 %) to 23 %. The measured EMF becomes appreciably lower than the calculated EMF, as the SO_2 gas content becomes less than 1000 ppm.

Na_2SO_4-$Gd_2(SO_4)_3$-SiO_2 systems: Figure 7 shows the variation of the EMF for the electrolyte sample consisting of Na_2SO_4-$Gd_2(SO_4)_3$-SiO_2 (51.0:8.2:40.8). There is almost no difference between the measured and the calculated EMF, for inlet SO_2 gas concentrations ranging from 3200 ppm(0.32 %) to 10 %. In the SO_2 gas concentration region lower than 0.32 % as well as higher than 10 %, the measured EMF becomes smaller than the calculated value. The response of this electrolyte is not as good as that observed for the Na_2SO_4-$Y_2(SO_4)_3$-SiO_2 systems. The Na_2SO_4-$Y_2(SO_4)_3$-SiO_2 and the Na_2SO_4-$Gd_2(SO_4)_3$-SiO_2 gave excellent response in the range of 0.1 %-23 % and 0.32 %-10 %, respectively. The sodium sulfate electrolytes mixed with $Y_2(SO_4)_3$ and SiO_2 are better solid electrolytes for the SO_2 gas detector.

The reason that the Na_2SO_4-$Y_2(SO_4)_3$-SiO_2 solid electrolyte can not detect adequately the SO_2 gas lower than 0.1 %, is attributed to the difficulties in the formation and decomposition of sodium sulfate on the electrolyte surfaces, and in obtaining the suitable contact between the Pt net electrode and the electrolyte.

In order to improve the EMF characteristics in the lower SO_2 gas content range(<1000 ppm), platinum sputtering on the center surface(about 5×10^{-3} m in diameter) of the electrolyte(1.3×10^{-2} m in diameter) was attempted.

Na_2SO_4 systems(Pt sputtering): In the EMF measurements for the electrolyte with Pt sputtering, the reference SO_2 and O_2 gas mixture was regulated with a commercial gas flow meter KOFLOG RK 1200 from Kojima Flow Instruments Co. and Pt mesh was installed instead of the quartz wool(B) shown in Figure 5. The sodium sulfate samples were prepared with platinum sputtering for 10 min on working and on both surfaces. Approximately 10 nm thickness of the Pt film was obtained by the 10 min Pt sputtering. Figure 8 presents the EMF variations

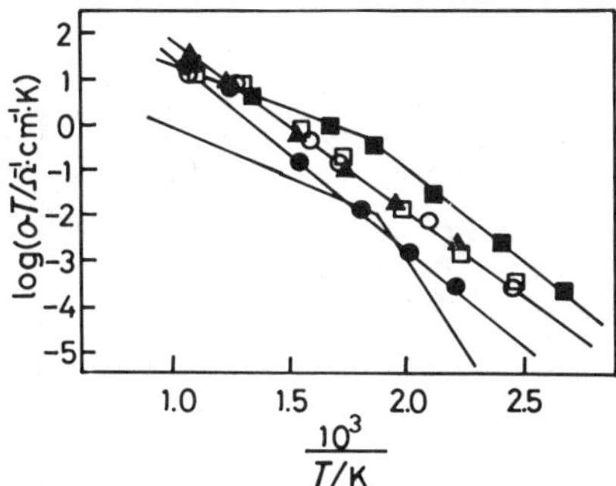

Figure 4. Temperature dependences of electrical conductivities
for the Na_2SO_4-$Gd_2(SO_4)_3$-SiO_2.

—□— Na_2SO_4:$Gd_2(SO_4)_3$:SiO_2 = 55.2: 4.7:40.1
—○— Na_2SO_4:$Gd_2(SO_4)_3$:SiO_2 = 52.0: 8.0:40.0
—▲— Na_2SO_4:$Gd_2(SO_4)_3$:SiO_2 = 51.0: 8.2:40.8
—●— Na_2SO_4:$Gd_2(SO_4)_3$:SiO_2 = 48.2:11.6:40.2
—■— Na_2SO_4:$Gd_2(SO_4)_3$:SiO_2 = 43.8:17.2:39.0
——— Na_2SO_4

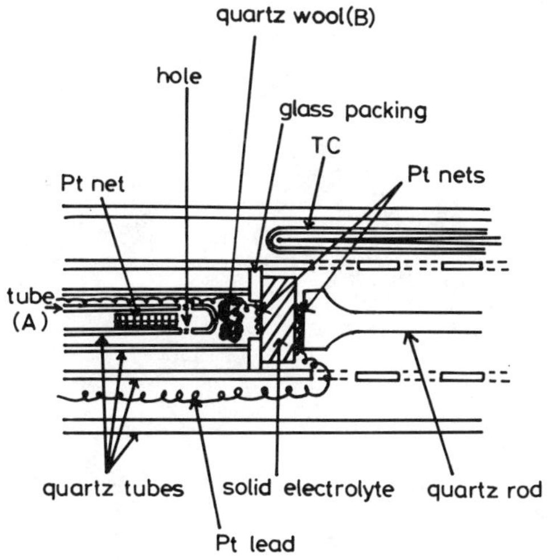

Figure 5. The apparatus for the electromotive force(EMF)
measurements(An SO_2 gas concentration cell method).

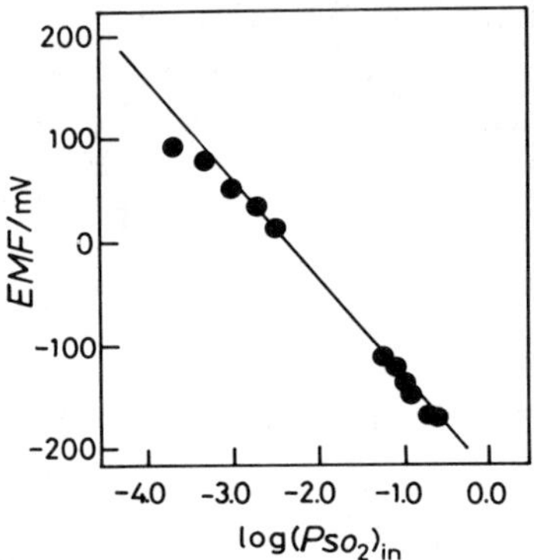

Figure 6.   Variation of the EMF for the concentration cell;
Pt|$O_2(p_1')$, SO_2 (p_2')|Na_2SO_4(9.9 mol % $Y_2(SO_4)_3$, 40.0 mol %
SiO_2)|$O_2(p_1'')$, $SO_2(p_2'')$,|Pt, with p_1''=0.996 and p_2''=0.004
at 973°K.
———————— is calculated EMF(1).

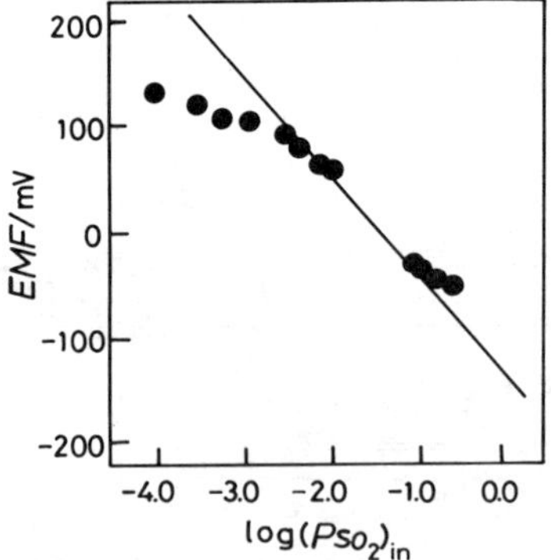

Figure 7.   Variation of the EMF for the concentration cell;
Pt|$O_2(p_1')$, $SO_2(p_2')$|Na_2SO_4(8.2 mol % $Gd_2(SO_4)_3$, 40.8 mol %
SiO_2)|$O_2(p_1'')$, $SO_2(p_2'')$|Pt, with p_1''=0.996 and p_2''=0.034
at 973°K.
———————— is calculated EMF(1).

together with sodium sulfate without sputtering. The measured EMF
was consistent with the calculated EMF in the inlet SO_2 gas concentration between 0.2 % and 1 %. The measured EMF of the sodium
sulfate without Pt sputtering becomes appreciably lower than the calculated value in the $\log(p_{SO_2})_{in}$ expression -2.7(2000 ppm). The EMF
characteristics of the Pt sputterred sample was improved in the lower
concentration(<0.2 %) compared with that of Na_2SO_4 without sputtering.
The suitable contact between the Pt net electrode and the electrolyte
can be obtained by the sputtering. No meaningful difference in the
EMF characteristics exists in the two Pt sputterred samples. The
difference between the measured and the calculated EMF becomes larger,
the SO_2 gas content smaller than 0.1 %($\log(p_{SO_2})_{in}$=-3.0), with Pt
sputterred samples. These may result from the permeation of the ambient gases in the electrolyte because of the cracks produced from
the III to I phase transformation.

Na_2SO_4-$Y_2(SO_4)_3$-SiO_2 systems(Pt sputtering): The results of the EMF
measurements as a function of SO_2 gas concentration is presented in
Figure 9. The samples with 5 min Pt sputtering and 10 min sputtering
on the working electrode side, were prepared in addition to the sample without sputtering. The measured EMF for the Na_2SO_4-$Y_2(SO_4)_3$-
SiO_2 without sputtering was in good accordance with the calculated
value in the $\log(p_{SO_2})_{in}$ range from -3.0 to -2.0. Compared with
the calculated EMF, the measured EMF suddenly decreases at an
SO_2 concentration lower than 1000 ppm. Five min sputtering of Pt
onto the electrolyte enables it to detect the SO_2 gas with good response from 500 ppm($\log(p_{SO_2})_{in}$=-3.3) to 10000 ppm(1 %). Sputtering
of Pt onto the electrolyte for 10 min lowered the SO_2 gas detection
limit to 200 ppm($\log(p_{SO_2})_{in}$=-3.7). As the platinum sputtering time
was increased, the measured EMF approaches closer to the calculated
value. The formation and the decomposition of the sodium sulfate on
the electrolyte surface come to occur easily by the increase of Pt sputtering time. The experimental results from sputtering Pt for 10, 20,
and 30 min on the working electrode surface are exhibited in Figure 10.
These is no significant differences in the EMF response from these
three samples. Only at 30 ppm($\log(p_{SO_2})_{in}$=-4.52), does the electrolyte
with 10 min Pt sputtering shows the nearest EMF value to that expected.
The dependence of the Pt sputtering time on the measured EMF/calculated
EMF ratio is shown in Figure 11. The measured EMF/calculated EMF proportion changes significantly from 0-10 min. Little change occurred
for sputtering times greater than 10 min. The optimum platinum sputtering time is determined to be 10 min. Because the 10 min is good
enough to sputter, both surfaces of the electrolyte has been sputterred for 10 min. The results for Na_2SO_4-$Y_2(SO_4)_3$-SiO_2 solid electrolyte with sputtering on both surfaces is presented in Figure 12
together with the result of Na_2SO_4. The EMF characteristics for
Na_2SO_4 remarkably decreases as a result of the penetration through
the cleavage appearing in the electrolyte, at an SO_2 gas concentration less than 0.1 %. On the other hand, the measured EMF for the
Na_2SO_4-$Y_2(SO_4)_3$-SiO_2 show excellent agreement with the calculated
EMF in the range of SO_2 concentration 30 ppm to 1 %.
     In the laboratory, the SO_2 gas concentration cell method using
the solid electrolyte is considered to be a good technique for detecting the SO_2 gas. However, in a practical utilization, the method is

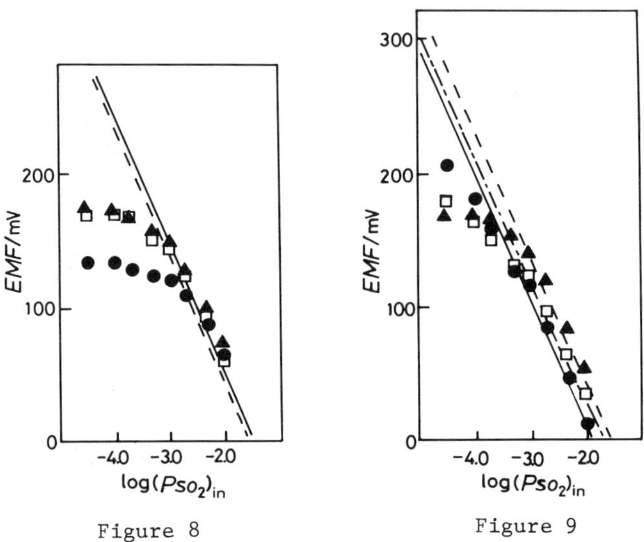

Figure 8   Figure 9

Figure 8.   The variation of the EMF for Na_2SO_4 solid electrolyte
with the SO_2 gas concentration

    ● Na_2SO_4 (without Pt sputtering)(————)

    □ Na_2SO_4 (10 min Pt sputtering on working
                 electrode surface only)(— —)

    ▲ Na_2SO_4 (10 min Pt sputtering on both
                 surfaces)(————)

———— and — — are calculated EMF($\underline{1}$), respectively.

Figure 9.   The variation of the EMF for $Na_2SO_4-Y_2(SO_4)_3-SiO_2$
(48.1:11.8:40.1) solid electrolyte with the SO_2 gas concentration
(Pt sputtering on working electrode surface only)

    ▲ $Na_2SO_4-Y_2(SO_4)_3-SiO_2$ (0 min)(————)

    □ $Na_2SO_4-Y_2(SO_4)_3-SiO_2$ (5 min)(—— - ——)

    ● $Na_2SO_4-Y_2(SO_4)_3-SiO_2$ (10 min)(————)

   — — —, —— - ——, and ———— are calculated EMF($\underline{1}$),
respectively.

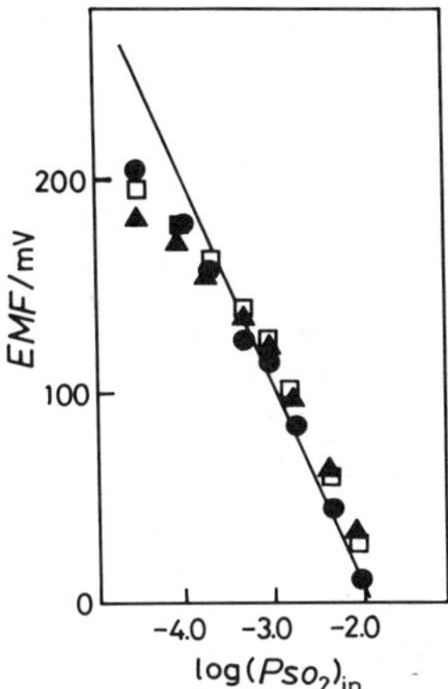

Figure 10. The variation of the EMF for the $Na_2SO_4-Y_2(SO_4)_3-SiO_2$ (48.1:11.8:40.1) solid electrolyte with the SO_2 gas concentration (Pt sputtering on working electrode surface only)
● $Na_2SO_4-Y_2(SO_4)_3-SiO_2$ (10 min)
□ $Na_2SO_4-Y_2(SO_4)_3-SiO_2$ (20 min)
▲ $Na_2SO_4-Y_2(SO_4)_3-SiO_2$ (30 min)
——— is calculated EMF(1).

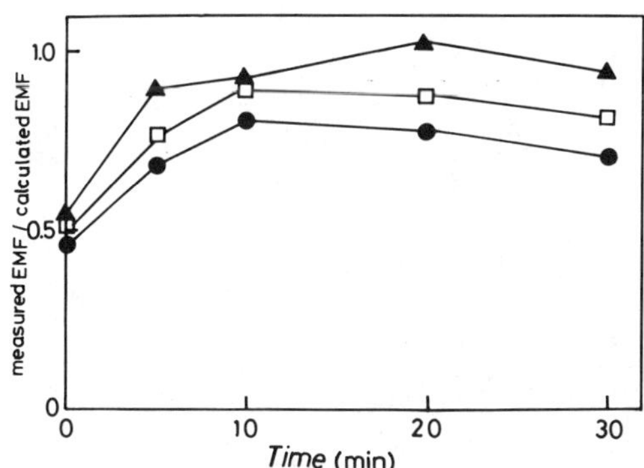

Figure 11. The Pt sputtering time dependences of measured EMF/calculated EMF with various test SO_2 gas concentration
●   30 ppm
□ 100 ppm
▲ 500 ppm

not suitable, particularly because the apparatus is expensive and complicated. The solid reference electrode method was examined in order to approach a more practical application.

The apparatus for the solid reference electrode method is depicted in Figure 13. The solid electrolyte was directly kept in contact with the solid reference electrode by fixing the reference platinum electrode between them. The sample is covered with a bonding agent(SUMICERAM from Sumitomo Chemical Industries Ltd.). As the solid reference electrode, the equimolar mixture of nickel sulfate and nickel oxide was applied. Figure 14 presents the EMF results of the Na_2SO_4-$Y_2(SO_4)_3$-SiO_2 solid electrolyte with the solid reference electrode method. In the case of the sample without Pt sputtering, the measured EMF was almost the same as the calculated EMF from 100 ppm ($\log(p_{SO_2})_{in}$=-4.0) to 1 %. The EMF at 30 ppm($\log(p_{SO_2})_{in}$=-4.52) was approximately 30 mV smaller than the calculated EMF. When the platinum was sputterred on both surfaces of the electrolyte, the EMF characteristics coincided very well with the calculated value for the measured SO_2 gas content in the range from 30 ppm to 1 %.

The solid reference electrode technique is able to obtain almost the same results as the SO_2 gas concentration cell method. The apparatus can be made more compact, simple and cheaper by using the solid reference electrode technique. The SO_2 gas detection with solid reference electrode method is a promising technique for practical applications.

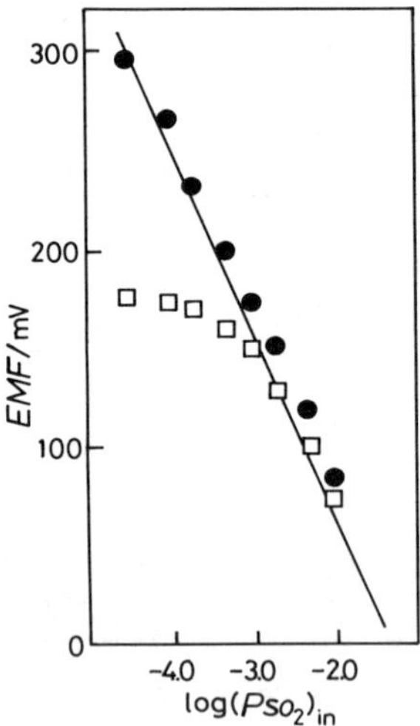

Figure 12. The variation of the EMF for Na_2SO_4 and Na_2SO_4-$Y_2(SO_4)_3$-SiO_2 solid electrolytes with the SO_2 gas concentration (Pt sputtering on both surfaces for 10 min)
● Na_2SO_4-$Y_2(SO_4)_3$-SiO_2
□ Na_2SO_4
———— is calculated EMF(1).

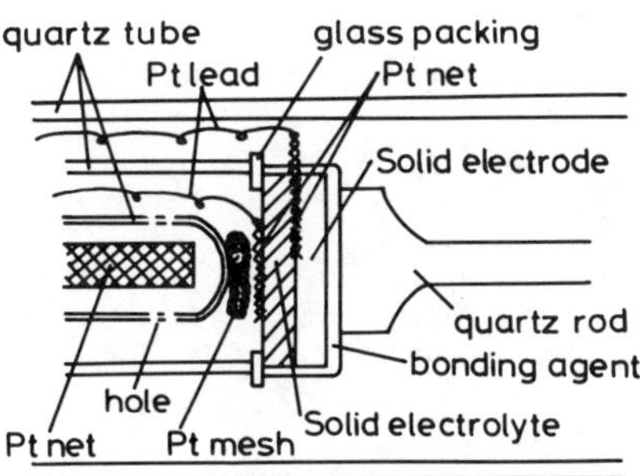

Figure 13. The apparatus for the EMF measurements
(A solid reference electrode method)

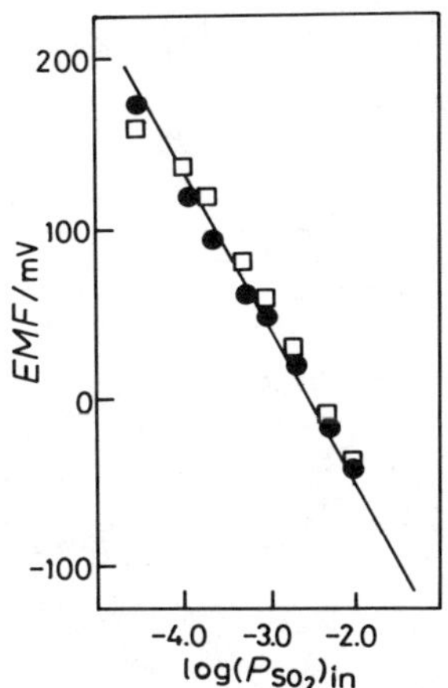

Figure 14. The variation of the EMF for $Na_2SO_4-Y_2(SO_4)_3-SiO_2$ (48.1:11.8:40.1) solid electrolyte with the solid reference electrode($NiSO_4$+NiO)

● 10 min Pt sputtering on both surfaces
□ without Pt sputtering
—— is calculated EMF(7).

In conclusion, the sodium sulfate mixed with rare earth sulfates (Ln=Y and Gd) and silicon dioxide exhibits high electrical conductivity and is more durable than the pure sodium sulfate. Furthermore, the $Na_2SO_4-Y_2(SO_4)_3-SiO_2$ solid electrolyte maintains a similar phase to Na_2SO_4-I, which is excellent in Na^+ cation conduction. The measured EMF was in excellent accordance with the calculated EMF, at SO_2 gas concentration in the range of 30 ppm to 1 %. In fact, the solid reference electrode method could be applicable as a practical SO_2 gas detector.

Acknowledgments

The present work was partially supported by a Grant-in-Aid for Developmental Scientific Research NO.57850250 from the Ministry of Education, Science and Culture.

Literature Cited

1. Jacob, K. T.; Rao, D. B. J. Electrochem. Soc., 1979, 126, 1842-7.
2. Worrell, W. L. Proc. the International Meeting on Chemical Sensors, Fukuoka, 1983, p. 332.
3. Worrell, W. L.; Liu, Q. G. J. Electroanal. Chem. Interfacial Electrochem., 1984, 168, 355-62.
4. Gauthier, M.; Chamberland, A. J. Electrochem. Soc., 1977, 124, 1579-83.
5. Gauthier, M.; Chamberland, A.; Bélanger, A.; Poirier, M. J. Electrochem Soc., 1977, 124, 1584-7.
6. Gauthier, M.; Bellemare, R.; Bélanger, A. J. Electrochem. Soc., 1981, 128, 371-8.
7. Gauthier, M.; Bale, C. W. Metall. Trans. B, 1983, 14B, 117-24.
8. Imanaka, N.; Adachi, G.; Shiokawa, J. Chem. Lett., 1983, 287-8.
9. Imanaka, N.; Adachi, G.; Shiokawa, J. Denki Kagaku, 1983, 51, 93-4.
10. Imanaka, N.; Adachi, G.; Shiokawa, J. Bull. Chem. Soc. Jpn., 1984, 57, 687-91.
11. Imanaka, N.; Adachi, G.; Shiokawa, J. Proc. the International Meeting on Chemical Sensors, Fukuoka, 1983, p. 348.
12. Itoh, M.; Sugimoto, E.; Kozuka, Z. Trans. Jpn. Inst. Met., 1984, 25, 504-10.
13. Saito, Y.; Maruyama, T.; Matsumoto, Y.; Yano, Y. Proc. the International Meeting on Chemical Sensors, Fukuoka, 1983, p. 326.
14. Saito, Y.; Maruyama, T.; Sasaki, S. Report of the Research Laboratory of Engineering Materials, Tokyo Institute of Technology, 1984, 9, 17-26.
15. Kreidl, E. L.; Simon, I. Nature, 1958, 181, 1529.
16. Saito, Y.; Kobayashi, K.; Maruyama, T. Solid State Ionics, 1981, 3/4, 393-6.
17. El-Kabbany, F. A. I. Phys. Stat. Sol., (a), 1980, 58, 373-8.
18. Kvist, A.; Lundén, A. Z. Naturforschg., 1965, 20a, 235-8.
19. Höfer, H. H.; Eysel, W.; Alpen, U. v. J. Solid State Chem., 1981, 36, 365-70.
20. Höfer, H. H.; Eysel, W.; Alpen, U. v. Mater. Res. Bull., 1978, 13, 265-70.
21. Murray, R. M.; Secco, E. A. Can. J. Chem., 1978, 56, 2616-9.

22. Keester, K. L.; Eysel, W.; Hahn, Th. <u>Acta Crystallogr., Sect. A,</u>
     1975, 31, S79.
23. Höfer, H. H.; Alpen, U. v.; Eysel, W. <u>Acta Crystallogr., Sect. A,</u>
     1978, 34, S358.
24. Imanaka, N.; Adachi, G.; Shiokawa, J. <u>Can. J. Chem.,</u> 1983, 61,
     1557-61.
25. Hong, H. Y-P. <u>Mater. Res. Bull.,</u> 1976, 11, 173-82.
26. Goodenough,J. B.; Hong, H. Y-P.; Kafalas, J. A. <u>Mater. Res. Bull,</u>
     1976, 11, 203-20.
27. Alpen, U. v.; Bell, M. F.; Wichelhaus, W. <u>Mater. Res. Bull.,</u> 1979,
     14, 1317-22.
28. Boilot, J. P.; Salanié, J. P.; Desplanches, G.; Potier, D. Le
     <u>Mater. Res. Bull.,</u> 1979, 14, 1469-77.
29. Cordon, R. S.; Miller, G. R.; McEntire, B. J.; Beck, E. D.;
     Rasmussen, J. R. <u>Solid State Ionics,</u> 1981, 3/4, 243-8.
30. Takahashi, T.; Kuwabara, K.; Shibata, M. <u>Solid State Ionics,</u>
     1980, 1, 163-75.
31. Bogusz, W.; Krok, F.; Jakubowski, W. <u>Solid State Ionics,</u> 1981, 2,
     171-4.
32. Alpen, U. v.; Bell, M. F.; Hofer, H. H. <u>Solid State Ionics,</u> 1981,
     3/4, 215-8.
33. Imanaka, N.; Yamaguchi, Y.; Adachi, G.; Shiokawa, J. <u>Bull. Chem.
     Soc. Jpn.,</u> 1985, 58, 5-8.
34. Imanaka, N.; Yamaguchi, Y.; Adachi, G.; Shiokawa, J. <u>Proc. the
     1984 International Chemical Congress of Pacific Basin Societies,</u>
     Honolulu, Hawaii, 1984, 03C14.

RECEIVED February 3, 1986

8

High-Temperature Oxygen Sensors Based on Electrochemical Oxygen Pumping

E. M. Logothetis and R. E. Hetrick

Scientific Research Laboratories, Ford Motor Company, Dearborn, MI 48121–2053

This paper reviews the area of high temperature oxygen sensors based on oxygen pumping. After an initial discussion of the principle of oxygen pumping, the various sensor designs developed over the years are described and their properties are analyzed and compared in detail. The main characteristic of all these sensors is that their signal output is linearly proportional to the oxygen partial pressure. Their other characteristics, however, depend on the specifics of the sensor design. It is shown that these sensors have several advantages over other oxygen sensors such as the Nernst concentration cell and the metal oxide resistive sensors including higher sensitivity to oxygen and weaker dependence on temperature.

High temperature oxygen sensors are finding an increasing use in a variety of applications including monitoring and control of industrial processes(1) and of internal combustion engine operation(2).

There are two types of high temperature solid state devices which have been developed and already extensively used for O_2 sensing. One type is an electrochemical Nernst oxygen concentration cell based on ZrO_2 solid electrolyte(3-4) which generates a voltage signal given by $EMF=(RT/4F)\ln(P_O/P_{O,R})$, where F and R are respectively the Faraday and ideal gas constants, T is the absolute temperature and P_O and $P_{O,R}$ are the oxygen partial pressures in the unknown and in a reference atmospheres. Sensors of the second type(5) consist of a metal oxide element such as TiO_2 having a resistance that depends on P_O according to the relationship $R=R_0 P_O^{\pm n}$, where R_0 depends on material properties and includes an exponential variation with temperature, while values of n are usually in the range 1/4 to 1/6. Although these sensors have been successfully used for automotive engine control as well as for other applications, their relatively low sensitivity to oxygen (as gauged by their logarithmic or fractional power dependence on P_O) restricts their use to cases where P_O shows large changes (e.g. detection of stoichiometric combustible gas mixtures) or to cases where parameters affecting device operation such as temperature and pressure are well-controlled (e.g. in sampling systems).

Another type of high temperature solid state O_2 sensor that has been developed is based on the principle of electrochemical pumping of oxygen with ZrO_2 electrolytes. These sensors have higher sensitivity (generally, a first power dependence on P_O) than the Nernst cell and the resistive device and possess a number of other characteristics that make them very promising for many new applications.

In this paper, we review the various types of sensors based on O_2 pumping and discuss and compare their characteristics. Since the commercialization of these devices is just beginning, some of the sensor performance characteristics (e.g. accuracy, reproducibility and durability) are not yet known and, consequently, will not be included in this discussion. It is pointed out that although our discussion is limited to O_2-sensing with oxygen-ion conducting solid electrolytes, most of the concepts and sensor designs discussed here are equally applicable to sensing other ions using the corresponding ion-conducting solid electrolytes.

<u>Oxygen Pumping</u>

Consider a slab of an oxygen-ion-conducting material such as yttrium-doped ZrO_2 with platinum electrodes on both sides separating two regions with different oxygen concentration (Fig.1). This difference in the oxygen chemical potential will drive oxygen from the high oxygen concentration region to the low concentration region through the ZrO_2 electrolyte. At the lower oxygen partial pressure (P_1) side, two oxygen ions combine to give an oxygen molecule to the gas phase leaving four electrons on the Pt electrode:

$$2 \ O^= \ (ZrO_2) \ ----> \ O_2 \ (gas) \ + \ 4 \ e^- \ (Pt) \tag{1}$$

The oxygen lost from the ZrO_2 material by reaction (1) is recovered by the reverse reaction occurring at the gas/Pt/ ZrO_2 interface in the higher oxygen partial pressure (P_2) side. The net result of these processes is the transfer of one oxygen molecule from the high to the low P_O side and of four electrons from electrode 2 to electrode 1. As a result of this electron transfer, an electric field develops within the ZrO_2 and exerts an opposing force on the oxygen ions in the electrolyte. At equilibrium, the net current through the ZrO_2 material is zero and the open-circuit EMF developed between the two Pt electrodes is given by the well-known Nernst equation(<u>4</u>).

$$EMF = (RT/4F)\ln(P_1/P_2) \tag{2}$$

where R, F and T have been previously defined.

If a load resistor R_L is connected across the ZrO_2 electrochemical cell of Fig.1, a current will continuously pass through the closed circuit given by

$$I = EMF/(R_L + R_i) \tag{3}$$

where R_i is the internal impedance of the cell; R_i is the sum of the resistance of the ZrO_2 slab and the resistance of the Pt electrodes (including any effective resistance relating to the transport of O_2 through the electrodes). In this case, the electrochemical cell operates as a <u>fuel cell.</u>

Suppose, next, that an external voltage V is applied across the ZrO_2 cell (Fig.2). The current, in this case, is given by

$$I = (V + EMF)/(R_L + R_i) \tag{4}$$

Depending on the relative magnitude and sign of V and EMF, the current will transfer (<u>pump</u>) oxygen from the high to the low or from the low to the high oxygen partial pressure sides. In this config- uration, the electrochemical cell operates as an <u>oxygen pump</u>.

Equation (4) is valid under the assumption that the oxygen pumping action does not change appreciably the values of P_1 and P_2 at the two gas/Pt/ZrO_2 interfaces. This, however, is not usually valid. In fact, as it will be seen shortly, in most oxygen sensing devices, one purposely configures the structure so that at least one of the P_1 and P_2 changes at the gas/Pt/ZrO_2 interface. Consider, for example, the structures in Fig.3. In Fig.3a, a porous layer is placed on top of electrode 2 and acts as a barrier to the diffusion of O_2 from the bulk of the gas to the Pt electrode 2. In Fig.3b, a similar situation is obtained by introducing adjacent to electrode 2 an enclosed volume v which communicates with the bulk of the gas through a restriction, an aperture C. In the cases illustrated in Fig.3, the pumping action can change the oxygen partial pressure at electrode 2 from the bulk value P_1 to a lower value P_1'. The difference between P_1 and P_1' will induce a diffusional flux G of O_2 from the bulk of the gas to the gas/Pt/ZrO_2 interface:

$$G = \sigma_L (P_1 - P_1') \tag{5}$$

where σ_L depends on the diffusion constant of O_2 in its carrier gas, D_0, and on the geometrical characteristics of the diffusion barrier. At steady state, the flux of O_2 pumped by the current I is equal to the diffusional flux G,

$$I/4e = G \tag{6}$$

Combining Eqs. (4), (5) and (6) and using the fact that $EMF = (RT/4F) \ln(P_1'/P_1)$, one obtains

$$I = \frac{V + (RT/4F)\ln(1 - I/4e\sigma_L P_1)}{R_L + R_i} \tag{7}$$

Equation (7) is plotted in Fig.4 (with the assumption that the electrode contribution to R_i remains constant with V) for $R_L = 0$, $R_i = 50$ ohms, $T = 1000°K$ and $4e\sigma_L P_1 = 10$ mA. As seen in Fig.4, for low pumping voltages, the pumping current is determined by the resistance of the circuit (cell resistance and external resistance). For high pumping voltages, the pumping current saturates at the value $I_S = 4e\sigma_L P_1$; this corresponds to the case where oxygen has been completely depleted at electrode 2. It is apparent that the satura- tion current I_S is linearly proportional to the oxygen partial pressure in the gas, P_1; therefore, the measurement of I_S provides a means for obtaining an oxygen sensor with higher sensitivity than the

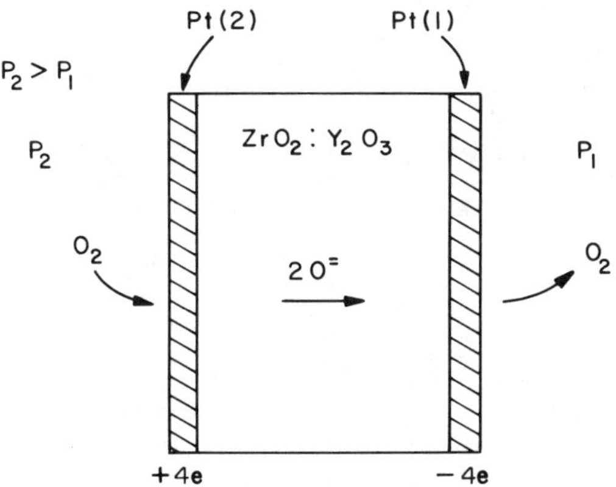

Fig. 1   Schematic of a ZrO_2 oxygen concentration cell.

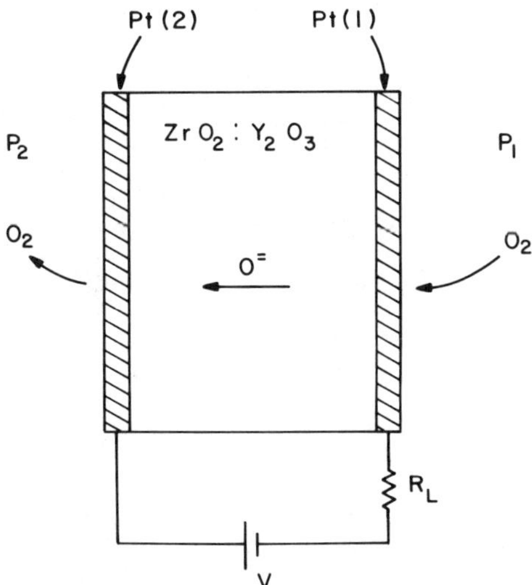

Fig. 2   Principle of oxygen pumping. A voltage V applied across
the ZrO_2 cell transfers oxygen from one side to the other
side of the cell.

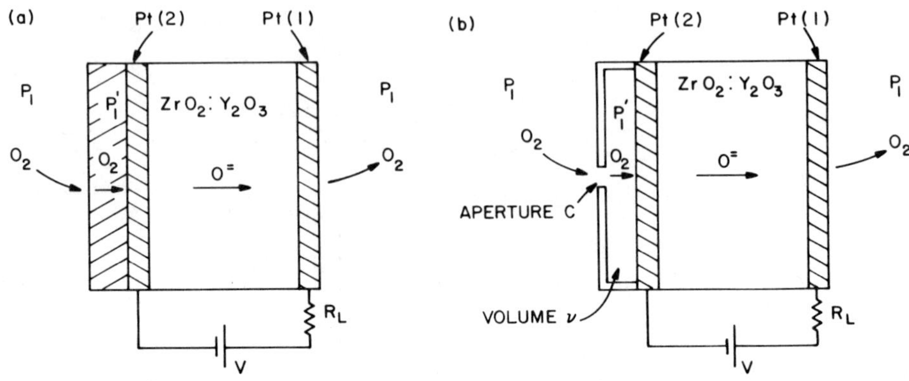

Fig. 3   Pumping cells with different diffusion barriers: (a) porous layer placed on top of one of the cell electrodes; (b) aperture C restricting the flow of O_2 from the bulk of the gas to the electrode.

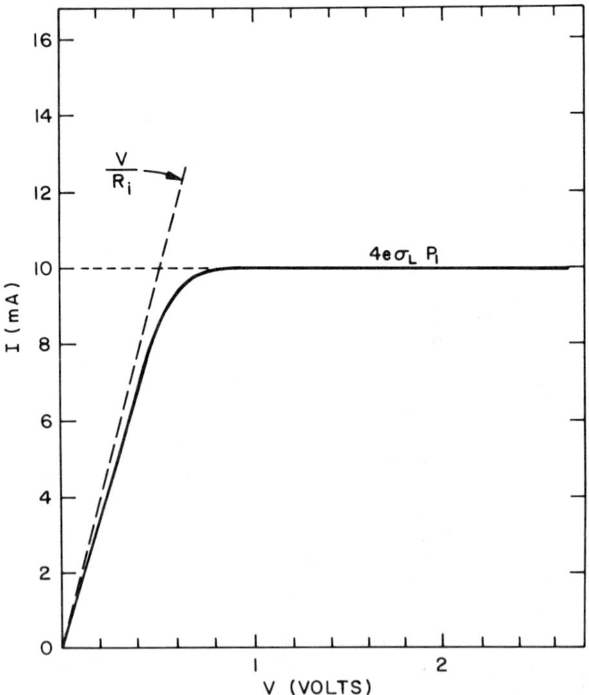

Fig. 4   Current voltage characteristic of an ideal ZrO_2 cell with $R_i=50$ ohms, $4e\sigma_L P_1=10$mA and $T=1000°$K. The broken and dotted lines represent the term V/R_i and the saturation current $I_s=4e\sigma_L P_1$ respectively; the solid line represents Eq. (7).

Nernst cell or the resistive device. Of course the minimum detectable O_2 concentration or difference in concentration depends on device design or signal processing electronics.

Figure 5 shows typical experimental current-voltage characteristics obtained with a device having essentially the configuration of Fig.3b. These results are in good agreement with the expected behavior of Fig.4 except at high voltages where the measured pumping current shows a sharp increase. Excess current in this region is frequently subject to drifts and can arise for a number of reasons (<u>6-9</u>). One is electronic conduction induced by partial decomposition of the ZrO_2 electrolyte which occurs when the pumping current lowers the oxygen activity at the cathode of the pump cell to very small values. Another reason is ionic current flow associated with the electrolytic decomposition of the electrolyte while yet another is the electrolytic decomposition of gaseous species containing oxygen such as CO_2 or H_2O.

<u>Sensor Designs</u>

In the last 20 years, several different types of oxygen sensors based on O_2-pumping have been developed(<u>10</u>). All these sensors have the common characteristic that their signal output is linearly proportional to the ambient oxygen partial pressure. Their other characteristics, however, such as response time, temperature and absolute pressure dependence, and effect of gas flow depend on the specific design and mode of operation of the sensor.

Most of the sensors developed to date incorporate in their structure some type of barrier to diffusion of oxygen as indicated in the previous section. In some sensors, however, diffusion of oxygen is not fundamental to the operation of the device. For the purpose of our discussion, we will classify the sensors based on oxygen pumping into two categories, namely, diffusion-limited and nondiffusion-limited sensor structures. Since devices of the first type have some advantages and appear to be the most developed, our discussion will concentrate on the characteristics of these sensors; however, some of the other types of devices will also be briefly described. In addition, some oxygen pumping devices capable of measuring combustible gas mixtures with excess of reducing species will also be discussed.

<u>Diffusion-Limited Sensor Structures</u>.   These sensors can be further classified according to the number of ZrO_2 electrochemical cells used in the structure and according to the type of the diffusion barrier employed. In the <u>single-cell</u> sensors, the same ZrO_2 cell is used for both, oxygen pumping and sensing. The diffusion barrier is formed either by a porous inactive layer covering one electrode (Fig.6a) or by a small cavity with one or more (macroscopic) apertures adjacent to one electrode (Fig.6b). In both cases, the device operates in the limiting-current mode (Figs.4, 5). If one applies a sufficiently high voltage across the cell (e.g. about 1 Volt), the limiting current I_S is proportional to P_O:

$$I_S = 4e\sigma_L P_O \tag{8}$$

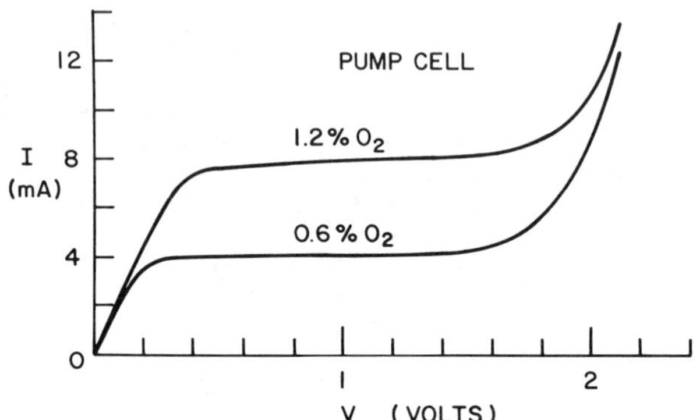

Fig. 5   Experimental current-voltage characteristic of a pumping cell having the structure shown in Fig. 3b for two different O_2 concentrations in N_2. The temperature of the device was 750°C.

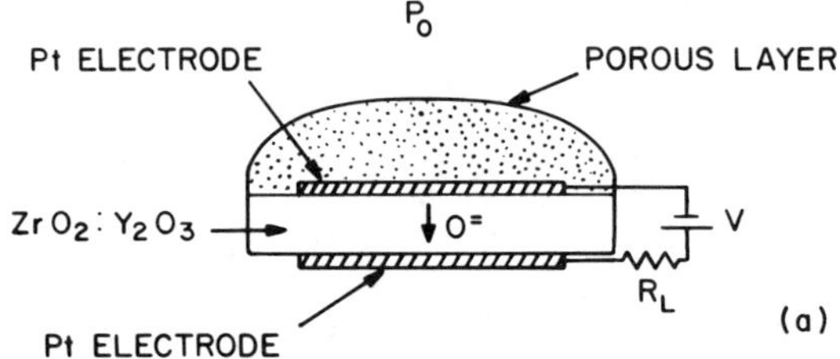

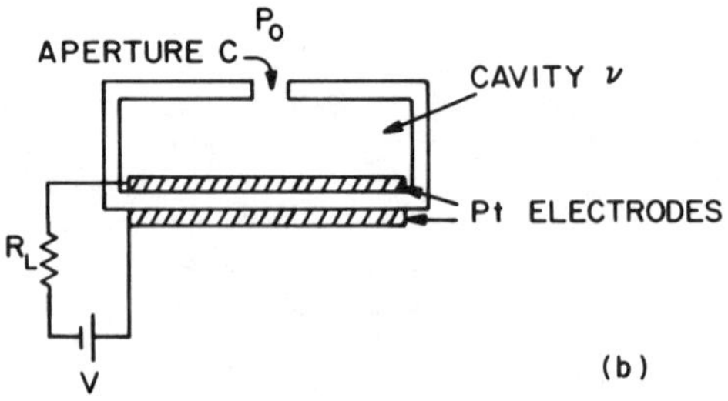

Fig. 6   Schematics of single-cell O_2-pumping sensors: (a) sensor with a porous layer; (b) sensor with a cavity v and an aperture C.

Under this condition, there is complete depletion of O_2 at the electrode next to the porous layer of Fig.6a or inside the cavity of Fig.6b. The constant σ_L depends on the diffusion constant of O_2 (in its particular carrier gas), D_O, and the geometrical characteristics of the diffusion barrier. In the device of Fig.6b (sensor with integral cavity), the diameter of the aperture C (usually greater than 50 microns) is much larger than the mean free path of the gas molecules at 1 atm (about 1 micron) and bulk diffusion dominates. In this case($\underline{11\text{-}12}$), $D_O = K_1 T^a/P$ and $\sigma_L = (D_O A)/(kTd)$, where K_1 is a constant, P is the absolute pressure, a is a constant having a value between 1.5 and 2 and A and d are the cross-sectional area and length of the aperture C. Representative values for D_O are about 1.5 cm^2/s at 700 °C and 0.15 cm^2/s at 20 °C. Since $P_O = cP$ with c the percentage of O_2 molecules in the gas, we have

$$I_S \sim T^b \, c \tag{9}$$

i.e. the sensor output is independent of the absolute gas pressure and has a weaker than first-power dependence on temperature (0.5<b<1). Eq.(8) with $\sigma_L = (D_O A)/(kTd)$ is derived under the assumption that there is a negligible gradient in the O_2 concentration inside the cavity during the pumping process; this approximation is increasingly valid as the parameter σ_L for the cavity ($\sigma_L' = (D_O A')/(kTd')$ with A' and d' an appropriate cross-sectional area and a length of the cavity) becomes larger than σ_L of aperture C. Also neglected are O_2 concentration gradients which can occur within cell electrodes; these effects can be reduced with thin, porous electrodes and elevated temperatures.

In the device of Fig.6a (sensor with distributed cavity), depending on the dimensions of the pores of the diffusion barrier, the oxygen diffusion can be bulk diffusion or Knudsen diffusion($\underline{11}$). In the former case, the sensor output (limiting current I_S) is given by Eq. (9). Knudsen diffusion occurs when the average pore diameter is much smaller than the mean free path of the gas molecules, in which case collisions between molecules and pore walls are the dominant events. In this case($\underline{12}$), $D_O' = K' T^{1/2}$, and

$$I_S \sim T^{-1/2} cP \tag{10}$$

i.e. I_S depends on the absolute pressure. If the average pore diameter is of the same order of magnitude as the mean free path of the gas molecules, the diffusion is a mixture of bulk and Knudsen diffusion and I_S has temperature and pressure dependences intermediate between those given by Eqs. (9) and (10). It is possible that by proper choice of pore dimensions, one could obtain a device with a very small temperature dependence. Details on the effect of the diffusion barrier microstructure on sensor characteristics can be found in a paper by Dietz($\underline{11}$).

A single-cell sensor based on the measurement of the limiting current can function properly only if the rates of the electrode reaction (1) and its inverse are higher than the rate by which O_2 molecules arrive at the electrode by diffusion. The ability of an electrode to support reaction (1) and its reverse is conventionally expressed($\underline{8}$) by the so-called exchange current density J_O, which is of the order of 1 A/cm^2 for thin porous Pt electrodes at T > 600 °C.

The condition $J_0 > J_S$ (J_S = I_S/(area of electrode)) is favored by small values of σ_L. However, small values of σ_L lead to long sensor response times (see below).

Another important point to emphasize is that the limiting-current type of sensor requires that the impedance of the device (i.e. the resistance of the ZrO_2 material and of the electrode processes) is sufficiently small so that

$$I_R > I_S \tag{11}$$

(where I_R = $V/(R_L + R_i)$ and I_S = $4e\sigma_L P_0$) for all P_0 and T of interest. Expressed differently, condition (11) states that the device impedance must be smaller than an equivalent "diffusion resistance" $V/(4e\sigma_L P_0)$. Condition (11) is again favored by small values of σ_L. Another point of importance for the single-cell device is that under the conditions of limiting current, the cathode is completely depleted of oxygen. This will lead to the onset of new processes(6-9) which could compromise effective device operation. For example, under these conditions the decomposition of oxygen containing gases such CO_2 or H_2O (which could be present in significant amounts) would occur as well as the decomposition of the electrolyte itself. These processes will produce larger currents and could result in significant errors.

For a sensor with good electrodes (high J_0), the response time is determined by the characteristics of the diffusion barrier and the sensor structure. For the sensor of Fig.6b, for example, a simple rate equation analysis for P_V, the oxygen pressure inside v, predicts that for a cylindrical aperture

$$\tau = vd/D_0 A \tag{12}$$

For values of the parameters consistent with convenient device fabrication, this relation gives a value for τ of the order of 50 ms.

A number of variations on the limiting current approach using a single cell have been proposed. For example, the anode of the pump cell can be blocked from easy communication with the gas while at the same time the solid electrolyte is made porous. Pumping then results in an increase (or decrease if the current direction is changed) of oxygen partial pressure at the anode. The EMF induced by pumping is modified and the I/V characteristic of the structure under appropriate conditions can be used to measure oxygen concentration over a wide range(13). Although the structure of this device has an appealing simplicity, practical devices would seem to require stringent materials control to maintain a consistent porosity.

Although several O_2-conducting solids exist, all reported devices were made from ZrO_2 doped with Y_2O_3. Electrodes are usually made from sputtered, e-beam or screen-printed platinum. The porous layer is made from spinel or aluminum oxide. The sensor configuration and fabrication methods vary. Some sensors are made by laminating ZrO_2 membranes(14); others are made by standard ceramic sintering techniques(15); film-type structures on appropriate substrates have also been reported(16). All sensors incorporate some type of heater to keep the device at some elevated temperature (usually higher than 600 °C).

In   the   <u>double-cell</u>   sensors, different ZrO_2 cells   are   used   for oxygen pumping and sensing.   The separation   of these   two functions allows   for   other   than   the   limiting   current   modes   of   operation. Figure 7 shows a   two-cell   device($\underline{17}$)   with   a   macroscopic enclosed volume v   which communicates with   the ambient through the aperture C. If the oxygen inside volume v is <u>partially</u> depleted by the passage of a current I through the pump cell, the resulting lower oxygen partial pressure P_v inside volume v causes   an EMF, $V_S=(RT/4F)\ \ln(P_v/P_0)$, to develop   across   the   sensor   cell.   Since,   at   <u>steady state</u>,   I=4e σ_L (P_0-P_v), one has

$$I = 4e\ \sigma_L\ P_0\ \{1\ -\ \exp(-4FV_S/RT)\} \tag{13}$$

If the current is controlled so that V_S is kept   constant, then   I is proportional to P_0.   This can be accomplished with a simple feedback circuit such as the   one shown   schematically in   Fig.7.   A feedback amplifier produces   a pump   current which   always acts to maintain V_S equal to a predetermined reference value V_R.   A known resistor in the pump cell   circuit can   be used   to measure I.   As in the case of the single-cell sensor, the pumping   current of   the   double-cell device with bulk diffusion is independent of the total gas pressure.   On the other hand, the temperature dependence of   I for   the double-cell can be weaker than that of the single cell device, Eq.(9).   The reason is that $\sigma_L \sim T^b$ (0.5<b<1),   but   {1   - $\exp(-4FV_S/RT)$}   decreases with temperature.   For a given application, V_S can be chosen so that I has only a very weak dependence on   T over   a certain   temperature range. Figure 7   shows experimental   results on   the dependence   of I on the percentage of O_2 in N_2.   The   predicted linear   dependence is accura- tely followed by the experimental data.

One   of   the main   advantages of   the double-cell device over the single-cell one is that all the problems associated with the applica- tion   of   appreciable   currents   and voltages to ZrO_2 electrochemical cells   (such   as   electrode   polarization   and   electrolyte decompo- sition)   are   avoided.   A   second   important   advantage   is that, in contrast   to   the   single-cell device,   the   condition   expressed by Eq.(11) does not have to be satisfied for the double-cell device; the only requirement is that the pumping voltage is sufficient to provide the   current   necessary   to   keep V_S constant;   in other words, the double-cell device can operate at any part of   the current   - voltage characteristic of   Fig.4.   Additionally, as explained in the previous paragraph, the double-cell device has a weaker temperature dependence than the single-cell device with bulk diffusion; although a very weak temperature dependence   could also   be achieved   with the single-cell sensor   by   using   a   diffusion   barrier with mixed, bulk and Knudsen diffusion, the resulting low value of the parameter σ_L would compro- mise the response time of the device (see Eq.(12)).

It is   apparent that   the double-cell   sensor can be operated in the limiting current mode (which corresponds   to large   values of V_S, of   the   order   of   600   to   800   mV).   The advantage of this mode of operation is that errors associated   with   the   measurement   of small values of V_S (e.g. less than 100 mV) are avoided.   However, limiting current   operation   is   generally   undesirable since   the problems associated with   complete depletion   of oxygen   at the cathode of the pump cell can seriously limit the performance of   the device.   It is

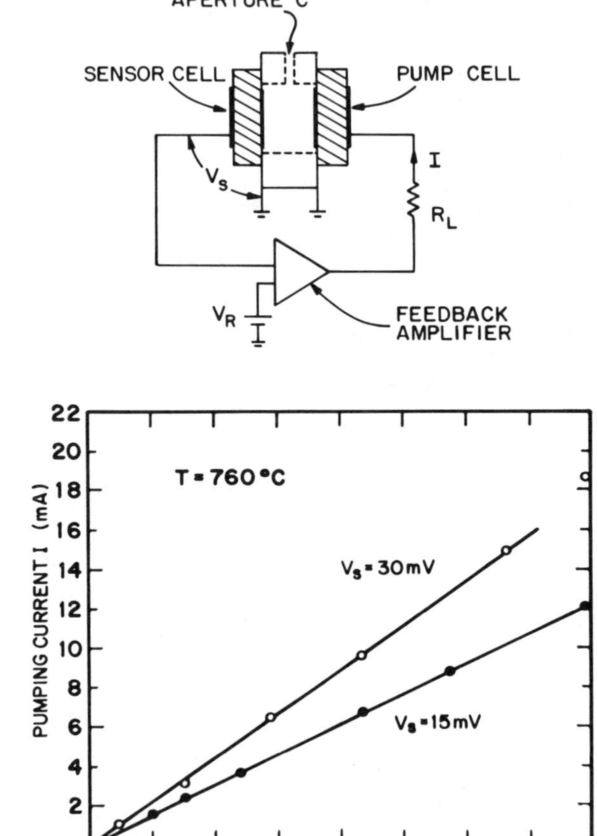

Fig. 7 Double-cell O_2-sensor with a cavity: (a) schematic of the sensor; (b) pumping current versus O_2 concentration in N_2 for two different values of the sensor voltage V_S. The temperature of the device was 760°C.

pointed out that in the double-cell sensor, an error in the measurement of V_S is not as important as a similar error in the case of a concentration Nernst cell. A simple analysis shows that the ratio of the errors in P_O as measured by the two devices corresponding to the same error in the EMF is given by the relation

$$\Delta P_O(\text{Nernst})/\Delta P_O(\text{pump}) = [1 - \exp(-4FV_S/RT)]/\exp(-4FV_S/RT) \qquad (14)$$

This ratio can be easily made larger than 10 by choosing values for V_S greater than 40-50 mV.

One important aspect of the double-cell device is that, for proper operation of the device, the two cells must be electrically decoupled. This can be accomplished by proper design of the device structure.

In addition to the steady-state mode, the double-cell device of Fig.7 can be operated in the so-called <u>transient</u> and <u>oscillatory</u> modes of operation(<u>18</u>). In the transient mode, one measures the slope dV_S/dt when the pumping current is changed in a stepwise fashion from zero to a constant value. A simple analysis(<u>18</u>) gives the following differential equations for V_S and P_v,

$$dV_S/dt = -(RT/4FP_v) \; (dP_v/dt) \qquad (15)$$

$$dP_v/dt = -(RT/4Fv)I + (P_O - P_v)/\tau \qquad (16)$$

The first term in the right-hand side of Eq.(16) is due to the pumping current whereas the second term is due to the diffusion of O_2 through the aperture C. If I is sufficiently large, this diffusion term can be neglected and one then has,

$$(dV_S/dt)_{t=0} = (RT/4F)^2 \; I/(vP_O) \qquad (17)$$

Thus an initial measurement of the slope of the function $V_S = f(t)$ allows the determination of P_O. Following this measurement, the current is turned off and the device is allowed to relax to the initial condition $P_v = P_O$. The advantage of this type of device is that the sensor output does not depend on the characteristics of the aperture C and thus it is immune to changes in these characteristics. On the other hand, the slope measurement proves to be tedious as discussed in Ref. 18.

In the oscillatory mode of operation(<u>18</u>), a step in I is applied to the pump cell to remove O_2 out of the cavity until an external control circuit detects that V_S has risen to a certain reference voltage V_R (in the range 5 to 50 mV). At this point, the circuit causes the sign but not the magnitude of I to change. Oxygen is now pumped into the cavity v from the ambient and V_S decreases and eventually changes sign. When $V_S = -V_R$, the control unit again reverses the sign of I so that the process of pumping O_2 in and out of v is repeated indefinitely. The period of this oscillatory operation is calculated(<u>18</u>) to be,

$$\tau_o = 4V_R/(dV_S/dt) = 4V_R vP_O/[(R^2T^2)/(16F^2I)] \qquad (18)$$

under the condition that the current I is sufficiently large so that the diffusion term in Eq.(16) can be neglected. As in the transient mode, the sensor output here again does not depend on the diffusion process through the aperture C; the latter process only serves to insure that the average value of P_V is equal to P_0 during the oscillatory process. Figure 8 shows a plot of τ_0 vs P_0 for O_2 in N_2 with the indicated values for the relevant parameters. The advantages and disadvantages of the oscillatory mode of operation are discussed in detail in Ref. 18.

The double-cell sensor of Fig.7 can also be made with a structure having a distributed cavity rather than an integral cavity. One example([19]) is shown in Fig.9, where the porous diffusion barrier is made from ZrO_2 and forms the pump cell (The sensor and pump cells are interchangeable). This structure has only three electrodes, the middle one being common to both, pump and sensor cells. Electrical decoupling of the two cells is accomplished by making the middle electrode large and maintaining it during operation at constant (ground) potential.

Several sensors representing variations of the various structures discussed above have appeared in the literature in the last 10 years. Interested readers can consult the relevant publications ([13-14,16,20-22]).

<u>Non-Diffusion-Limited Sensor Structures</u>. In this section, we will discuss a number of devices which do not require for their operation the existence of a barrier to the diffusion of oxygen.

One device of this type was discussed by Hickam and Witkowski([23]). In the diffusion limited devices that have been discussed, the flow rate of the gas is not normally an important factor. In the Hickam device, however, the gas flow is of paramount importance and new possibilities or complications arise. The structure consists of pump (upstream) and sensor (downstream) cells cylindrically surrounding a flowing stream of gas containing oxygen. The sensor cell EMF is fedback to the pump so that oxygen is either added to or subtracted from the stream in the amount required to keep the sensor EMF at a constant value. For a calibrated device, the amount of pump current required measures the oxygen content of the gas at the inlet of the structure provided the flow rate is held constant. Alternately, if a gas of constant composition were employed, the structure could be used to measure flow rate.

Another device([20]) has essentially the structure of Fig.6b but is operated in a cyclic mode. At the start of the measuring cycle, a constant current is passed through the cell to pump oxygen out of the cavity with a rate much higher than the rate by which oxygen diffuses in the cavity through the aperture C. The result is that after a time interval t short compared to the time τ (given, for example, by Eq.(12)), practically all oxygen is pumped out of the cavity. This condition is indicated by a sharp increase (of the order of 1 Volt) in the voltage across the cell. It is apparent that the time interval t is proportional to the original concentration of oxygen in the cavity and provides a means of measuring the oxygen concentration in the ambient gas atmosphere. After all oxygen has been removed from the cavity, the pumping current is turned off and the oxygen is allowed to diffuse back in the cavity. After several time intervals τ have elapsed, the oxygen concentration in the cavity has become equal to that in the ambient gas and the device is ready for a new

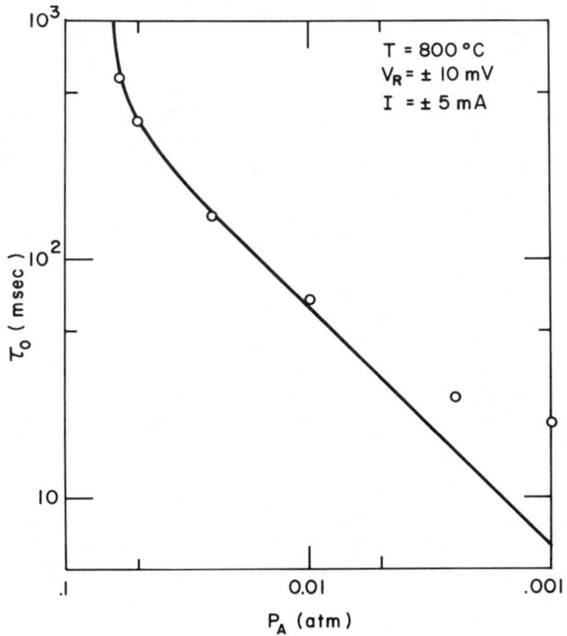

Fig. 8   Plot of the oscillatory period τ_0 versus the partial pressure of O_2 in N_2.

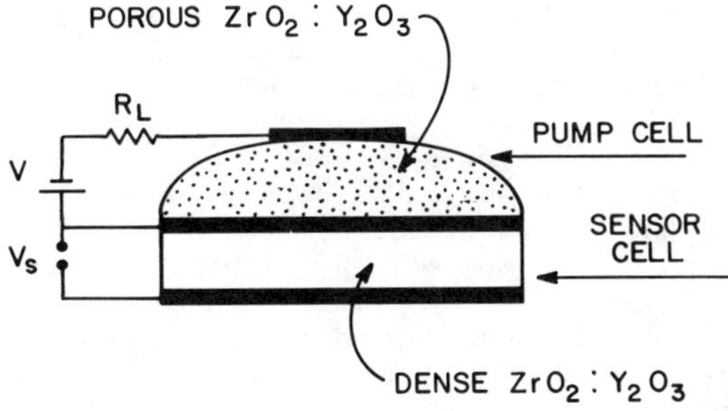

Fig. 9   Schematic of a double-cell O_2-sensor with a distributed cavity (porous ZrO_2 layer).

measuring cycle. One advantage of this device is that the calibration does not depend on the properties of the diffusion barrier; on the other hand, the response time of the device tends to be long (several τ).

Another device([24]) has essentially the structure of Fig. 7 except that the cavity is completely sealed from the ambient gas atmosphere. Initially, all oxygen is pumped out of the cavity by passing a current through the pump cell. This condition is indicated by a large voltage ($\sim$ 1 Volt) developed across the sensor cell. The pumping current is then reversed and oxygen is pumped back into the cavity until the oxygen partial pressures inside and outside the cavity are equal; this is indicated by a zero EMF across the sensor cell. The time integral of the current required to achieve this condition provides a measure of the amount of oxygen in the ambient gas atmosphere.

<u>Sensors for Combustible Gas Mixtures</u>. The oxygen sensors described so far operate in gas mixtures containing oxygen and non-reacting gases (e.g. N_2, Ar). The same sensors will also operate in reacting gas mixtures (e.g. O_2 with CO or H_2) provided that there is excess oxygen; in this case, under thermodynamic equilibrium conditions, the sensors will measure the (excess) oxygen concentration after the gases have been reacted to their equilibrium concentrations (within or near the sensor structure).

There are, however, applications where the oxygen concentration is smaller than that corresponding to a stoichiometric mixture, for example an internal combustion engine operated in the fuel-rich region. In this case, the concentrations of the combustible species (e.g. CO, H_2 or hydrocarbons) at thermodynamic equilibrium sufficiently characterize the combustible mixture. Although the devices described above do not function properly under these conditions, other oxygen pumping devices have been developed which can measure combustible mixture concentrations with high sensitivity.

One pumping device for combustibles([25]) is shown in Fig. 10. It is a double-cell device with an integral cavity and one electrode of the pump cell exposed to air as a reference atmosphere. In environments containing excess oxygen, the device is operated in the same manner as the double-cell device of Fig.7. For monitoring and control of stoichiometric gas mixtures, only cell B is used to provide the conventional single Nernst cell. In oxygen-deficient combustible mixtures, the device is operated by pumping oxygen into the cavity v from the air reference by the application of a reverse current I_p through the pump cell B. The pumped oxygen reacts with the reducing species inside the cavity v and causes a decrease in their concentration. The corresponding increase in the equilibrium oxygen partial pressure inside v induces across the sensor cell A an EMF V_C with a sign opposite to that of the EMF induced in the case of oxygen-rich mixtures. Under certain conditions([25]), a simple relation exists between I_p, V_C and the concentration of combustibles. For a $O_2/CO/CO_2$ mixture, for example, one has

$$I_p = 2e\sigma_{CO}P_{CO}\left[1 - \exp(-2FV_C/RT)\right]\left[1 + (\sigma_{CO}/\sigma_{CO_2})(P_{CO}/P_{CO_2})\right.$$
$$\left.\exp(-2FV_C/RT)\right]^{-1} \tag{19}$$

where σ_{CO} and σ_{CO_2} are the diffusion parameters of the structure for CO and CO_2, and P_{CO_2}, P_{CO_2} are the (equilibrium) partial pressures of CO and CO_2. Figure 11 shows some results obtained with such a device in a propane burner(<u>25</u>).

<u>Concluding Remarks</u>

In this paper, the properties of various oxygen sensors based on oxygen pumping were discussed. It is apparent from this discussion that these sensors possess several advantages over other oxygen sensors such as the Nernst cell or the resistive device. These advantages are higher sensitivity, weaker temperature dependence, and weaker or zero dependence on the absolute gas pressure. Compared to the Nernst cell, the oxygen pumping sensors have the additional advantages of higher signal level (volts compared to millivolts for the Nernst cell) and generally lower sensitivity to electrode properties. On the other hand, the pumping devices need calibration and require somewhat more complicated electronics.

Questions of accuracy, reproducibility and durability have only been touched on slightly. They depend critically on such factors as fabrication techniques, materials properties, etc. which are still in a state of development for the specific device designs.

Of the various types of oxygen pumping sensors, the double-cell devices appear to have the best overall performance characteristics. These devices can be constructed to achieve the best combination of response time, independence of temperature and absolute pressure, insensitivity to electrode properties and operation not limited by the resistance of the ZrO_2 material.

Although oxygen sensors based on oxygen pumping have not yet been used extensively, it seems certain that this will change in the near future as new applications appear which require high temperature oxygen sensors with the properties possessed by the pumping devices. One application of this kind is control of internal combustion engines operated in the fuel-lean region. It is also emphasized that the concepts and the sensor designs discussed here are also appli-

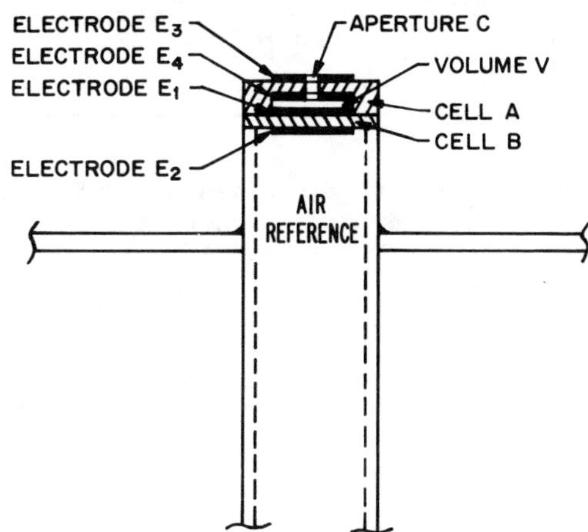

Fig. 10   Schematic of a sensor for combustible gas mixtures.

cable to sensing other chemical species using the corresponding ion
conducting electrolytes.

     Finally, it is noted that the O_2-pumping process can be
utilized to carry out other physical measurements and functions. For
example, some of the structures described above can be used to
control the oxygen content in a gaseous environment(<u>9</u>), electrolyze
water(<u>26</u>), measure gas flow(<u>23</u>) or absolute pressure(<u>27</u>) and realize
an ionic transistor(<u>28</u>).

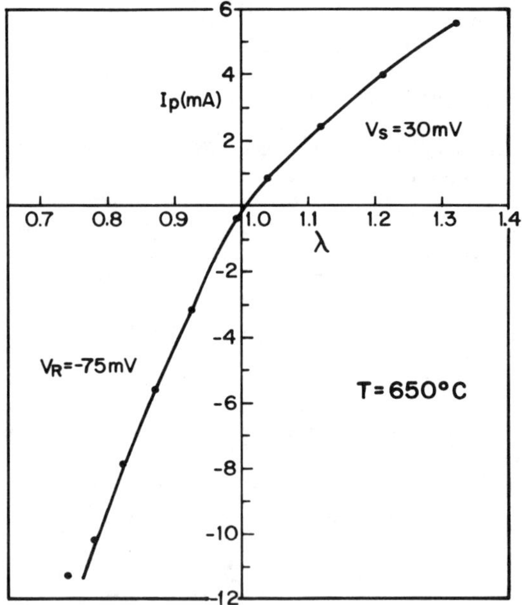

Fig. 11   Pumping current of the sensor of Fig. 10 as a function of
             λ in a propane burner.   λ is defined as λ =
             (A/F)/(A/F)$_{stoich}$, where A/F is the ratio of the mass
             of air to the mass of fuel (propane) in the gas mixture
             and (A/F)$_{stoich}$ is the stoichiometric value of this
             mixture.   The solid lines are fits to the experimental
             data (circles) of an equation similar to Eq. (19) but
             modified[25] to describe an air/propane mixture.

Literature Cited

1. B. C. H. Steele, J. Drennan, R. K. Slotwinski, N. Bonanos and E. P. Butler, Advances in Ceramics $\underline{3}$, p. 286, Eds. A. H. Heuer and L. W. Hobbs, Amer. Ceramic Soc., 1981.

2. E. M. Logothetis, Advances in Ceramics $\underline{3}$, p. 388, Eds. A. H. Heuer and L. W. Hobbs, Amer. Ceramic Soc., 1981.

3. H. Dietz, W. Haecker and H. Jahnke, in Advances in Electrochemistry and Electrochemical Engineering, Vol. 10, p. 1. Edited by H. Gerischer and C. Tobias, Wiley, New York, 1977.

4. P. Kofstad, Nonstoichiometry, Diffusion, and Electrical Conductivity in Binary Metal Oxides. Wiley Interscience, New York, 1972; Chapters 6 and 8.

5. E. M. Logothetis, Ceram. Eng. Sci. Proc., 8th Automotive Mater. Conf., $\underline{1}$, 281 (1980).

6. D. Yuan and F. A. Kroger, J. Electrochem. Soc. $\underline{116}$, 594 (1969).

7. H. Yanagida, R. J. Brook and F. A. Kroger, J. Electrochem. Soc. $\underline{117}$, 593 (1970).

8. T. H. Etsell and S. N. Flengas, J. Electrochem. Soc. $\underline{118}$, 1890 (1971).

9. J. Fouletier, G. Vitter and M. Kleitz, J. Appl. Electrochem. $\underline{5}$, 111 (1975).

10. Some of the first ideas and work on O_2-pumping and sensing with ZrO_2 cells were reported by: J. Weissbart et al (Electrochem. Soc. Meeting, Oct. 1964, Ext. Abstr. No.31, and U.S. Patent 3,400,054, Sept. 1968); O. Anonsen et al (Rev. Brown-Boveri $\underline{53}$, 21 (1966)); W. M. Hickam and J. F. Zamaria (Inst. Contr. System $\underline{40}$, 87 (1967)); N. M. Beekmans and L. Heyne (Philips Tech. Rev. $\underline{31}$, 51 (1971)); M. Kleitz et al (Fr. 1,580,819, Aug. 1969 and "Reactions d' electrode dans les oxydes electrolytes solids," Thesis, Grenoble 1968).

11. H. Dietz, Solid State Ionics $\underline{6}$, 175 (1982).

12. W. Jost, Diffusion in Solids, Liquids, Gases, p. 425, Academy Press, New York, 1960; R. Jackson, Transport in Porous Catalysts, p. 9, Elsevier, Amsterdam, 1977.

13. S. Kimura, H. Tokas, S. Ishitani, K. Ikezawa and H. Aoki, U.S. Patent 4,224,113.

14. S. Suzuki, T. Sasayama, M. Miki, H. Yokono, S. Iwenaga, and S. Ueno, Paper no. 850379, SAE Congress, Detroit, Michigan, February, 1985.

15.   T. Kamo, Y. Chujo, T. Akasuka, J. Nakano and M. Suzuki, Paper No. 850380, SAE Congress, Detroit, Michigan, February, 1985.

16.   G. Velasco, J. P. Schnell and M. Croset, Sensors and Actuators $\underline{2}$, 371 (1982).

17.   R. E. Hetrick, W. A. Fate and W. C. Vassell, Appl. Phys. Letters $\underline{38}$, 390 (1981).

18.   R. E. Hetrick, W. A. Fate and W. C. Vassell, IEEE Trans. Electron Dev. $\underline{ED-29}$, 129 (1982).

19.   E. M. Logothetis and W. C. Vassell, U.S. Patent 4,504,522, March, 1985.

20.   N. M. Beekmans and L. Heyne, Philips Techn. Rev. $\underline{31}$, 112 (1970).

21.   S. Soejima and S. Mase, Paper No. 850378, SAE Congess, Detroit, February, 1985.

22.   K. Saji, H. Kondo, H. Takahashi, T. Takeuchi and I. Igarashi, 3rd Intern. Conf. Solid State Transducers, Proceed. p. 336, Philadelphia, June, 1981.

23.   W. Hickam and R. Witkowski, U.S. Patent 3,650,934, March, 1972.

24.   D. Haaland, Analytical Chemistry $\underline{49}$, 1813 (1977).

25.   W. C. Vassell, E. M. Logothetis and R. E. Hetrick, Paper No. 841250, SAE Passenger Car Meeting, Dearborn, October, 1984.

26.   G. B. Barbi and C. M. Mari, Solid State Ionics $\underline{6}$, 341 (1982).

27.   W. A. Fate, R. E. Hetrick and W. C. Vassell, Appl. Phys. Lett. $\underline{39}$, 924 (1981).

28.   R. E. Hetrick and W. C. Vassell, J. Electrochemical Soc. $\underline{128}$, 2529 (1981).

RECEIVED February 26, 1986

Microsensor Vapor Detectors Based on Coating Films of Phthalocyanine and Several of Its Metal Complexes

W. R. Barger, Hank Wohltjen[1], Arthur W. Snow, John Lint[2], and Neldon L. Jarvis[3]

Chemistry Division, U.S. Naval Research Laboratory, Washington, DC 20375-5000

Experiments were conducted to assess relative sensitivities of phthalocyanine and its Fe, Co, Ni, Cu, and Pb complexes to vapors that cause changes in the resistance of thin films of these materials. When sulfur dioxide, benzene, air, ethanol, water, dimethyl methylphosphonate, and ammonia vapors were introduced into a stream of helium flowing over thin layers of phthalocyanines that were sublimed onto the surface of interdigital electrodes, the conductivities of the coatings increased or decreased depending on the type of phthalocyanine and on the type and concentration of vapor. Behavior of the sublimed coatings is briefly contrasted to the behavior of coatings of an organic derivative of copper phthalocyanine deposited by the Langmuir-Blodgett technique.

A program of research directed at the eventual development of small, low-cost gas sensors that take advantage of modern microfabrication technology and advances in microcomputing capabilities is presently under way at our laboratory. Specifically, thin, chemically selective coating films are being placed on the surfaces of micro interdigital electrodes, and the usefulness of these devices as vapor detectors is being investigated. Figure 1 illustrates the size and typical design of such devices. A significant part of the effort is directed at fundamental studies of the materials to be used as coating films for the microsensors and at techniques to deposit these films on sensor surfaces. There are numerous coating techniques that might be used, such as casting from a volatile solvent, polymerization after contact with the surface, spin coating, sublimation, dipping, or deposition of one monomolecular layer at a time by the Langmuir-Blodgett technique.

This paper describes studies in which phthalocyanine and metal complexes of phthalocyanine were sublimed onto the surfaces of

---

[1] Current address: Microsensor Systems, Inc., P.O. Box 90, Fairfax, VA 22030
[2] Current address: 5536 Fillmore Ave., Alexandria, VA 22311
[3] Current address: Chemical Research and Development Center, U.S. Army, Aberdeen Proving Grounds, MD 21010

interdigital electrodes in order to measure the change in the
conductivity when the coating films were exposed to a series of test
gases. We are currently developing vapor detectors based on organic
derivatives of phthalocyanine and its metal complexes which can be
deposited on electrode surfaces by the Langmuir-Blodgett (L-B)
technique (1). The L-B work will be the subject of a future report,
but some preliminary data is included here for comparison of the
general characteristics of sublimed films and those of the L-B films.

Phthalocyanines were chosen for these experiments because they
are electronic semiconductors and because they are quite stable
materials -- an important consideration in fabricating any practical
gas-detecting device. A considerable body of literature exists
describing the physical and chemical properties of the phthalocyanines. A review of the work prior to 1965 is contained in the chapter
by A. B. P. Lever in Volume 7 of Advances in Inorganic Chemistry and
Radiochemistry (2). Electrical properties of phthalocyanines have
been receiving increased attention in recent years. The photocon-
ductivity of metal-free phthalocyanine has been studied in detail
(3,4). Electrical properties of lead phthalocyanine have been
studied extensively, especially by Japanese workers (5,6,7,8). They
have also studied the alteration of the conductivity of this material
upon exposure to oxygen (9,10). The effects of a series of adsorbed
gases (O$_2$, H$_2$, CO, and NO) on the conductivity of iron phthalocyanine have been recently reported (11). A study of general
electrical properties of metal phthalocyanines of the first transition period with some assessments of the effects of oxygen has been
prepared by Beales et al. (12). The use of thin Cu, Ni, and metal-
free phthalocyanine films as the sensing material in a gas sensor has
been described by Sadaoka et al. (13). These workers studied the
effect of NO$_2$, NO, SO$_2$, O$_2$, N$_2$, and CO on the electrical
conductivity of the phthalocyanine films. A quantitative study of
the effects of a series of gases adsorbed on the surface of single
crystals of Mn, Co, Ni, Cu, Zn, Pb, and metal-free phthalocyanines
and of other semiconducting materials has been done by van Ewyk et
al. (14). Gases studied included NO$_2$ + NO, O$_2$, BF$_3$, NH$_3$,
ethylene, CO, and other organic vapors. Recently, Jones and Bott
studied gas sensors made by sublimation of several phthalocyanines
onto interdigital platinum electrodes and reported on the influence
of temperature on performance (15). They also used a multisensor
array that included PbPc as a sensor for NO$_2$ and NO$_x$ in hazardous
atmospheres (16).

The preparation of L-B films from metal-free phthalocyanine and
from tetra tert-butyl phthalocyanine was reported by Baker et al. in
1983 (17). Since then the preparation of a series of tetracumylphenoxy derivatives suitable for preparing L-B coating films has been
reported by Snow and Jarvis (18), and the results of tests of gas
sensors made with these coatings have been reported by Barger et al.
(19).

Experimental

Chemicals. The phthalocyanine compounds used to prepare sublimed
films were prepared, purified, and characterized in our laboratory.
The preparation technique for the Cu tetracumylphenoxy phthalocyanine

applied to sensors by the L-B technique is described in the paper by
Snow and Jarvis (18).

Measurement System. For the studies involving the exposure of coating films to various gases, the phthalocyanines were sublimed onto
the surface of interdigital electrodes which were microfabricated on
a quartz substrate. Two types of devices that were used for these
experiments are shown in Figure 1. The interdigital electrodes at
each end of the larger device consisted of 50 gold finger pairs, each
25 micrometers wide with a gap of 25 micrometers between the fingers.
Finger overlap distance was 7.250 mm. The smaller microelectrode
had 300 niobium finger pairs, each 2.3 micrometers wide and spaced
0.7 micrometers apart. A direct current was passed through the
coating film which covered the electrode. Only one end of the larger
device with electrodes at both ends was used for the measurement of
conductivity changes in the presence of a test gas. The larger
device was housed in a rectangular, stainless-steel chamber with an
entrance and an exit for a flowing gas stream. The smaller microelectrode was coated with a multilayer film deposited by the Langmuir
Blodgett technique. Another device of the same small design was
coated by sublimation for comparison. A low-volume cell was
fabricated to hold the small devices and is shown in Figure 2. A
cylindrical insert of Teflon supported the electrode and surrounded
all sides except the top. The remainder of the block was fabricated
from polyvinyl chloride except for the stainless-steel tubing and the
gold-plated electrodes embedded in the block.

Electrical measurements with both types of cells were made by
using two devices as the gain-controlling elements of an inverting
operational amplifier circuit which was driven with a constant input
voltage (E_{in}). The ratio of the resistance of the reference device,
R_2, to the resistance of the measuring device, R_1, determined
the measured voltage, E_{out}:

$$E_{out} = -\left[R_2 \middle/ R_1 \right] E_{in} \tag{1}$$

The E_{out} voltage was passed to an A/D converter and recorded by
a microcomputer. The apparatus was arranged as shown in Figure 3. A
helium carrier gas flowing at 100 ml/min passed over the reference
electrode (REF), into a 500 ml dilution flask, and then over the
sample-coated electrode (SAM). The gas stream then passed through a
six-port valve to waste. By periodically switching the valve with an
automatic actuator, 5 ml samples of the gas mixture that had just
passed over the measuring film could be injected into a gas chromato-
graph (GC) in order to monitor the change in concentration with time
due to dilution. From this data the equation describing the dilution
could be determined, and the concentration as a function of time
could be extrapolated to values below the detection limit of the
thermal conductivity detector. The per cent of initial concentration
vs. time in this arrangement of experimental apparatus is shown in
the lower right quadrant of Figure 5. A sufficient amount of test
compound was injected into the dilution flask to produce an initial
concentration of either 100 or 1000 ppm by volume in the vapor phase.

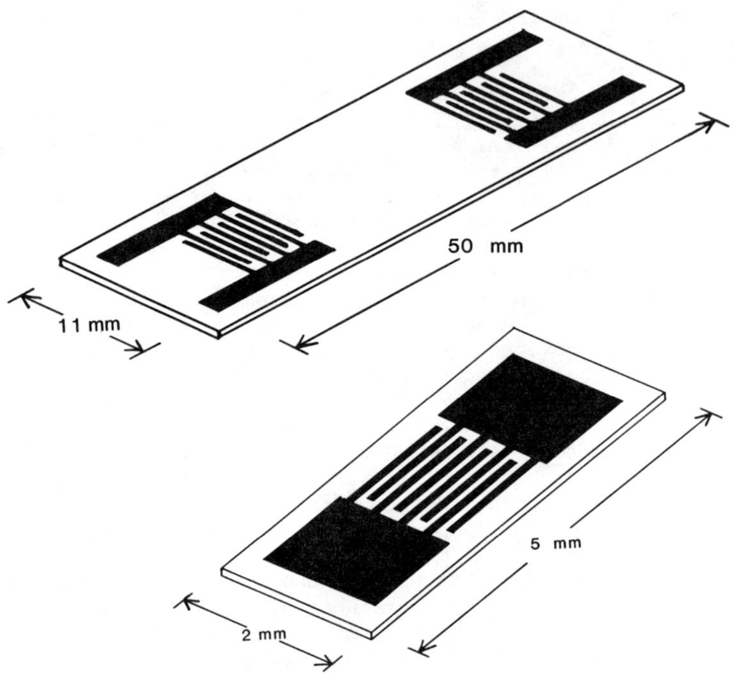

Figure 1.   Schematic representations and dimensions of interdigital microelectrodes that were coated with phthalocyanines.

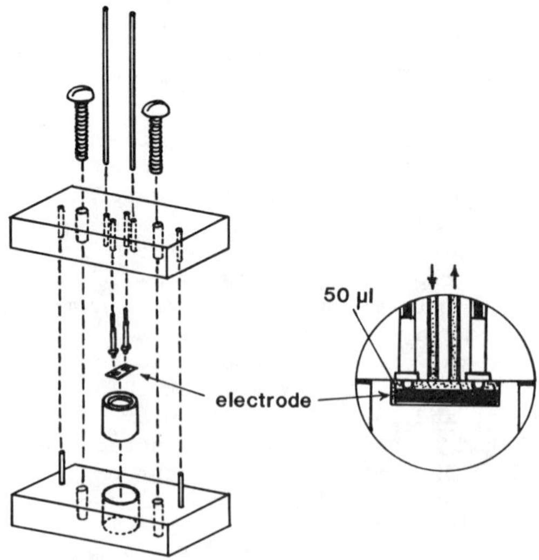

Figure 2.   Low dead-volume microelectrode detector cell.

The more reactive vapors were introduced into the apparatus at the lower concentration.

Results and Discussion

Sublimed Films Exposed to Vapors.   The six different types of phthalocyanine films that were sublimed onto the interdigital electrodes were exposed to each of a series of seven vapors.   Two devices, one for reference and one for measurement, were coated by sublimation simultaneously.   Metal-free, Fe, Co, Ni, Cu, and Pb phthalocyanines were applied to the interdigital electrodes in this manner.   Vapors studied were ammonia, dimethyl methylphosphonate (DMMP), water, air, benzene, and sulfur dioxide.   The relative change in conductivity, S, was computed from the measured voltage by the microcomputer-controlled apparatus using the following relationship:

$$S = \left(\sigma_2 - \sigma_1\right)\big/\sigma_1 = \left(E_2 - E_1\right)\big/E_1 \tag{2}$$

Here conductivity of the film during exposure to a test vapor is represented by σ_2 and initial conductivity by σ_1. E_1 is the initially measured voltage value prior to the vapor exposure, and E_2 is the measured voltage value during the exposure to a test vapor.   Figures 4 and 5 present the conductivity change data for the first 25 minutes of exposure to the test vapors which were decreasing exponentially in concentration with time.   Note the scale changes in Figures 4 and 5.   Each tick mark on the vertical axes of the figures represents a 5 per cent change in relative conductivity.

The maximum per cent change in relative conductivity of each phthalocyanine is listed for each vapor in Table I.   For the response values of metal-free phthalocyanine enclosed in parentheses in Table I, the magnitude of the signal was extremely variable--up to approximately 75% of the maximum.   For this reason the data for metal-free phthalocyanine were omitted from Figures 4 and 5 in the runs with water, ethanol, air, and benzene.

Table I.   Maximum Per Cent Change in Relative Conductivity
for Sublimed Phthalocyanine Films Exposed to Vapors

Vapor	Conc. (ppm)	H_2	Fe	Co	Ni	Cu	Pb
Ammonia	100	−41	−36	−72	−51	−81	−24
DMMP	100	+23	−10	−7	−5	−21	−51
Sulfur Dioxide	100	+10	+28	−9	+11	+3	−8
Ethanol	1000	(+18)	−11	−5	−1	−15	−4
Water	1000	(+21)	−11	−5	−5	−15	−6
Air	1000	(+25)	−4	−5	+2	0	−8
Benzene	1000	(+11)	−1	−2	+6	−2	+4

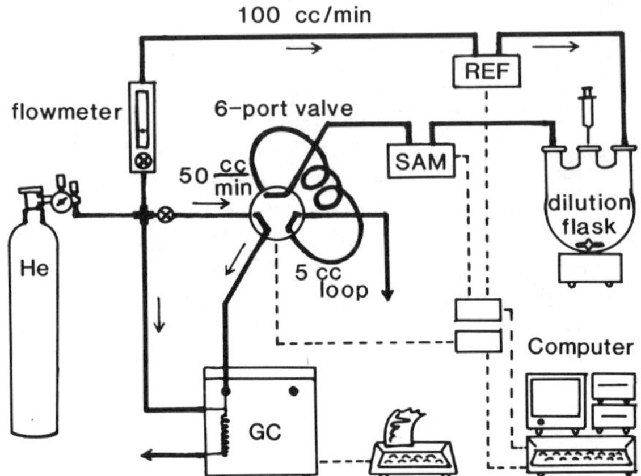

Figure 3. Vapor dilution apparatus to measure responses of coated electrodes.

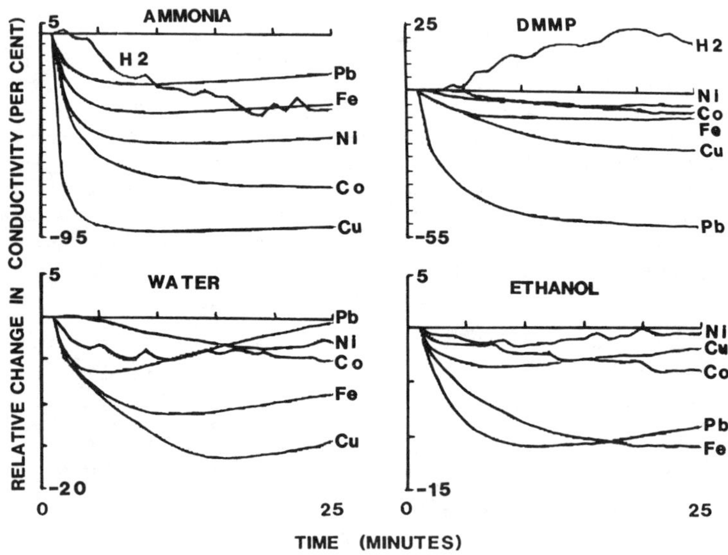

Figure 4. Change in conductivity with time of various phthalo-cyanine-coated electrodes exposed to exponentially diluted vapors. Initial concentration was 100 ppm for ammonia and DMMP but 1000 ppm for water and ethanol. The central atom of the phthalocyanine ring is indicated near each curve.

A detailed examination of the mechanisms of reaction of these vapors with the various types of phthalocyanines is beyond the scope of this work. The reader is directed to the papers by van Ewyk, Chadwick and Wright (14) or Langton and Day (11) for further information in this area. Experiments described here were conducted primarily to survey a series of materials under identical conditions in order to get a relative ranking of candidate materials for practical gas sensors, and to aid decisions about which monolayer-forming derivatives to synthesize.

A general characteristic of all the data shown in Figures 4 and 5 is the relatively slow initial response time and even slower return to baseline as the concentration of the test vapor was reduced. Because of the exponential dilution of vapor flowing to the sensor, the concentration dropped to 8.5% of its initial value after 25 minutes and to 0.3% of the initial value by the end of an hour. The sublimed films obviously do not equilibrate rapidly with the test vapors. This is a major reason for the interest in being able to deposit ultra-thin, ordered films reproducibly by the L-B technique.

In spite of the slow response times, the sublimed films might be used in an alarm that responds to specific vapors. Since the electrodes being coated with these materials are so small, an array of many electrodes could be incorporated into a small sensor. Assuming an array of electrodes coated with the six materials shown in Table I, it is apparent that a distinct recognizable pattern of maximum responses is produced by each of the test vapors. This is illustrated in Figure 6 by presenting the data of Table I as a bar graph. A microcomputer could be programmed to recognize these patterns.

Monolayer Films of Phthalocyanine Derivatives. A series of organic derivatives of phthalocyanines were prepared that have two important characteristics of materials to be deposited by the Langmuir-Blodgett technique: (1) they are soluble in volatile organic solvents, and (2) they form monomolecular films on the surface of water. Further study of deposited films of these phthalocyanine derivatives will be necessary in order to determine the exact orientations on the surface, but regardless of their orientations, they offer interesting possibilities for construction of thin films of ordered arrays of molecules on the surface of gas sensors.

Studies of the gas-sensing abilities of these films have just begun. To conduct such studies it is first necessary to transfer the material to a sensor surface. Early attempts to transfer the compounds from this group of surface-active phthalocyanine derivatives to a solid surface (glass) produced rather non-uniform films compared to classical L-B film materials such as stearic acid. Since the phthalocyanine derivatives are highly colored, irregularities in a many-monolayer L-B film may be observed visually. However, a mixed film of copper tetracumylphenoxy phthalocyanine and stearyl alcohol (abbreviated as CUPCCP/C18OH) produced a uniform multilayer film when transferred to glass, so this 1:1 mole ratio mixture was also transferred to a microelectrode for electrical testing. Similar tests for other phthalocyanine derivatives were conducted.

A rough comparison of the response of a L-B film of CUPCCP/C18OH to the response of a sublimed film of copper phthalocyanine (CUPC) is

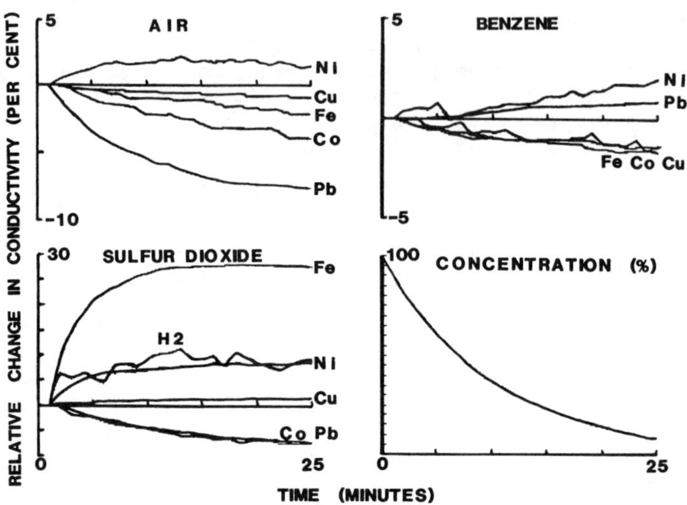

Figure 5. Change in conductivity with time of various phthalo-
cyanine-coated electrodes exposed to exponentially diluted vapors.
Initial concentration was 100 ppm for sulfur dioxide but 1000 ppm
for air and benzene. The dilution function is shown at the lower
right.

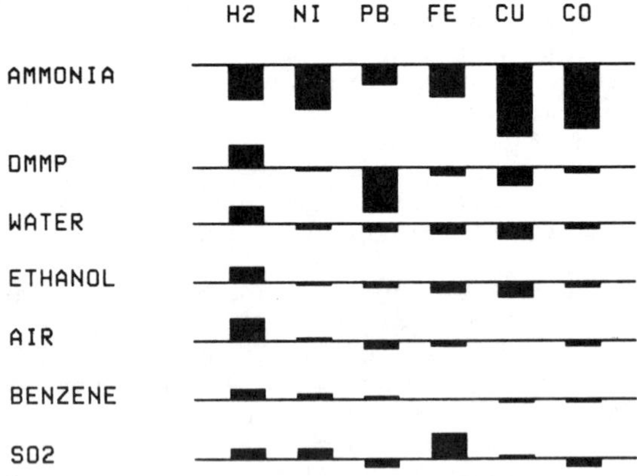

Figure 6. Characteristic patterns of response of an array of six
sensors coated with different phthalocyanines to various vapor exposures. The maximum concentration was 100 ppm for ammonia,
sulfur dioxide, and DMMP and 1000 ppm for water, ethanol, air and
benzene. The central metal of each phthalocyanine is shown at the
top.

shown in Figure 7. The thickness of the 47-monolayer film was
approximately 1100 Ångstroms (estimated from the length of octadec-
anol and the number of layers), while the thickness of the sublimed
CUPC was estimated to be about 6000 Angstroms (based on mass
increase of the substrate). The ammonia injection resulting in an
initial concentration of 500 ppm was placed into the dilution flask
after tracking the baseline for two minutes. Although the two
coatings are difficult to compare directly since the morphology is
strikingly different, some general features are readily apparent.
There was a greater change in conductivity for the CUPCCP/C18OH film,
although less coating material was present. The conductivity of
CUPCCP/C18OH increases on exposure to ammonia, but the conductivity
of CUPC decreases. The time to reach maximum response is less than
for the CUPC. This might be expected because the CUPC film is
thicker. The L-B film containing CUPCCP shows much greater reversibility. The fact that the signals for both coatings lag behind the
ideal dilution curves (shown by dotted lines in the figure) indicates
that the test vapor diffuses into the bulk of the coating and must
diffuse back out when the concentration above the electrode decreases. Greater reversibility is desirable in a practical detection
device.

Summary and Conclusions

A survey of interactions of a series of vapors with coatings of metal-
free and metal-substituted phthalocyanines sublimed onto the surface
of microelectrodes indicated, as others have observed, that the conductivity of these materials was altered significantly in the presence of certain vapors. The central metal atom in the phthalocyanine molecule strongly influences the response of the compound to
the challenging vapor. Although several of the sublimed coatings of
the phthalocyanines exhibit an easily measured change in conductivity
when exposed to test vapors, in general, the signal is very slow to
return to the preexposure baseline. It has been possible to create a
series of organic derivatives of the phthalocyanines that are soluble
in volatile organic solvents and that form monolayer films on water
surfaces. These materials can be transferred to solid surfaces, such
as microelectrodes, by the Langmuir-Blodgett technique of passing the
microelectrode repeatedly through the monolayer film. The ability to
deposit multiple monolayer films of phthalocyanine derivatives
offers the possibility of constructing films of precisely controlled
thicknesses with ordered arrays of molecules. These films can be
placed on the surfaces of microsensors to make improved vapor-
sensitive devices. Such films may have faster response and greater
reversibility than the less well controlled sublimed coatings.

Acknowledgments

This work was supported by the Office of Naval Research.

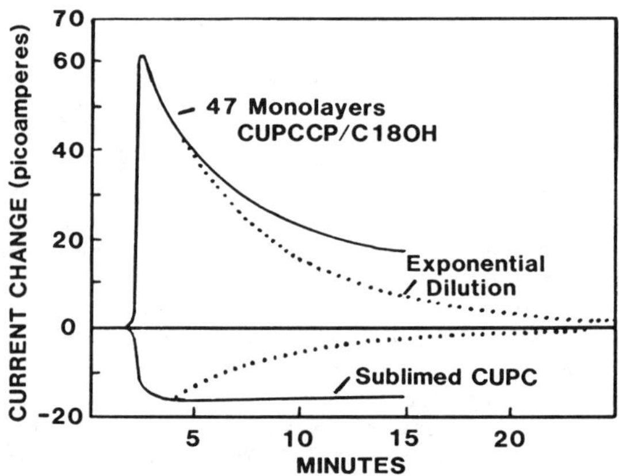

Figure 7. Comparison of response of a 47-layer L-B film coating
of Cu tetracumylphenoxy phthalocyanine to a sublimed coating of Cu
phthalocyanine. The signal expected if the electrodes tracked the
exponential dilution of the 500 ppm initial concentration of
ammonia is shown by the dotted lines.

Literature Cited

1. Gaines, G. L.   Thin Solid Films 1983, 99, ix-xii.
2. Lever, A. B. P.   In "Advances in Inorganic Chemistry and
    Radiochemistry"; Academic: New York, 1965; Vol. VII, pp. 27-114.
3. Tollen, G.; Kearns, D. R.; Calvin, M.   J. Chem. Phys. 1960, 32,
    1013-1019.
4. Kearns, D. R.; Tollen, G; Calvin, M.   J. Chem. Phys. 1960, 32,
    1020-1025.
5. Ukei;, K.; Takamoto, K.; Kanda, E.   Physics Letters 1973, 45A,
    345-346.
6. Ukei, K.   J. Physical Soc. Jap. 1976, 40, 140-143.
7. Hamann, C.; Hohne, H.-J.; Kersten, F.; Muller, M.; Przyborowski,
    F.; Starke, M.   Phys. Stat. Sol. 1978, (a), 50, K189-K192.
8. Yasunaga, H.; Shintaku, H.   J. Appl. Phys. 1980, 51, 2149-2152.
9. Yasunaga, H.; Kojima, K.; Yohda, H; Takeya, K.   J. Phys. Soc.
    Jap. 1974, 37, 1024-1030.
10. Yasunaga, H.; Kasai, K; Kenichi, T.   J. Phys. Soc. Jap. 1979,
    46, 839-845.
11. Langton, J; Day, P.   J. Chem. Soc., Faraday Trans. 2; 1982, 78,
    1675-1685.
12. Beales, K. J.; Eley, D. D.; Hazeldine, D. J.; Palmer, T. F.
    In "Katalyse an Phthalocyaninen"; Kropf, H.; Steinbach, F. Eds.;
    Georg Thieme Verlag: Stuttgart, 1973; pp. 1-32.
13. Sadaoka, Y.; Yamazoe, N.; Seiyama, T.   Denki Kagaku Oyobi Kogyo
    Butsuri Kagaku 1978, 46, 597-602.
14. van Ewyk, R. L.; Chadwick, A. V.; Wright, J. D.;   J. C. S.
    Faraday I 1980, 76, 2194-2205.
15. Jones, T. A.; Bott, B.   In "Transducers '85"; IEEE : New York,
    1985; IEEE No. 85CH2127-9; pp. 414-417.
16. Bott, B.; Jones, T. A.   In "Transducers '85"; IEEE : New York,
    1985; IEEE No. 85CH2127-9; pp. 128-130.
17. Baker, S.; Petty, M. C.; Roberts, G. G.; Twigg, M. V.   Thin Soid
    Films 1983, 99, 53-59.
18. Snow, A. W.; Jarvis, N. L.   J. Amer. Chem. Soc. 1984, 106, 4706-
    4711.
19. Barger, W. R.; Wohltjen, H.; Snow, A. W.   In "Transducers '85";
    IEEE: New York, 1985; IEEE No. 85CH2127-9; pp. 410-413.

RECEIVED November 13, 1985

10

Chemical Microsensors Based on Surface Impedance Changes

Stephen D. Senturia

Department of Electrical Engineering and Computer Science, Center for Material Science and Engineering, Massachusetts Institute of Technology, Cambridge, MA 02139

Sensing chemical species is a much more difficult task than the measurement of mechanical variables such as pressure, temperature, and flow, because in addition to requirements of accuracy, stability, and sensitivity, there is the requirement of specificity. In the search for chemically-specific interactions that an serve as the basis for a chemical sensor, investigators should be aware of a variety of possible sensor structures and transduction principles. This paper adresses one such structure, the charge-flow transistor, and its associated transductive principle, measurement of electrical surface impedance. The basic device and measurement are explained, and are then illustrated with data from moisture sensors based on thin films of hydrated aluminum oxide. Application of the technique to other sensing problems is discussed.

<u>Background</u>. The term <u>microsensor</u> denotes a transducer that, in some fashion, exploits advanced miniaturization technology, whether an adaptation of integrated circuit technology, or some other microfabrication technique. Within the past decade, a myriad of microsensors have been developed, with capabilities for measurement of temperature, pressure, flow, position, force, acceleration, chemical reactions, and the concentrations of chemical species. The latter measurements, of chemical species, are intrinsically more difficult than the measurement of mechanical variables because in addition to requirements of accuracy, stability, and sensitivity, there is a requirement for specificity.

Chemical sensors can be of the type that sense the species directly, such as ion-selective electrodes, or the type that sense the species indirectly through the change in another physical property produced by the species. An example of this latter type is the so-called "chemiresistor", in which the presence of the species being sensed modifies the electrical resistance of a transducer

0097–6156/86/0309–0166$06.00/0

material. This paper is directed toward the chemiresistor family of
sensors, and specifically toward applications of sensing gaseous
chemical species in gaseous ambient atmosphere. The domain of such
applications includes routine monitoring of moisture or atmospheric
pollutants, life-safety applications such as smoke and/or gas sensing, and pesticide or other chemical-agent detection. In all such
cases, one has a sensor material, usually in the form of a thin
film, and a suitable set of electrodes with which the electrical
resistance of the film can be monitored. This paper illustrates a
particular type of microsensor structure, called the charge-flow
transistor, that is ideally suited for chemiresistor application.

<u>Historical Perspective</u>. Development of the charge-flow transistor
was stimulated by the report by Byrd <u>1</u> that polymer derivatives of
poly(phenylacetylene) could be used in thin-film form to sense the
presence of NH_3 and SO_2. This led to several NASA-sponsored programs to explore in a more systematic way the application of such
polymers to life-safety applications <u>2-4</u>, including the possible
development of advanced sensor structures that could be used with
weakly conducting thin-film chemical-sensor materials. In 1977, the
first charge-flow transistor device was reported <u>5</u>. This device
consisted of a metal-oxide-semiconductor field-effect transistor
(MOSFET) in which a portion of the gate metal was replaced by the
thin-film material. Because the thin film was weakly conducting
compared to the metal gate, turn-on of the device was limited by how
quickly charge could flow along the sheet of thin-film material,
charging up the gate-to-semiconductor capacitance. Since the conduction in the sheet of thin film depended on the presence of suitable gases in the ambient, the turn-on time of the device could be
used to measure the presence of chemical species <u>6</u>.
     The dielectric constant, like index of refraction, has been used
to analyze binary mixtures for industrial process control since the
fifties <u>7</u> and used for humidity sensors shortly thereafter <u>8-14</u>. In
these humidity sensors, interdigital metal elements deposited on
ceramic substrates were used to increase the sensitivity. These
sensors used polymeric, inorganic or ceramic coatings on the interdigital structure to selectively adsorb water vapor. The idea that
these structures could be miniturized by forming them on a silicon
chip along with associated measuring circuitry was reported in 1977
as, first, discrete charge-flow transistors <u>5,6</u>, then, discrete
sense circuits <u>15</u>, and finally, two fully integrated monolithic
sense circuits <u>16</u>. In the work cited above, in which the thin-film
material actually served as the gate of a portion of the FET, there
were problems both in stability of the measurement, and in the quantitative interpretation of data. In the hope of finding improved
measurement accuracy, an alternative structure, the floating-gate
charge-flow transistor, was developed <u>17</u>, and has been applied to a
variety of measurment problems, including (a) the study of moisture
measurement both with thin polymer films <u>17</u> and thin films of
hydrated aluminum oxide <u>18</u>, and (b) in-situ chemical reaction monitoring in polymers <u>19,20</u>. The goal of the present paper is to
review both the charge-flow transistor measurement concept and its
applications to chemical sensing problems.

The Floating-Gate Charge-Flow Transistor (CFT)

The floating-gate charge-flow transistor (referred to hereafter as
CFT) is actually a small-scale silicon-integrated circuit. It
combines a pair of interdigitated electrodes with two field-effect
transistors to yield a transducer for the calibrated measurement of
low-frequency sheet resistance and sheet suspectance (hence the term
"surface impedance") of thin films. The heart of the device is the
electrode pair and transistor illustrated schematically in Figure 1.
One of the electrodes (the Driven Gate) has a signal applied to it;
the other electrode (the Floating Gate) extends over the channel
region of a field-effect transistor but has no explicit external
connection. The only path for charge to reach the floating gate is
across the spaces between the two electrodes, and it is here that
the thin film is placed, as illustrated in Figure 2.

The electrical equivalent circuit for the driven-gate-to-float-
ing-gate transfer function can be understood with reference to
Figures 2 and 3. The capacitance between the floating gate electrode and the conducting silicon substrate is denoted by C_L. It
includes the input capacitance of the FET. The total charge reaching C_L, or, equivalently, the total voltage across C_L, serves as an
input signal to the FET, modifying the conduction in the channel of
the FET. The coupling between the driven and floating gates consists of two parts. There is a stray capacitive coupling through
the air above the thin film, denoted by C_X, and there is a distributed R-C transmission line, denoted by R_S, C_S, and C_T, which represents the combined effect of sheet conduction in the thin film (R_S),
the film-to-substrate capacitance that must be charged by conduction
in the film (C_T), and a distributed capacitance that accounts for
dielectric polarization of the film (C_S). Mathematical analysis of
the transfer function is a straightforward problem in linear circuit
theory. In the sinusoidal steady state at frequency f, if V_{DG} is
the voltage phasor of the sinusoid applied to the driven gate, and
V_{FG} is the voltage phasor of the sinusoidal voltage appearing on the
floating gate, then the transfer function has the following form:

$$\frac{V_{FG}}{V_{DG}} = \frac{1 + (C_X/C_T)\alpha \sinh\alpha}{\cosh\alpha + (C_L/C_T + C_X/C_T)\alpha \sinh\alpha} \tag{1}$$

where

$$\alpha = \left[\frac{j(C_T/C_S)}{j + (2\pi f R_S C_S)^{-1}} \right]^{1/2} \tag{2}$$

The important point to note here is that the capacitive terms determine the overall shape of the transfer function, whereas the frequency f and the sheet resistance R_S occur only within the parameter
α, and only as a product. Therefore, an increase in frequency at
fixed sheet resistance can cause the same change in transfer function as an increase in sheet resistance at fixed frequency. This
property can be exploited in visualizing the transfer function by
plotting the amplitude of Equation 1 against its phase shift, using

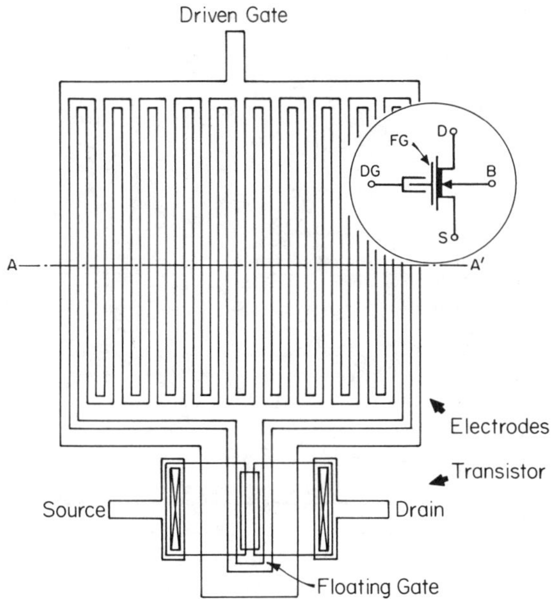

Figure 1.   Schematic top view of CFT device, with circuit symbol shown in insert.   Reproduced with permission from reference 17. Copyright 1982 Institute of Electrical and Electronics Engineers.

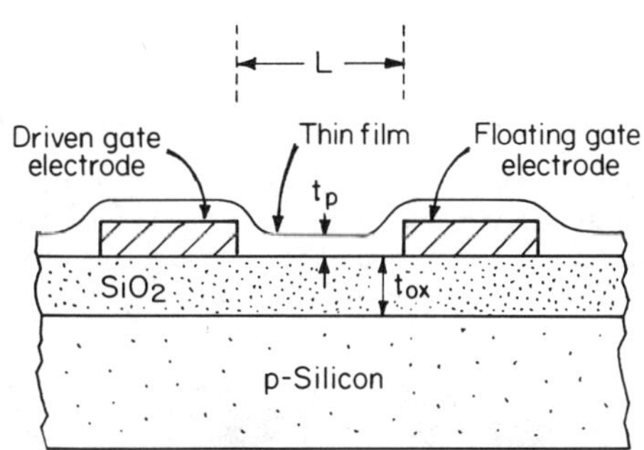

Figure 2.   Schematic cross section of CFT device through A–A'. Film thickness is exagerated relative to other dimensions. Reproduced with permission from reference 17. Copyright 1982 Institute of Electrical and Electronics Engineers.

the product fR_S as a parameter.   Figure 4 illustrates two different transfer functions for different assumed values of C_S, using actual device geometry to determine C_X and C_T.

A microphotograph of an actual CFT is shown in Figure 5.   The overall chip size is 2mm x 2mm, and the spacing between fingers of the interdigitated electrodes is 12 μm.   Note the additional FET on the chip, called the reference FET.   Its function is to compensate for process variations in the FET and for the temperature dependence of the transistor properties.   A feedback circuit in which the reference FET is used in conjunction with the sensing FET is shown in Figure 6.   Because the circuit requires the two transistor cur-

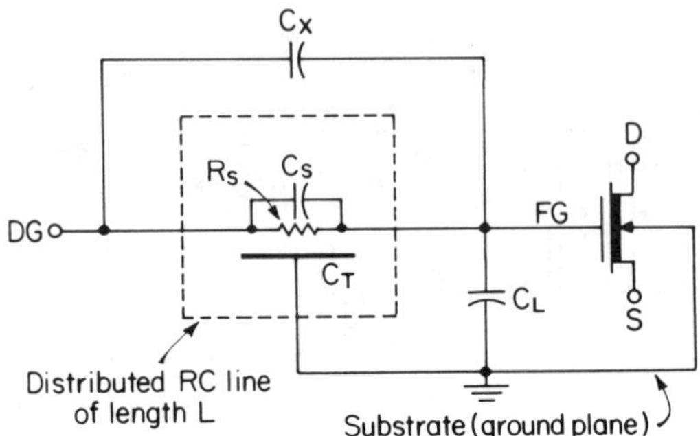

Figure 3. Electric circuit model for CFT device. Reproduced with permission from reference 18. Copyright 1982 Institute of Electrical and Electronics Engineers.

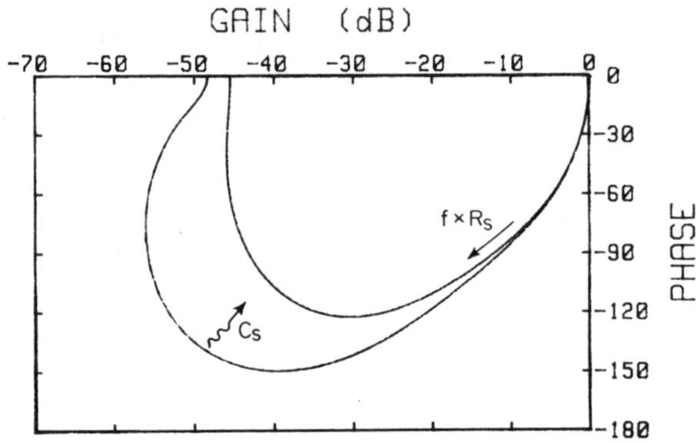

Figure 4. Calculated transfer function for CFT device illustrating effect of sheet capacitance on shape of calibration curve. Reproduced with permission from reference 18. Copyright 1982 Institute of Electrical and Electronics Engineers.

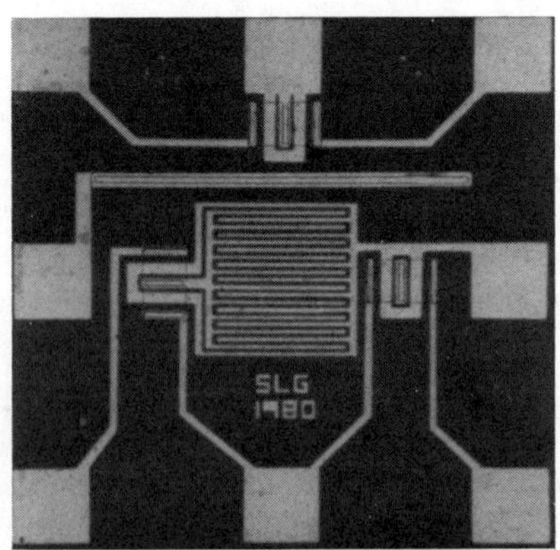

Figure 5. Microphotograph of CFT chip. Reproduced with permission from reference 17. Copyright 1982 Institute of Electrical and Electronics Engineers.

rents to be equal (via the feedback), and because the reference FET
is electrically identical to the sensing FET, the voltage applied to
the reference FET gate must be equal to the voltage that apears on
the floating gate through charge-flow processes. Thus, the entire
measurement process consists of applying a sinusoid to the driven
gate, and making an amplitude and phase measurement of the voltage
that the feedback circuit applies to the gate of the reference FET.
With the use of calibration curves of the type illustrated in Figure
4 (which, in practice, are stored as look-up tables in our data-
logging system), on-line real-time measurement of sheet resistance
and sheet capacitance is routinely possible. The range of frequencies presently in use is 0.1 to 10,000 Hz. The sheet resistance
range accessible to this measurement technique is between 10^9 and
10^{16} Ohms/square. Thus, this sensor becomes a useful replacement
for an electrometer when working with high impedance films, and has
the added advantages of a wide measurement bandwidth and the routine
use of AC measurement techniques to avoid electrode polarization
effects.

The sensitivity of the sheet-resistance measurement is a few
percent. Absolute accuracy is estimated at 10%. The sensitivity of
the sheet capacitance measurement is much lower, and depends
directly on film thickness. Our experience has been that sheet
capacitance must be known to within 50% in order to provide an
acceptably accurate calibration for sheet resistance, but that pre-
cisions much better than 50% are difficult to achieve. The temperature range over which the chip can operate is very wide. We have
made measurements between −200 and 300°C, although we do not yet
know whether there is a loss of calibration accuracy as temperature
varies.

Application to Moisture Measurement with Hydrated Aluminum Oxide
Films

The use of thin films of hydrated aluminum oxide to measure moisture
in the ambient is well known 21, and serves as the basis for a
number of commercial moisture-measurement products. The usual
device configuration consists of a parallel-plate structure in which
an aluminum electrode is in contact with a hydrated aluminum oxide
film (see Figure 7a), over which a porous gold electrode is placed.
As moisture reaches the oxide layer, the conductivity of the
hydrated oxide is changed. This is conventionally monitored by
measuring the effective capacitance and/or conductance between the
gold and aluminum electrodes. In the CFT approach, samples are
prepared by evaporating a very thin aluminum film (400−800 Å) over a
silicon wafer on which CFT device fabrication has been carried out.
The devices are then bonded to headers and wire bonded, after which
immersion in heated water serves to convert the thin aluminum film
to a moisture-sensitive hydrated oxide 22, yielding the cross-
section illustrated in Figure 7b.
      That the aluminum-oxide CFT obeys the transfer function of
Equations 1 and 2 is illustrated in Figure 8, in which the crosses
represent actual room temperature measurements in an ambient dew
point of −16.9°C at a a variety of frequencies, and the circles
represent values calculated from the model for $R_S = 1.7$ x
10^{14} Ohms/square, and $C_S = 1.6$ pF-square. Thus, by measuring the

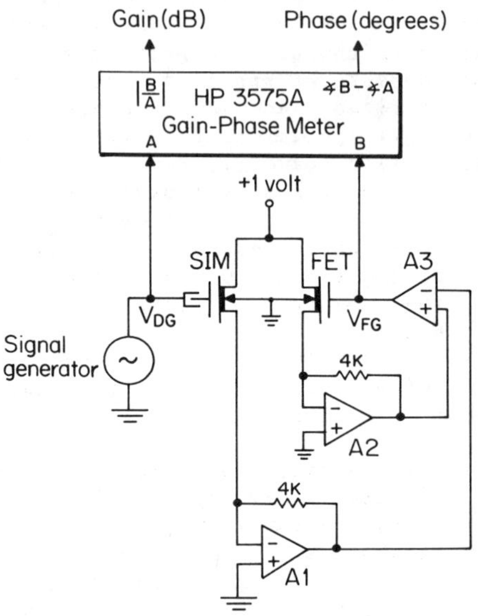

Figure 6. CFT feedback circuit and instrumentation for AC
measurements. Reproduced with permission from reference 17.
Copyright 1982 Institute of Electrical and Electronics Engineers.

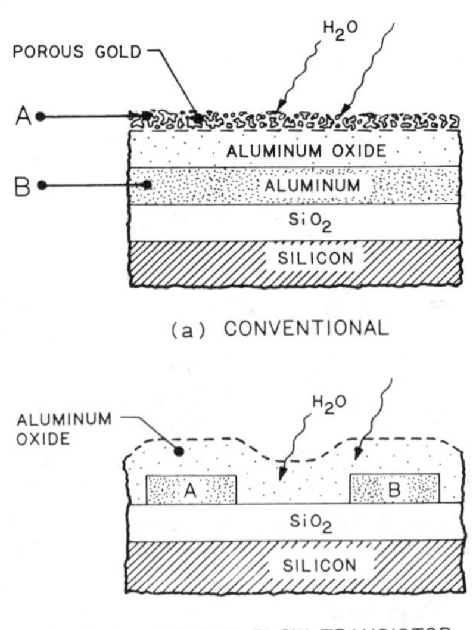

Figure 7. Schematic cross sections of (a) conventional aluminum
oxide moisture sensor, and (b) CFT aluminum oxide moisture sensor.

transfer function for different ambient dew points, one can follow
changes in sheet resistance. Both aging (Figure 9) and hysteresis
effects (Figure 10) have been observed with hydrated aluminum oxides
18,23.

In-Situ Measurement of Polymer Curing

Dielectric measurements have been established for nearly three decades as a technique for monitoring the cure of polymeric resins.
Dramatic changes in the dielectric properties of the material accom-
pany the transformation of the resin from a viscous liquid to a
solid.

The standard method for making measurements of dielectric properties is to place a sample between closely spaced parallel conducting plates, and to monitor the AC equivalent capacitance and
dissipation factor of the resulting capacitor. The capacitance is
proportional to the dielectric permittivity (ε') at the measurement
frequency, and the dissipation factor in combination with the value
can be used to extract the dielectric loss factor (ε'').

The CFT device shown in Figure 1 has been used to monitor
dielectric properties of polymeric materials. The medium to be
studied is placed over the electrodes, either by application of a
small sample or by embedding the entire integrated circuit in the
curing medium (see Figure 2 for a schematic cross-section through
the electrode region). The resulting device and associated electronics is called a microdielectrometer.

The microdielectrometer has been used to monitor polymer curing
in-situ 22,23. The CFT device can make measurements at lower frequencies than could be achieved by conventional dielectric measurement techniques. Measurements at multiple frequencies can be
made in real-time. A Fourier transform equivalent of the microdielectrometer has been developed to extend the frequency range to as
low as 0.005 Hz 24.

Discussion

The results illustrated above show that the CFT method is suitable
for making chemical-sensor measurements using both bulk polymers
and, in particular, thin film materials that are intrinsically weak
conductors. Therefore, the CFT looks promising for such materials
as poly(phenylacetylene) derivatives 24, for which carefully
shielded electrometer measurements have been required in the past
because of current levels at the threshold of detectability. Furthermore, the fact that the CFT always makes AC measurements reduces
the problem of DC polarization of electrodes. In addition, the CFT
approach should be suitable for other "chemiresistor" applications,
such as the metal-substituted phthalocyanines proposed by Jarvis et.
al. 25 and for Langmuir-Blodgett films 26, which, because they are
so thin, may prove impossible to use in parallel-plate form, but
which can be routinely used with the high-sensitivity interdigi-
tated-electrode approach provided by the CFT.

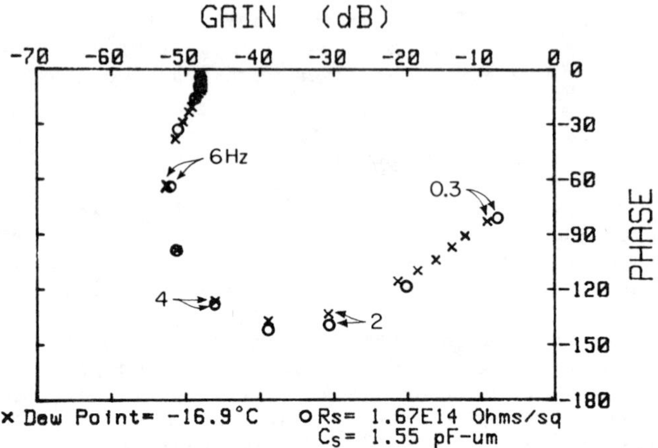

Figure 8. Typical comparison between experimental data (crosses)
and calculated transfer function (circles) for CFT coated with
hydrated aluminum oxide, measured at room temperature in an ambient
dew point of -16.9 ^{O}C. Reproduced with permission from reference 18.
Copyright 1982 Institute of Electrical and Electronics Engineers.

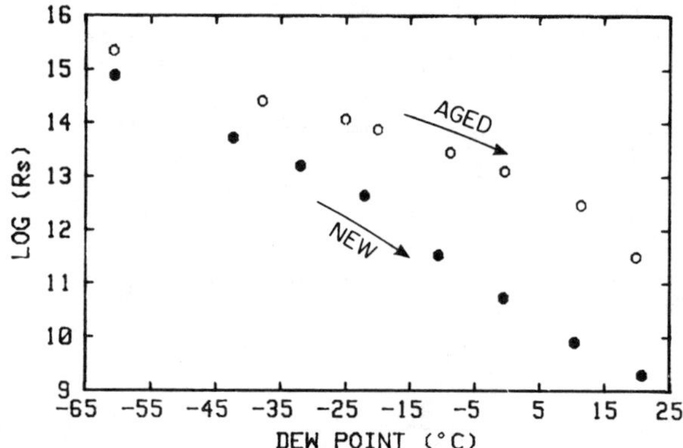

Figure 9. Sheet resistance data illustrating the effect of aging on
aluminum oxide moisture sensor. Reproduced with permission from
reference 18. Copyright 1982 Institute of Electrical and Electronics
Engineers.

Acknowledgments

This paper is based on an invited presentation at the Symposium on
Microsensors for Chemical Detection held as part of the 17th Mid-
Atlantic Regional Meeting of the American Chemical Society (MARM-
ACS), April 1983. The work was supported by the Thermal Processes
Division of the National Bureau of Standards under Grant NB80-DADA-
1004 and by the National Science Foundation under Grant ECS-8114781.
Devices used in this work were fabricated in the MIT Microelec-
tronics Laboratory, a Central Facility of the Center for Materials
Science and Engineering which is sponsored in part by the National
Science Foundation under contract DMR-81-19295. Some of the meas-
urement instrumentation was purchased under NSF Contract ENG-7717219
and under programs sponsored by the Office of Naval Research.

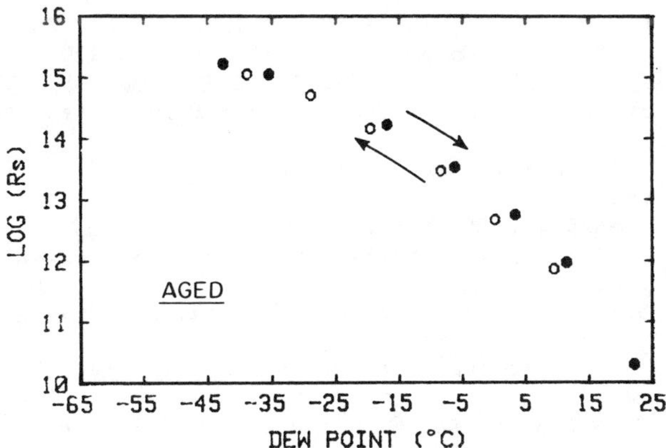

Figure 10. Sheet resistance data illustrating the hysteresis in aged aluminum oxide moisture sensor. Reproduced with permission from reference 18. Copyright 1982 Institute of Electrical and Electronics Engineers.

Literature Cited

1. Byrd, N. R. "Space Cabin Atmosphere Contaminat Detection Techniques"; Douglas Report SM-48446-F (Contract NAS 21-15); NTIS: Springfield, VA, 1968.

2. Byrd, N. R.; Sheratte, M. B. "Synthesis and Evaluation of Polymers"; NASA CR-134693 (Contract NAS 3-17515); NTIS: Springfield, VA, 1975.

3. Byrd, N. R.; Sheratte, M. B. "Semiconducting Polymers for Gas Detection"; NASA CR-134885 (Contract NAS 3-18919); NTIS: Springfield, VA, 1975.

4. Senturia, S. D. "Fabrication and Evaluation of Polymer Early-Warning Fire Detection Devices"; NASA CR-134764 (Contract NAS 3-17534); NTIS: Springfield, VA, 1975.

5. Senturia, S. D.; Sechen, C. M.; Wishneusky, J. A. Appl. Phys. Lett., (1977) 30, 106.

6. Senturia, S. D.; Huberman, M. G.; Van der Kloot, R. Proc. ARPA/NBS Workshop on Moisture Measurement in Integrated Circuit Packages, (1978) NBS Special Publication 400-69, p. 108.

7. Dee Snell, F.; Hilton, C. L. "Encyclopedia of Industrial Chemical Analysis", Section on Capacity and Dielectric Constant, Interscience Publishers, 1966, Vol. 1.

8. Musa, R. C.; Schnable, G. L. "Polyelectrolyte Electrical Resistance Humidity Elements", Paper 353 in Humidity and Moisture, Exler, A., Ed., Reinhold Publishing Corp., New York, 1965, Vol. 1.

9. Amdur, E. J.; Nelson, D. E. "A Ceramic Relative Humidity Sensor", Paper 38, ibid.

10. Regtien, P. P. L. Sensors and Actuators, 1981, 2, 85.

11. Wakabayashi, K.; Ohta, S.; Takemori, D.; shirae, K. "Non-Linear Behavior of Glass Substrate in High Humidity", in "Chemical Sensors", Seiyama, T.; Feuki, K.; Shiokawa, J.; Suzuki, S., Editors, Proceedings of the International Meeting on Chemical Sensors, Fukuoka, Japan, Analytical Chemistry Symposia Series - Volume 17, Elsevier, New York, pp. 439-444.

12. Uchikawa, F.; Miyao, K.; Horii, H.; Shimamoto, K. "Humidity Sensing with Silicone Composite Films", pp. 445-450, ibid.
13. Miyoshi, S.; Sugihara, T.; Jinda, A.; Hijikigama, M. "Thin Film Humidity Sensor Composed of Cross-Linked Polyelectrolyte", pp. 451-456, ibid.
14. Fleming, W. J. "A Physical Understanding of Solid State Humidity Sensors", Society of Automotive Engineers Paper No. 810432, Warrensdale, PA, 1981.
15. Senturia, S. D.; Fertsch, M. T. IEEE J. Sol. St. Circuits, SC-14 (1979) 753.
16. Senturia, S. D.; Garverick, S. L.; Togashi, K. Sensors and Actuators, 1981/82, 2, 59.
17. Garverick, S. L.; Senturia, S. D. IEEE Trans. Elec. Dev., ED-29, 1982, 90.
18. Davidson, T. M.; Senturia, S. D.; Proc. IEEE Int. Rel. Phys. Symp., 1982, pp. 249-252.
19. Sheppard, Jr., N. F.; Coln, M. C. W.; Senturia, S. D. "Microdielectrometry: A New Method of In-Situ Cure Monitoring", Proc. 26th SAMPE Symposium, Los Angeles, CA, 1981, 26, 65-76.
20. Senturia, S. D.; Sheppard, N. F.; Lee, Jr., H. L.; Day, D. R. J. Adhesion, 1982, 15, 69.
21. Kovac, M. G., Chleck, D.; Goodman, P. Proc. IEEE Int. Rel. Phys. Symp., 1977, pp. 85-91.
22. Vedder, W.; Vermilyea, D. A. Trans. Faraday Soc., 1969, 65, 561.
23. Lin, C.-H.; Senturia, S. D. Proc. Solid State Transducers, 83, Delft, June, 1983; Sensors and Actuators, 1983, 4, 497.
24. Coln, M. C.; Senturia, S. D. "The Application of Linear System Theory to Parametric Microsensors", Proc. Third Int'l Conf on Solid-State Sensors and Actuators, Philadelphia, June 1985 (in press).
25. Wentworth, S. E.; Libby, J. B.; Bergquist, P. R. Symp. on Microsensors for Chemical Detection, 17th MARM-ACS, 1983.
26. Wohltjen, H.; Jarvis, N. L.; Snow, A. Symp. on Microsensors for Chemical Detection, Chapter 20 in this book.
27. Roberts, G. Sensors and Actuators, 1983, 4, 131.

RECEIVED February 19, 1986

Schottky–Barrier Diode and Metal-Oxide-Semiconductor Capacitor Gas Sensors
Comparison and Performance

S. J. Fonash and Zheng Li

Engineering Science Program, Pennsylvania State University, University Park, PA 16802

Schottky-barrier diode and metal-oxide-semiconductor
(MOS) capacitor gas sensors have established themselves
as extremely sensitive, versatile solid state sensors.
In this review the basis for the chemical sensitivity
of these devices will be explored and the various
device structures used for these sensors will be discussed. A survey of the performance of the diode-type
and capacitor-type structures will be presented and a
comparison of characteristics of these two classes of
solid state gas sensors will be given.

Schottky-barrier type diodes, in the form of metal-semiconductor
(M-S), metal-interfacial layer-semiconductor (M-I-S), or degenerate
semiconductor-interfacial layer-semiconductor structures, can be
configured to be extremely sensitive solid state gas detectors. If
these chemically sensitive diode structures are used as a gate, one
can fabricate chemically sensitive MESFET (metal-semiconductor field
effect transistor) sensors. Metal-oxide-semiconductor (M-O-S) or,
to be more general, metal-insulator-semiconductor (M-I-S) capacitor
structures, as well as degenerate semiconductor-insulator-semicon-
ductor capacitor structures, can also be configured to be extremely
sensitive solid state gas detectors. In this case, if these chemically sensitive capacitor structures are used as a gate, one can
fabricate chemically sensitive MOSFET (metal-oxide-semiconductor
field effect transistor) sensors. The chemical sensitivity of MESFET
sensors is due to the chemical sensitivity of the diode-type structure and the chemical sensitivity of MOSFET sensors is due to the
chemical sensitivity of the capacitor-type structure. Consequently,
this review will focus its attention on the basic physics and chemistry of sensing taking place in diode-type or capacitor-type configurations.
  Since both metals and degenerate semiconductors have been used
as the counter-electrode to the semiconductor in both diode and
capacitor-type devices, a more general notation than that usually
found in the literature will be employed in this review. This more
generalized notation will refer to the counter-electrode as the conductor (c). Hence, M-S, M-I-S, and degenerate semiconductor-interfacial layer-semiconductor diode devices all become C-S or C-I-S

0097–6156/86/0309–0177$07.50/0

diode sensors, as appropriate. Following this scheme, all the capacitor-type structures become C-I-S capacitor sensors. As we will see,
gas detection occurs when the presence of a gaseous species modifies
the current-carrying capability of a C-S or C-I-S diode or the
charge-storage capability of a C-I-S capacitor structure.

<u>Structures and Principles of Operation</u>

<u>Diode Sensors</u>. In general, C-S and C-I-S Schottky diode-type gas
sensors have the configurations indicated in Figures 1 and 2. As
seen in Figure 1A, the conductor-semiconductor (C-S) structure is
the simpler configuration. The essential feature of this structure
is the conducting film/semiconductor interface and the concomitant
barrier (space charge) region in the semiconductor. We note that
whenever dissimilar materials are joined together (in this case a
conductor and a semiconductor), a double layer (i.e., space charge)
is created at the interface (1). Since a semiconductor has very few
free carriers to use to establish its part of the double layer, the
space charge (barrier) region in a semiconductor extends relatively
far into its bulk as seen in the figure. It is this extended barrier
region in the semiconductor which controls transport across the
conductor/semiconductor interface as seen schematically in Figure 1B.
Any change in the double layer at the conductor/semiconductor interface would reflect into a change in the extent and height of the barrier seen by charge carriers in the semiconductor. Any change in the
barrier or, more generally, any change in transport across the barrier region will produce a change in the current-carrying capabilities of the C-S junction seen in Figure 1.

The barrier (space charge) region in the semiconductor can be
modified by applying a bias to the structure as seen in Figure 1B.
This causes the current flow to change. The semiconductor barrier
can also be modified due to the presence of some chemical species in
the environment if that chemical species can modify the double layer
at the conductor/semiconductor interface. If a chemical species can
effect a change in the double layer, then the barrier in the semiconductor is correspondingly changed and consequently the current flowing across the conductor/semiconductor interface, at some voltage,
is changed. Such a change in the current flowing across the structure at some bias due to the presence of a chemical species consti-
tutes detection of that species.

We note that the presence of a chemical species, if it affects
the double layer, can cause an increase or decrease in the barrier
height ϕ_B in the semiconductor depending on the interaction between
the species and the C-S structure. Those species that cause the barrier to increase would enhance rectifying current-voltage behavior
in the C-S structure; those species that cause the barrier to
decrease would enhance ohmic (lack of rectification) current-voltage
behavior in the C-S structure.

A chemical species in the environment of a C-S structure can
interact with the device and thereby modify its double layer in a
number of ways. These are the following: (1) the species can modify
the work function (electrochemical potential) of the C-layer which
consequently results in a modification of the double layer (2,3),
(2) the species can introduce dipoles at the conductor/semiconductor
interface thereby modifying the charge distribution of the double

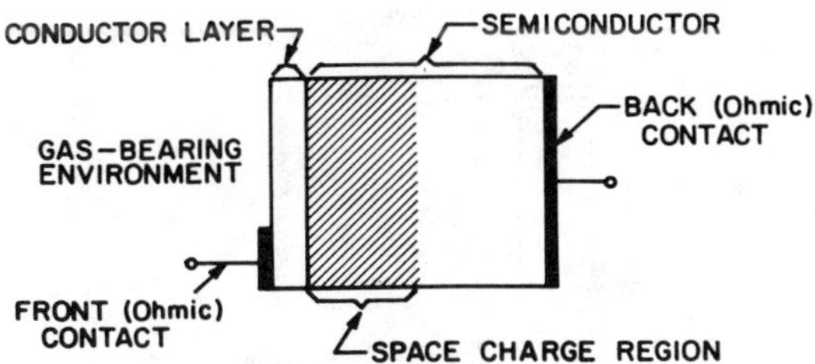

(A) C-S DIODE CONFIGURATION

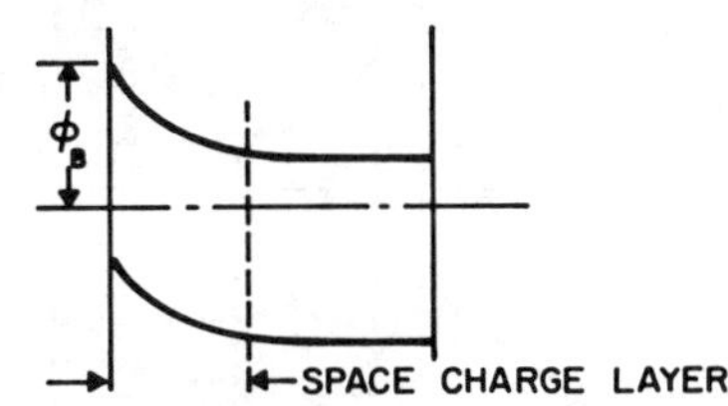

THERMODYNAMIC EQUILIBRIUM

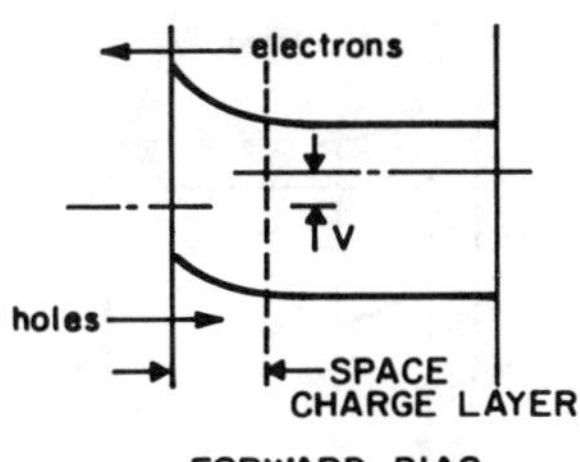

FORWARD BIAS

(B) C-S DIODE ENERGY BAND DIAGRAM

Figure 1.   Schematic of a C-S diode.   (A) physical configuration;
(B) energy band diagram in thermodynamic equilibrium (zero bias)
and under forward bias.

layer (4,5), (3) the species can significantly change the interface
state density in the semiconductor interfacial region (pinning or
unpinning the Fermi level) thereby changing the charge, and hence the
barrier, in the semiconductor (6), or (4) the species can diffuse
into the semiconductor where it may modify the doping density of the
semiconductor thereby changing the charge density in the barrier
region and, consequently, the barrier (7).
    In principle, if this fourth possible mechanism were to occur in
a particular sensing situation, it could occur because (1) the diffusing species acted as a donor or acceptor level in the semiconductor resulting in augmentation or compensation of the doping density
or (2) the diffusing species de-activated the semiconductor dopant
(7). We note that, in general, the species to be detected need not
be, by itself, the entity that causes the work function change in the
conductor, produces the interface state density change, introduces
the dipoles, or diffuses into the semiconductor. Rather, it may be

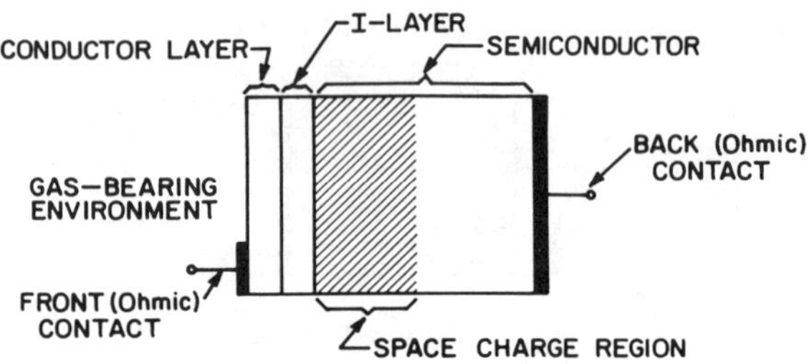

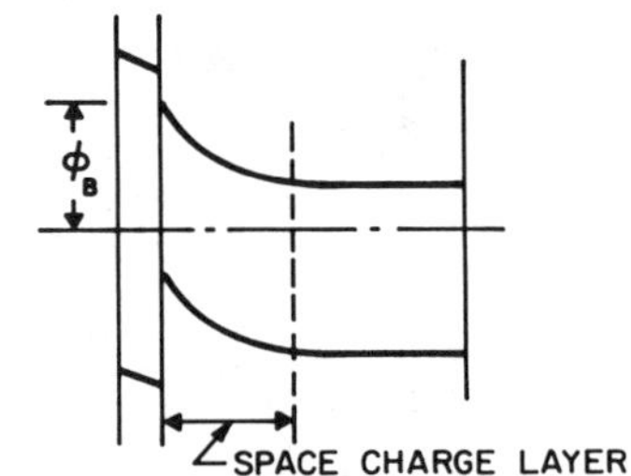

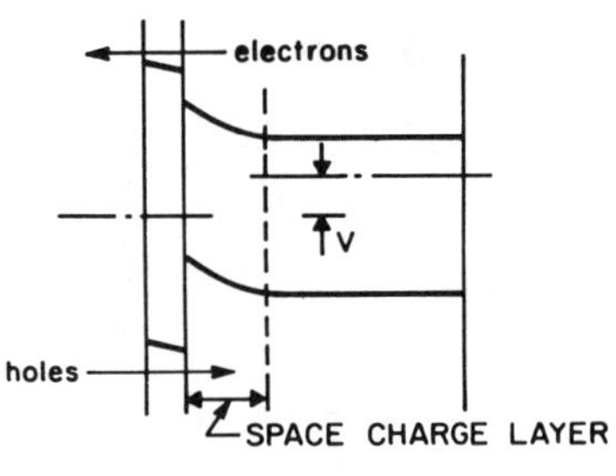

(B) C-I-S DIODE ENERGY BAND DIAGRAM

Figure 2.   Schematic of a C-I-S diode.   (A) physical configuration; (B) energy band diagram in thermodynamic equilibrium (zero bias) and under forward bias.

some reaction product generated, perhaps catalytically, at the conductor surface which interacts with the sensor structure.

The four mechanisms we have just enumerated give rise to detection due to modification of the double layer. Each of these four mechanisms leads to detection by changing the carrier transport across that barrier region due to a change in the barrier height ϕ_B. As we have noted, a direct change in the transport properties of the barrier region will also give detection. Hence, a fifth mechanism for detection has been proposed whereby a change in the interface state density directly changes transport (6). In this fifth mechanism it is assumed that carriers are by-passing over-the-barrier transport at the interface by flowing via interface state recombination (1,2). Thus, in this proposed mechanism changes in the interface state density due to the presence of some chemical species directly affects the efficacy of transport.

In all of these five potential detection mechanisms it is the semiconductor which is the key element of the C-S sensor device (as it is in the C-I-S diode and capacitor structures). The conductor layer is, however, also a very important part of the device. In addition to serving as the counter electrode, it must be capable of one or more of the following: (1) absorbing the chemical species itself or some product resulting from a reaction, perhaps catalytic, of that species at the conductor surface, (2) interacting with the absorbed species or its reaction product to give a conductor work function change, or (3) serving as an efficient bulk, or grain boundary, diffusion conduit allowing the chemical species or some reaction product of the chemical species to reach the semiconductor surface.

The C-I-S variant of the diode sensor is shown in Figure 2A. The essential features of this structure are the conductor/purposefully inserted interfacial layer/semiconductor configuration and the concomitant barrier (space charge) region in the semiconductor. As with the C-S diode, it is this extended barrier region in the semiconductor which controls current transport across the C-I-S diode as noted in Figure 2B. Since current flows in the structure, the I-layer must be capable of supporting transport of electrons or holes or both carriers. If the I-layer is an insulator or a cavity, then transport of carriers must be by tunneling. In that case, a readily established upper bound exists on the thickness of the I-layer; viz, thickness $\leq$ 50 Angstroms ($\underline{1},\underline{2}$).

The mechanisms of detection and the functions of the conductor layer and of the semiconductor are the same in a C-I-S diode sensor as they are in a C-S diode sensor. The only difference between these two structures is the presence of the purposefully inserted interfacial layer (I-layer) between the conductor and the semiconductor in the C-I-S devices. In general, this I-layer is employed in the C-I-S sensor configuration for one of two reasons: (1) either it is used to block chemical reactions between the conductor and the semiconductor or (2) it is used to augment or reduce the role of the interface in establishing the double layer or controlling transport.

If a conductor material undergoes a work function change when exposed to a certain chemical species, then clearly one has the foundations of a C-S diode sensor. However, this sensor cannot be made to function if the conductor chemically reacts with the semiconductor. This loss of sensitivity occurs because the new material resulting from the reaction, in general, will not have the same work function sensitivity to the chemical species as the conductor has. The C-I-S configuration solves this problem since a properly chosen I-layer, capable of supporting an electrical current, can be inserted between the conductor and semiconductor to prevent their reaction. The resultant C-I-S structure is able to respond to the effects of the gas species on the conductor.

We continue with this hypothetical case of a conductor, whose work function changes on exposure to a certain chemical species, to discuss the second reason we listed for employing I-layers. In this example, we assume that there is no chemical reaction between the conductor and the semiconductor. Further, we assume here that this chemical species has only the effect of modifying the conductor work-function. For this demonstration of the second use of I-layers, it is assumed that there is a very high density of interface states at the conductor/semiconductor junction in the C-S configuration. Such a high density of states would pin the barrier height (the quantity

ϕ_B seen in Figure 1B) causing the barrier (space charge) layer in the semiconductor to be unresponsive to work function changes in the metal. Put succinctly, in this situation of barrier height (Fermi level) pinning, there is such a large amount of charge possible in the interface states that the semiconductor barrier region becomes electrostatically shielded from any work function changes in the conductor layer. In this case, a properly chosen interfacial layer, grown or deposited on the semiconductor surface, can remove the large density of interface states thereby unpinning the barrier (Fermi level). With the barrier unpinned, the semiconductor space charge layer in the resulting C-I-S structure can now respond to changes in the double layer.

This discussion of the second reason for using I-layers in diode sensor configurations has focused so far on employing the I-layer to reduce the role of interface states in establishing the double layer. As noted in our listing of the second reason for using I-layers, such layers may also be designed to augment the role of the interface in establishing the double layer or controlling transport. Hence, one can envisage an entirely different approach to the use of interface layers employing them to develop interface states on the introduction of some chemical species. The charge residing in these states created by the presence of some chemical species could for example, then cause the charge in the semiconductor to change. This would produce a barrier modification in the semiconductor and chemical detection due to the shift in current passed for a given voltage. Using this approach, the barrier would ultimately be pinned by the interface states resulting from the presence of the detected chemical species. This would lead to an eventual saturation of the response of the C-I-S diode.

In a variation of this use of the I-layer we note that it may also be designed to augment the role of the interface by being tailored to support dipoles. That is, in such a situation detection would be based on diffusion to the I-layer of the detected chemical species, or some product resulting from the detected chemical species, and the setting-up of dipoles in the I-layer by these species. The presence of these dipoles in the I-layer would produce a modification of the device double layer. In sensors employing this approach, the interface state modification approach, or the semiconductor effective doping modification approach for the detection mechanism, one can easily envisage structures where the C-layer need not cover the entire sensor surface.

We note that, in the case of both C-S and C-I-S diode sensors, the current density J passed at a bias voltage V by these devices is given by ($\underline{1},\underline{2}$)

$$J = J_0 \left(e^{V/nkT} - 1\right) \tag{1}$$

where kT is Boltzmann's constant times the absolute temperature and n and J_0 are diode parameters. The quantity n is termed the diode ideality factor. In general, $n \geq 1$ and it may vary with bias and temperature depending on the details of electron and hole transport across the C-S or C-I-S junction ($\underline{1},\underline{2}$). The quantity J_0 is termed the diode saturation current. The value of J_0 also depends on the transport mechanism dominating across the C-S or C-I-S junction. If thermionic emission of majority carriers is dominating, then J_0 is modelled by ($\underline{1},\underline{2}$)

$$J_o = A^* T^2 e^{-\phi_B/kT} \tag{2}$$

where A^* is the so-called effective Richardson constant and ϕ_B is the
semiconductor barrier height seen in Figures 1B and 2B. Obviously,
changes in the double layer of a diode can exponentially affect the
current through ϕ_B, if thermionic emission is the dominant transport
mechanism.

<u>Capacitor Sensors</u>. The C-I-S capacitor-type sensor has, in general,
the configuration shown in Figure 3. Figure 3A is the physical
structure and Figure 3B is its energy band diagram showing the barrier region in the semiconductor. Since this structure is designed
to store charge in the space charge region of the semiconductor, the
I-layer must block electron and hole transport and, hence, must be
an insulator or cavity. From our previous discussion it follows
that this layer must be greater than at least 50 Angstroms in thickness in the capacitor structure.

The ability of a C-I-S structure to store charge is measured by
its capacitance C_T (per area) which is given by (<u>8</u>)

$$\frac{1}{C_T} = \frac{1}{C_I} + \frac{1}{C_S} \tag{3}$$

where C_I is the geometrical capacitance per area of the insulator
layer and C_S is the capacitance per area of the semiconductor space
charge region. These two capacitances are expressed as (<u>8</u>)

$$C_I = \varepsilon_I/t \tag{4}$$

and

$$C_S = \varepsilon_S/W \tag{5}$$

where ε_I and t are the I-layer permittivity and thickness, respectively, and ε_S and W are the corresponding quantities for the semiconductor space charge layer. It is the semiconductor surface potential ψ_S (related to the barrier height ϕ_B as seen in Figure 3B) and
space charge layer thickness W which will change in response to modifications of the device double layer caused by the presence of some
chemical species. Consequently, it follows that it is C_S which we
wish to monitor in C-I-S capacitor structures. Since W in Equation
5, for typical semiconductor doping levels, is of the order of
microns and since it is C_T and not C_S that is directly measured, it
follows from Equation 3 that the I-layer thickness cannot exceed
several thousand Angstroms in thickness in these devices. Here we
assume the dielectric constant of SiO_2 for the I-layer; this thickness must be appropriately adjusted for other I-layers. Hence, the
thickness of the I-layer is constrained to 50 Å < t $\lesssim$ several thousand Angstroms in C-I-S capacitor sensors. As a practical matter the
lower bound on the I-layer thickness is usually hundreds of Angstroms
to overcome inhomogeneities.

A chemical species in the environment of a C-I-S capacitor sensor structure can interact with the device and thereby modify its
double layer in the same four ways it can interact with the double
layer of a C-S or C-I-S diode sensor. These four mechanisms need to

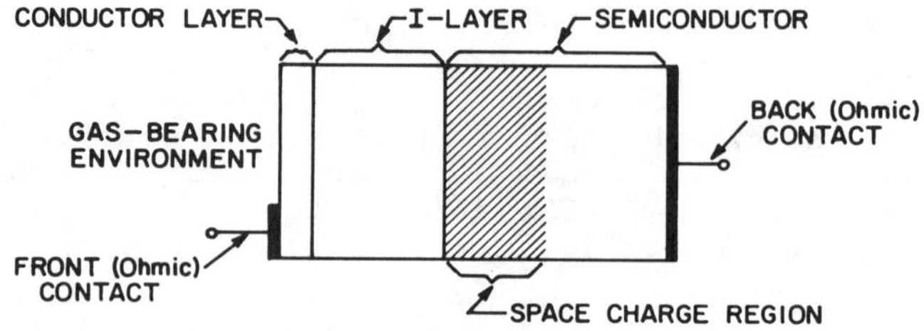

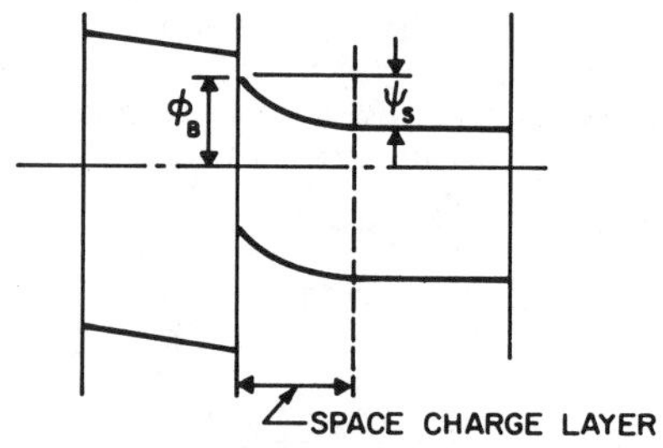

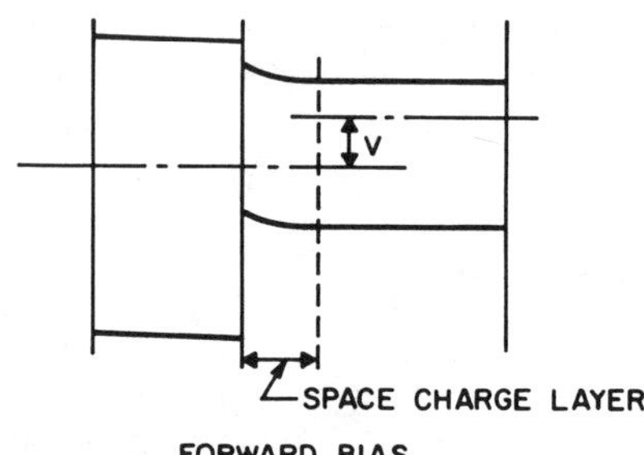

Figure 3. Schematic of a C-I-S capacitor. (A) physical configuration; (B) energy band diagram in thermodynamic equilibrium (zero bias) and under forward bias.

be generalized somewhat due to the presence of the thicker I-layer of the capacitor structure; hence, we re-list them in this more generalized form. The four mechanisms are the following: (1) the species can modify the work function of the C-layer, (2) the species can introduce dipoles at the conductor/I-layer interface, in the I-layer, or at the I-layer/semiconductor interface, (3) the species can change localized state densities in the I-layer or at the I-layer/semiconductor interface, or (4) the species can diffuse into the semiconductor where it may modify the doping density of the semiconductor. In addition, there is a fifth mechanism that can give rise to detection in C-I-S capacitor structures: the detected species can cause charged impurity redistribution in the I-layer thereby modifying the double layer and giving detection. As may be deduced from the above

listing of the sensing mechanisms, all the layers of a C-I-S capacitor have the same roles in sensing as they do in diode structures.

Performance of Diode-type Structures

General Behavior. The use of C-S and C-I-S diode structures to sense
hydrogen or hydrogen-bearing gases such as NH_3 and SiH_4 (2-7,9,10)
has been the most prevalent application of these devices to-date.
Hence this section will focus on the performance of these devices as
hydrogen detectors. The development of C-S and C-I-S devices for
hydrogen detection has occurred because of the ability of Pd, Pt, Ir,
Ni, and other transition metals and their alloys to perform, for
hydrogen-bearing gases, one or more of the three functions (viz,
absorbing the species (or its products) to be detected, undergoing a
work function change due to the presence of the species (or its products), or serving as a diffusion conduit) generally required of the
conductor layer in the sensor structure.

Figure 4 shows the response of two different C-S devices to mixtures of hydrogen in air and N_2. For both of these sensors the C-
layer is palladium. In the device of Figure 4a the semiconductor is
single crystal TiO_2 (5); in the device of Figure 4b the semiconductor
is single crystal CdS (3). The C-S structure can be employed for
these two semiconductors since the Fermi level is not pinned at their
surfaces (1). Because these devices are diodes, the detection of
hydrogen is shown in Figure 4 by giving the device current-voltage
(I-V) data in air and in the hydrogen bearing ambients. If the
devices are operated at constant voltage V, then the sensitivity is
given by $\Delta I/I|_V$. If the devices are operated at constant current I,
then the sensitivity is given by $\Delta V/V|_I$. Figure 5 gives I-V data for
three C-I-S sensor structures. Figure 5a shows the response in air
and in parts per million (ppm) mixtures of H_2 in air of a Pd/SiO_x
(20-40 Å)/(n)Si/(p)Si switching diode (11,12), Figure 5b shows the
response in air and in hydrogen mixtures of a Pd/SiO_x (40-50 Å)/(n)Si
diode, and Figure 5c shows the response in 100% O_2 and in 2% H_2 in N_2
of a Pd/($\sim$1100 Å) TiO_x/(n)Si diode (13). In all three of these C-I-S
devices the C-layer is palladium whereas the semiconductor is single
crystal Si. The C-I-S configuration must be employed when using Pd
with Si since palladium silicide is formed if the metal is in direct
contact with the Si and the resulting junction is not sensitive to
hydrogen (2). An I-layer is also generally needed with Si due to the
tendency of the Fermi level to be pinned at the surface of this
semiconductor (1).

The C-I-S structures of Figures 5b and 5c are simple diodes; one
employs SiO_x for the I-layer and the other uses TiO_x. The evolution
of their I-V characteristics on exposure to hydrogen is similar to
that seen for the simple C-S diodes. The structure of Figure 5a is
more involved since it consists of a C-I-S junction in series with a
n/p junction. Consequently, the evolution of its I-V characteristic
on exposure to hydrogen is different than that of the single diodes.
When the presence of hydrogen modifies the transport of current
across the C-I-S junction, the device I-V characteristic switches as
seen in Figure 5a. This switching is explained in terms of a
thyristor-like model (11). As may be noted in Figure 5a, the structure switches to essentially a short-circuited configuration on exposure to hydrogen with the switching threshold voltage V_{TH} being a
function of hydrogen concentration. The on and off nature of the I-V
characteristic of this C-I-S sensor structure allows it to also serve
as an actuator (11).

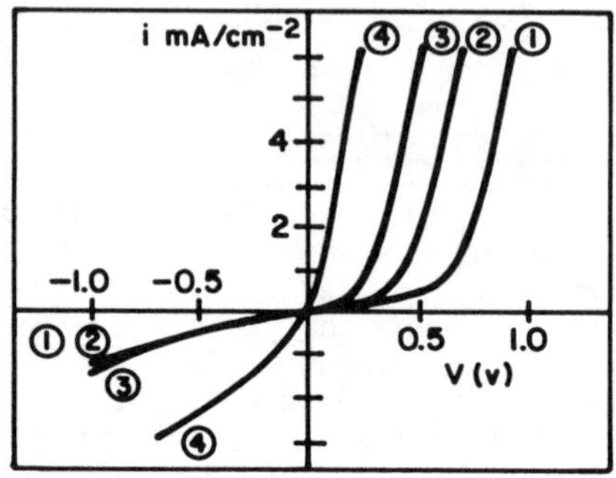

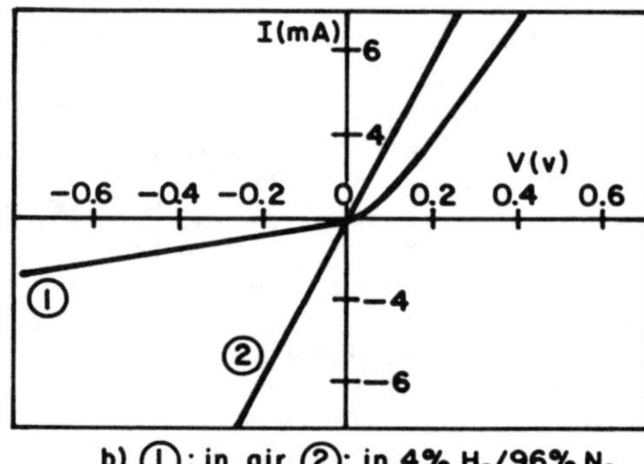

Figure 4. I-V characteristics at RT for a) a Pd/TiO$_2$ (C-S) diode
in air and in various mixtures of H$_2$ and air (after Reference 5,
with permission; and b) a Pd/CdS (C-S) diode in air and in 4% H$_2$/
96%N$_2$. (Reproduced with permission from Ref. 3. Copyright 1976
J. Appl. Phys.)

Response and Recovery Times. Typical response-time and recovery-time
behavior for Pd/SiO$_x$/(n)Si C-I-S hydrogen sensors is given in Figures
6, 7, 8, and 9. The data are for several temperatures and hydrogen
mixtures as indicated in the figures. On the basis of these typical
data, a number of observations can be made. For example, Figure 6
shows that high concentrations of oxygen give a fast response time.
However, we note that for SiO$_x$ I-layer structures high concentrations
of oxygen decrease the sensitivity to hydrogen. This latter point of
the influence of oxygen concentration in the ambient and of I-layers
will be discussed further in the section on interferences. Water
vapor is seen from Figure 7 and from Table I to also speed-up the response and recovery of diode sensors to hydrogen; however, it too
lowers the sensitivity. Figure 8 shows that the presence of CO
slows down device response for SiO$_x$ based structures but CO does not
significantly suppress sensitivity.
     Figure 9 shows the typical result that higher temperatures gener-
ally give faster C-S and C-I-S diode sensor response. However, as

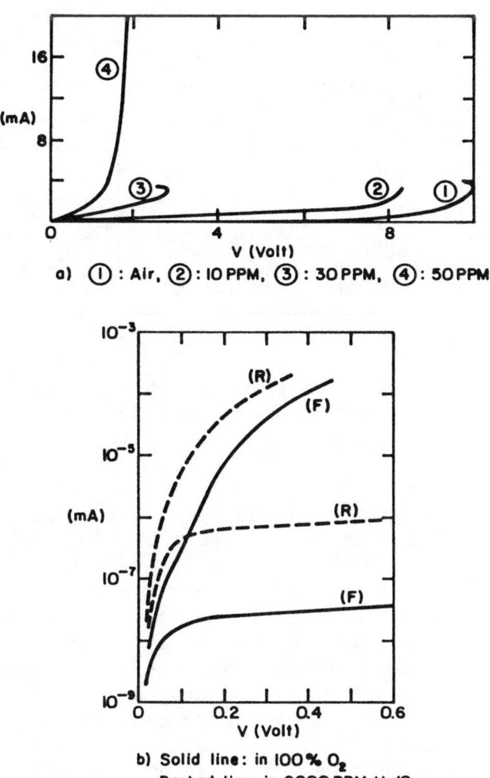

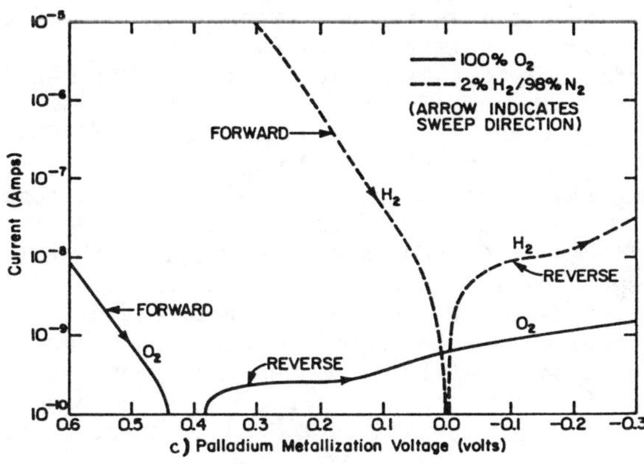

Figure 5.   I-V characteristics for a) a Pd/SiO$_x$ (20-40 Å)/n-Si/ p$^+$-Si (C-I-N-p$^+$) switching diode in air and in various mixtures of H$_2$ and air at 100°C (after Ref. 11, © 1985 IEEE); b) a Pd/ SiO$_x$ (∿45 Å/n-s (C-I-S) diode in 100% O$_2$ and in 2000 PPM H$_2$ in O$_2$ at RT; and c) a Pd/TiO$_x$ (∿1100 Å)/n-Si (C-I-S) diode in 100% O$_2$ and in 2% H$_2$/n N$_2$ at RT.

Table I. Effect of Water Vapor on the Device Hydrogen Response
Time (τ_r) and Recovery Time (τ_d) at RT

Device Structure	H_2 Partial Pressure	Ambient	τ_r (min.)	τ_d (min.)
Pd/SiO$_x$/Si	6000 ppm	Dry O$_2$	9.3	25
Diode	H$_2$ in	Wet O$_2$	3.3	13
Pd/SiO$_2$/Si	6000 ppm	Dry Air	9.3	12
Capacitor	H$_2$ in	Wet Air	4.2	5 $\sim$ 6

will be further explored under temperature effects, increased operating temperature also improves the lower detection limit, reduces
the upper detection limit, and generally gives a reduced sensitivity
at saturation.

Interferences. Solid state C-S and C-I-S gas sensing diodes, as well
as C-I-S capacitor-based structures, can be used in conjunction with
selectively permeable membranes to isolate the devices from interferences from other gases (9). Without such membranes, interferences
can be a problem. Interferences to detection of hydrogen, for example, can come from gases such as O$_2$ and H$_2$O whereas some gases such
as CO have no significant effect. As noted, high concentrations of
oxygen decrease the sensitivity to hydrogen of Pd/SiO$_x$/Si device configurations apparently due to the scavenging of hydrogen by the oxygen. On the other hand, Pd/TiO$_x$/Si device configurations have increased sensitivity to hydrogen in oxygen rich ambients due to the
interaction of the O$_2$ with the TiO$_x$ layer (13). For these TiO$_x$ I-
layer sensors, lower concentrations of oxygen lead to a lower detection, lower-bound and also to a lower detection, upper-bound. The
effects of oxygen on the hydrogen detection ability of these two
C-I-S diode sensors is summarized in Table II.
     The interference to the hydrogen detection of C-I-S structures
caused by varying amounts of water vapor is also summarized in Table
II. As seen in that table, high concentrations of H$_2$O vapor lower
the sensitivity of Pd/SiO$_x$/Si diodes whereas water vapor, in general,
lowers the sensitivity of Pd/TiO$_x$/Si diodes at room temperature.

Temperature Effects. The general observation is that temperature
enhances the speed of response of C-S and C-I-S diode structures to
hydrogen as noted earlier. It also generally leads to a lower detection limit but also to a lower upper detection bound due to saturation of device response at lower ambient hydrogen concentrations.
The maximum sensitivity to hydrogen decreases for diode structures
at elevated operating temperatures. Many of these points are seen in
Table III.
     The effect of elevated operating temperatures on the interfer-
ence to hydrogen detection arising from O$_2$ for Pd/SiO$_x$/Si diodes is
also seen in Table III as a function of H$_2$ ppm levels. As may be
noted from this table, the reduced sensitivity to hydrogen, especially at low concentrations, caused by the interference from oxygen
is not significant at elevated device operating temperatures; i.e.,
temperature effects dominate.

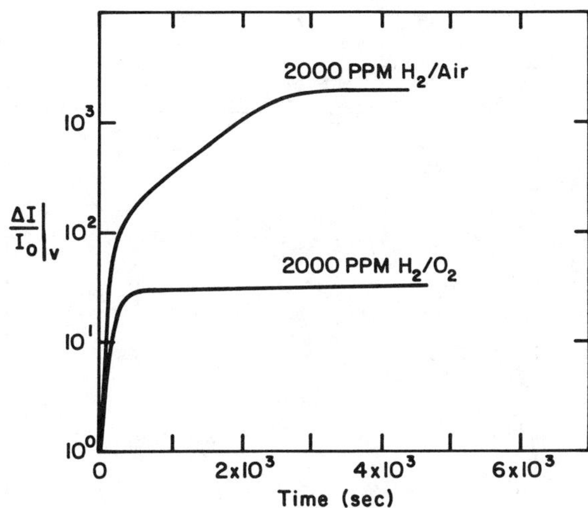

Figure 6.   Device kinetic responses to 2000 PPM H$_2$ in air ambient and in O$_2$ ambient at RT.   Device used here is a Pd/SiO$_x$(∿45 Å)/ n-Si (C-I-S) diode and the initial ambients are air and O$_2$, respectively.

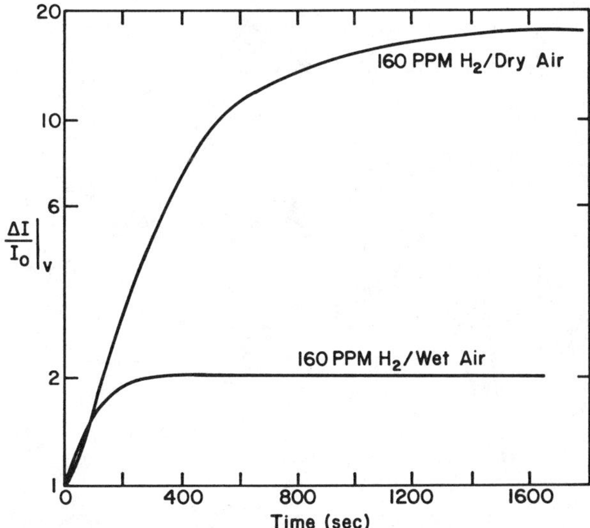

Figure 7.   Device kinetic response to 160 PPM H$_2$ in wet air (saturated with water vapor) ambient and in dry air ambient at RT. Device used here is a Pd/SiO$_x$ (∿45 Å)/n-Si diode and wet air and dry air are initial ambients, respectively.

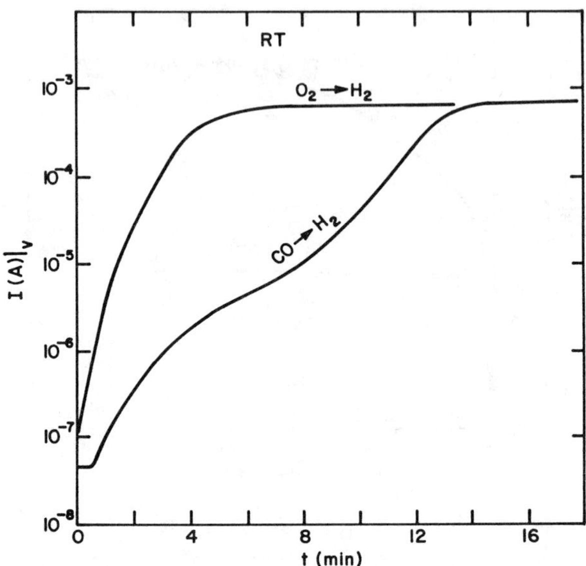

Figure 8. Device kinetic responses to 2% H_2/96%N_2 at RT. Device used here is a Pd/SiO$_x$($\sim$45 Å)/n-Si diode and initial ambients are 100% O_2 and 4% CO/96% N_2, respectively.

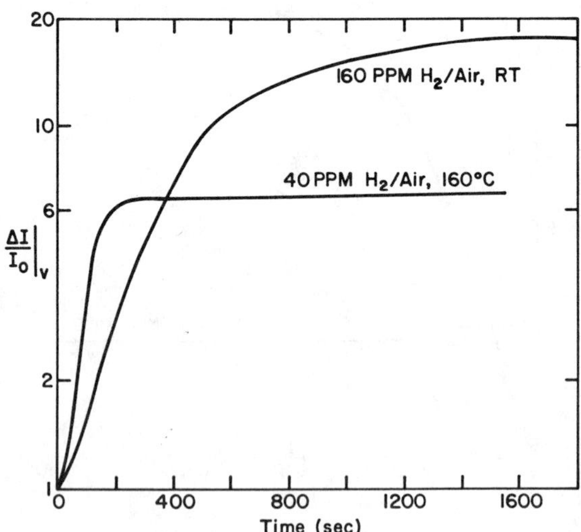

Figure 9. Device kinetic response to 160 PPM H_2 in air at RT and 40 PPM H_2 in air at 160°C. Pd/SiO$_x$ ($\sim$45 Å)/n-Si diode is used and initial ambients are air.

Table II. Interference to the Device Hydrogen Sensitivity at RT Due to Water Vapor (H_2O) and O_2

Interference / H_2 in	Structure	TiO_x I-layer C-I-S Diode	SiO_x I-layer C-I-S Diode	SiO_2 I-layer C-I-S Capacitor
Effect of High Conc. of O_2	Ambient	100%, Dry Air	100% O_2	100% O_2
	H_2 Sensitivity	$\frac{\Delta I}{I_o}\big\vert 4\ ppm \simeq 50$ $\frac{\Delta I}{I_o}\big\vert 4000\ ppm \simeq 1 \times 10^5$	$\frac{\Delta I}{I_o}\big\vert 600\ ppm \simeq 1$ $\frac{\Delta I}{I_o}\big\vert 2000\ ppm \simeq 27$	$\Delta V_{FB}\big\vert 40\ ppm \simeq -0.36\ V$ $\Delta V_{FB}\big\vert 2000\ ppm \simeq -0.77\ V$
Effect of Low Conc. of O_2	Ambient	Ar, N_2	Dry Air	Ar, N_2
	H_2 Sensitivity	$\frac{\Delta I}{I_o}\big\vert 40\ ppm \simeq 6.7 \times 10^3$ $\frac{\Delta I}{I_o}\big\vert 4000\ ppm \simeq 1.8 \times 10^5$	$\frac{\Delta I}{I_o}\big\vert 100\ ppm \simeq 18$ $\frac{\Delta I}{I_o}\big\vert 2000\ ppm \simeq 1.9 \times 10^3$	$\Delta V_{FB}\big\vert 40\ ppm \simeq -0.73\ V$ $\Delta V_{FB}\big\vert 2000\ ppm \simeq -0.86\ V$
Effect of High Conc. of H_2O	Ambient	Wet Air (saturated with water vapor $\sim$ 2.4%)		
	H_2 Sensitivity	$\frac{\Delta I}{I_o}\big\vert 40\ ppm \simeq 1.1 \times 10^2$ $\frac{\Delta I}{I_o}\big\vert 4000\ ppm \simeq 2.2 \times 10^3$	$\frac{\Delta I}{I_o}\big\vert 100\ ppm \simeq 1$ $\frac{\Delta I}{I_o}\big\vert 2000\ ppm \simeq 4$	$\Delta V_{FB}\big\vert 600\ ppm \simeq -0.24\ V$ $\Delta V_{FB}\big\vert 1200\ ppm \simeq -0.33\ V$
Effect of Low Conc. of H_2O	Ambient	Wet Air (but not saturated with water vapor, $\lesssim$ 1%)		
	H_2 Sensitivity	$\frac{\Delta I}{I_o}\big\vert 40\ ppm \simeq 1.1 \times 10^2$ $\frac{\Delta I}{I_o}\big\vert 4000\ ppm \simeq 2.2 \times 10^3$	about the same as in dry air	about the same as in 100% O_2

Table III.   Performance of the Pd/SiO$_x$/Si C-I-S Diode H$_2$ Sensor in Air and O$_2$ at RT and 160°C

Temp. / Ambient	Room Temperature		160°C	
	H$_2$ Concentrations	$\Delta I/I_o\|V$	H$_2$ Concentrations	$\Delta I/I_o\|V$
Dry Air	100 ppm	1.8×10^1	40 ppm	6.1×10^0
	600 ppm	4.9×10^1		
	1130 ppm	1.5×10^3	80 ppm	1.5×10^1
	1820 ppm	1.9×10^3		
	3300 ppm	2.5×10^3	160 ppm	3.2×10^1
	4600 ppm	2.7×10^3		
	6670 ppm	3.5×10^3		
Dry 100% O$_2$	1820 ppm	2.7×10^1	40 ppm	9.0×10^0
	3330 ppm	4.9×10^1	80 ppm	1.5×10^1
	4620 ppm	1.0×10^3	160 ppm	1.7×10^1
	6670 ppm	2.6×10^3		

The influence of elevated operating temperatures on the interference to hydrogen detection arising from water vapor is generally advantageous. That is, water vapor does not have a detrimental effect on SiO$_x$-based C-I-S devices for T $\geq$ 100°C. This is apparently due to water   leaving the sensor surface for T $\geq$ 100°C. We also note that, as at room temperature, CO does not interfere with hydrogen detection in diode sensors at elevated temperatures (14).

Performance of Capacitor-type Structures

General Behavior. For the same reasons discussed in the diode performance section, most of the work on capacitor-type C-I-S sensors has focused on structures for detecting hydrogen or hydrogen-bearing gases. Consequently, this section will mainly examine the performance of C-I-S hydrogen sensors; however, detection of CO will be discussed.

Figure 10 shows the capacitance-voltage (C-V) characteristic of a Pd/1000 Å SiO$_2$/(n) Si capacitor sensor in 100% O$_2$ and in 600 ppm H$_2$ in O$_2$ at room temperature. As seen, the presence of hydrogen produces a shift in the charge storage capabilities of the device at a given voltage. From the C-V characteristic presented in Figure 10, it is apparent that sensitivity can be defined in terms of the change in capacitance ΔC at a fixed voltage V; i.e., sensitivity can be given by $\Delta C/C\|_V$. It is also apparent that it could be defined as $\Delta V/V\|_C$. Often, however, sensitivity for C-I-S capacitors is simply given as ΔV_{FB} where V_{FB} is the flat band potential. That is, V_{FB} is the particular bias needed to force the surface potential ψ_S in Figure 3b to zero (8). This flat band bias shifts ΔV_{FB} on exposure to hydrogen. We define ΔV_{FB} to be negative, if the C-V characteristic shifts in the sense seen in Figure 10.

In addition to the expected sensitivity to hydrogen, Table IV shows that the Pd/SiO$_2$/Si capacitor, unlike the corresponding diode structure, is also sensitive to CO in oxygen or air. However, as

seen in the table, an elevated operating temperature is needed. This
unexpected sensitivity to a non-hydrogen-bearing gas is believed to
be actually due to the removal of O_2 from the structure by the CO at
elevated temperatures (15).

Table IV. CO in O_2 Response of Pd (600 Å)-C-I-S Device at 160°C

CO conc. in O_2	Sensitivity $-\Delta V_{FB}$ (mV)	τ_r (min.)	τ_d (min.)
1%	∿5	∿6	--
1.4%	26	20	13
2.3%	85	31	15
3.0%	130	19	2.3
4.5%	173	17	3.0
10%	195	17	3.0

<u>Response and Recovery Times</u>. Figure 11A gives ΔV_{FB} for a Pd/SiO$_2$/
(n) Si capacitor device as a function of exposure time to 600 ppm H_2
in 100% O_2 at room temperature. Table I summarizes the effects of
water vapor on Pd/SiO$_2$/(n)Si capacitor response and recovery times.
As may be noted, water vapor speeds up device response to hydrogen.
Figure 11b gives V_{TH} (where V_{TH} is the turn on voltage for transport
down the channel created by the stored charge) versus time for a
MOSFET based on the Pd/SiO$_2$/Si structure. These data are for an exposure to 180 ppm H_2 in air at 150°C. In this case, both response and
recovery behavior are shown. Figure 12 clearly shows that Pd/SiO$_2$/
Si capacitor sensor response is enhanced by raising the operating
temperature. Here data are shown for room temperature and 160°C.
    The effect of several ambients on the speed of hydrogen detec-
tion is seen for room temperature in Figure 13. The hydrogen level
is 600 ppm and the ambients are dry air, dry 100% O_2 and dry 100% Ar.
As may be noted, device response to hydrogen is the fastest in the O_2
ambient and the longest in the Ar ambient. Figure 14 shows the
effect of CO in device response to hydrogen. As may be seen, CO
slows down the room temperature response.
    Figure 15 presents response and sensitivity data for two device
structures which differ from the Pd/SiO$_2$/Si configuration. In Figure
15b ΔV_{FB} is plotted versus time for an ultra thin-Pt/Pd/SiO$_2$/(p)Si
capacitor for ppm levels of NH_3 in air at T = 150°C. The response
and recovery times for NH_3 detection for this device are seen to be
rapid even though detection probably requires generation of hydrogen.
In Figure 15a ΔV_{FB} is plotted versus time for Pd/I-layer/Si capacitors for devices with various I-layer compositions. As may be seen
from these data, the response time is longest for the SiO$_2$ I-layer
and shortest for the Si$_3$N$_4$ I-layer (16). The origin of the longer
response time (and the larger sensitivity) for hydrogen seen for the
I-layers other than Si$_3$N$_4$ is attributed to a phenomenon termed hydrogen induced drift (HID) (16). Hydrogen induced drift is probably due
to mechanisms (2), (3), or (5), or some combination, of our sensing
mechanisms list for C-I-S capacitors in the Structures and Principles
of Operation Section.
<u>Interferences</u>. Figures 13, 16, 17, and 18 together with Table II
provide information on O_2, CO, or H_2O interferences to hydrogen
detection observed when using Pd/SiO$_2$/Si capacitor devices. As may

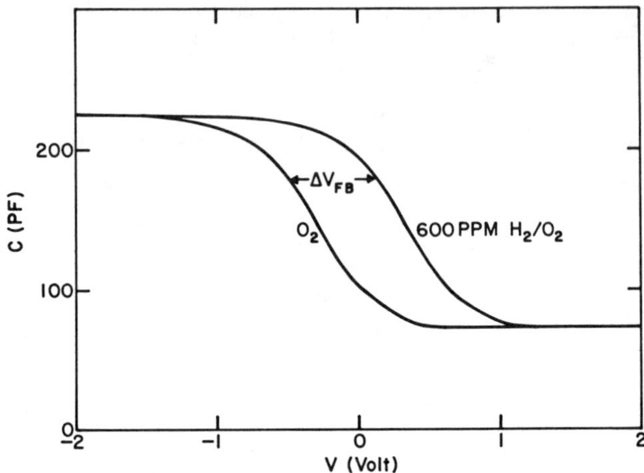

Figure 10.   C-V characteristics of a Pd/SiO_2 ($\sim$1000 Å)/n-Si
(C-I-S) capacitor in 100% O_2 and in 600 PPM H_2 in O_2 ambients at
RT.

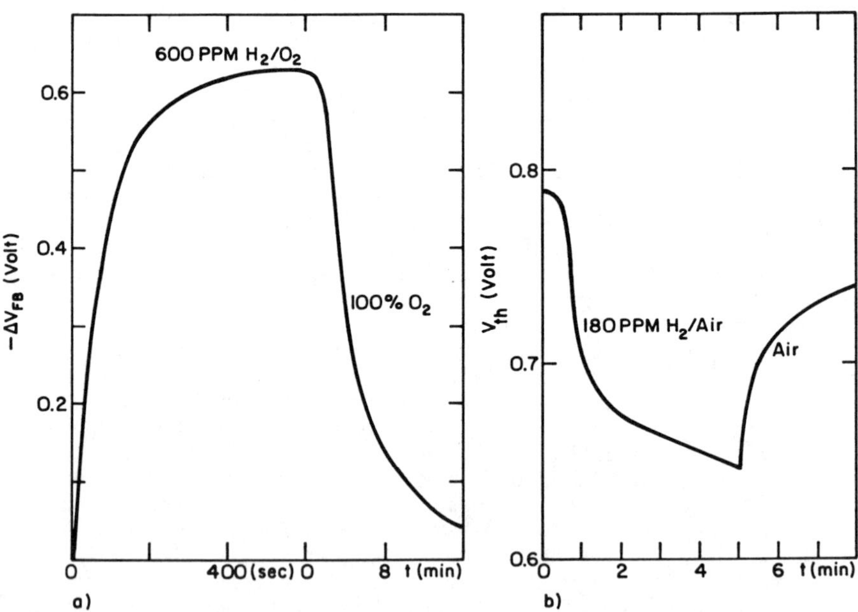

Figure 11.   Device kinetic response to a) 600 PPM H_2 in O_2 at RT.
Pd/SiO_2 ($\sim$1000 Å)/n-Si capacitor is used here and initial ambient
is 100% O_2; and b) 180 PPM H_2 in air at 150°C.   Device here is a
Pd-gate MOSFET and initial ambient is air (after Ref. 4, with
permission).   The recoveries in 100% O_2 and air ambients are
also shown.

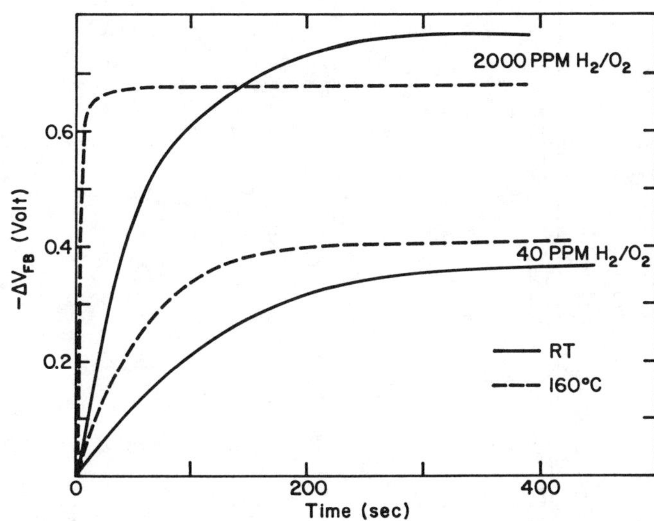

Figure 12. Device kinetic response to 40 PPM H_2 in O_2 and 2000 PPM H_2 in O_2 at RT and at 160°C. Pd/SiO$_2$ ($\sim$1000 Å)/n-Si capacitor is used here and 100% O_2 is initial ambients.

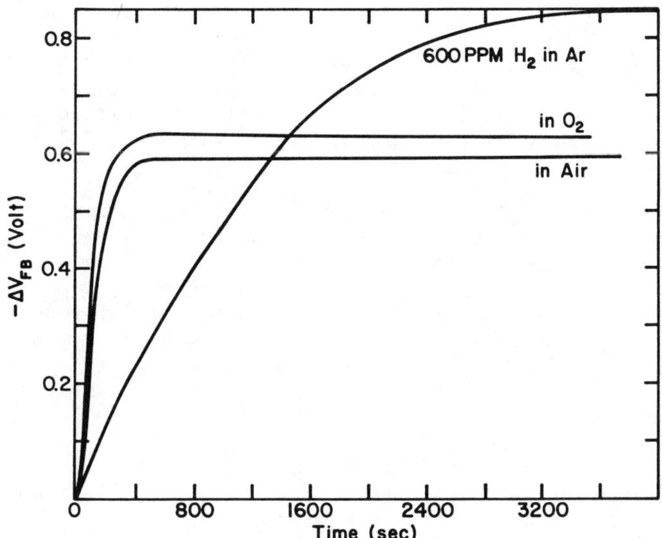

Figure 13. Pd/SiO$_2$ (1000 Å)/n-Si capacitor kinetic response to 600 PPM H_2 in Ar, O_2, and air ambients at RT. Initial ambients are corresponding 100% Ar, 100% O_2, and air ambients.

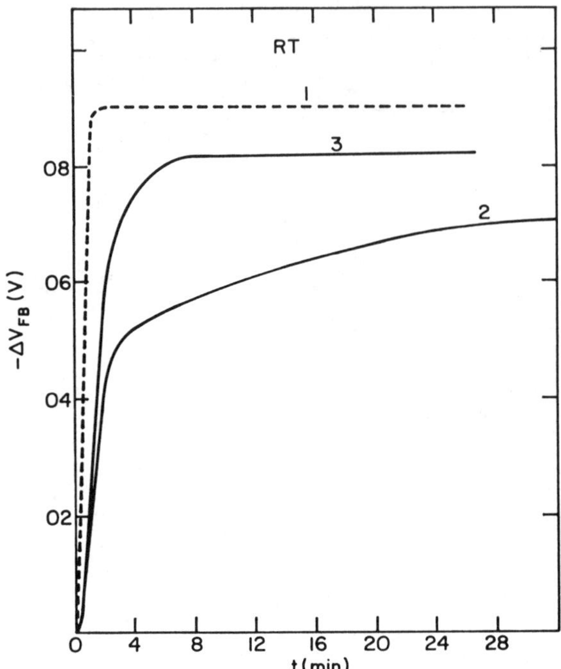

Figure 14. Effect of CO on the hydrogen sensitivity and response
time of a Pd/SiO$_2$ ($\sim$1000 Å)/n-Si capacitor. Device kinetic response to 2% H$_2$/96% N$_2$ at RT. Curve 1, the device is not exposed
to 4% CO/96% N$_2$ and initial ambient is 100% O$_2$; curve 2, the
device is exposed to 4% CO/96% N$_2$ prior to the initial ambient
100% O$_2$; and curve 3) the initial ambient is 4% CO/96% N$_2$.

be seen from Figure 13, the presence of high concentrations oxygen
relative to that found in air tends to increase somewhat the capacitor sensitivity to hydrogen at room temperature. This same trend is
also found at 160°C. However, a total absence of O$_2$ in the ambient
(i.e., using Ar as the ambient) is seen in the Ar environment data of
Figure 13 to give a much larger sensitivity to hydrogen, but this
large sensitivity to H$_2$ injected into an Ar ambient is found to have
a weak H$_2$ ppm level dependence. These results are also seen in the
data of Table II.
     The data of Figure 13 and Table II suggest that oxygen is invol-
ved in several interactions with the C-I-S capacitor device structure. The comparison between an O$_2$ ambient and an O$_2$-free (i.e., Ar)
ambient shows that O$_2$ most certainly competes for hydrogen reducing
the supply into the C-I-S device. However, the increase of the sensitivity for O$_2$ concentrations exceeding that found in air indicates
that O$_2$ may also sensitize the Pd, perhaps by removing other adsorbates on the Pd surface or in Pd grain boundaries (15).
     The effect of the presence of water vapor on hydrogen detection
at room temperature for C-I-S capacitor sensors is seen in Figure 16.
The data show that water vapor tends to reduce the sensitivity to

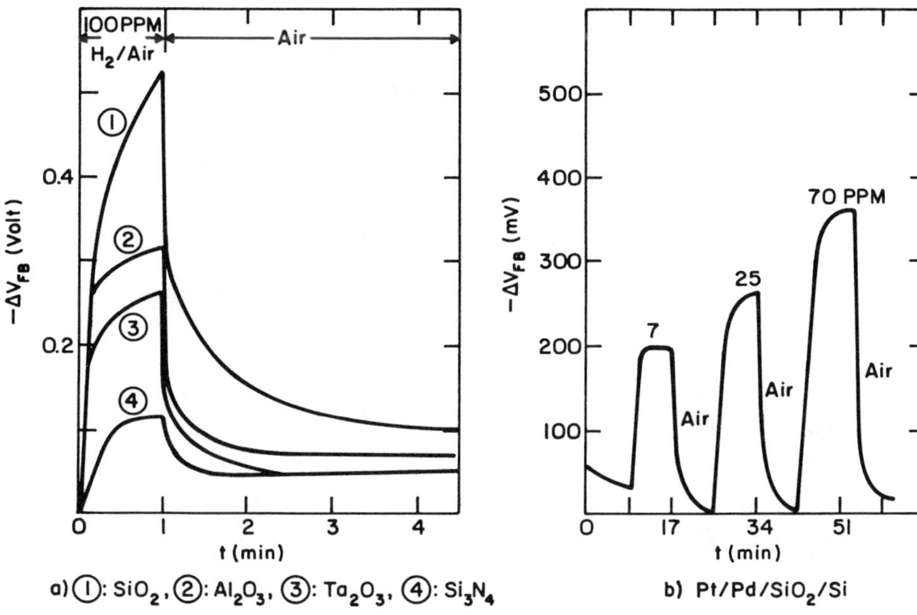

Figure 15.   Effect of different I-layer and C-layer on the device sensitivity and response time to a) 1000 PPM H_2 in air at 150°C. The device is a Pd/Top I-layer (SiO_2, Al_2O_3, Ta_2O_3, Si_3N_4)/ SiO_2/p-Si capacitor and the initial ambient is air (after Ref. 16, ©1984 IEEE) and b) 7 PPM, 25 PPM, and 70 PPM NH_3 in air at 150°C, device used here is an ultra thin Pt-layer/Pd/SiO₂/p-Si capacitor (after Ref. 9, with permission).

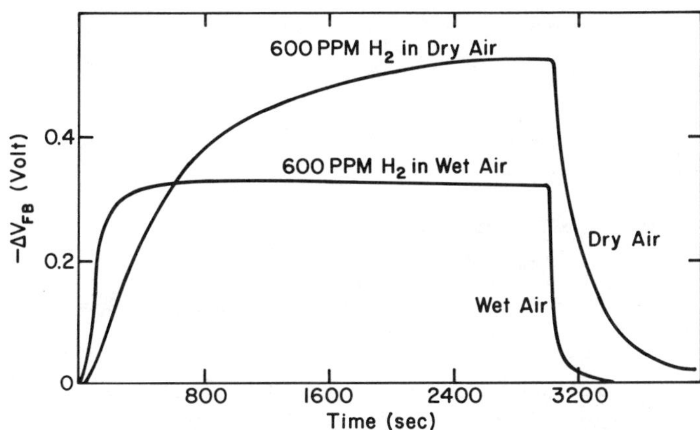

Figure 16.   Effect of water vapor on the device hydrogen response in air and recovery in air at RT.   Pd/SiO₂ (∼1000 Å)/n-Si capacitor is used.   Dry air and wet air (saturated with water vapor are initial ambients, respectively.

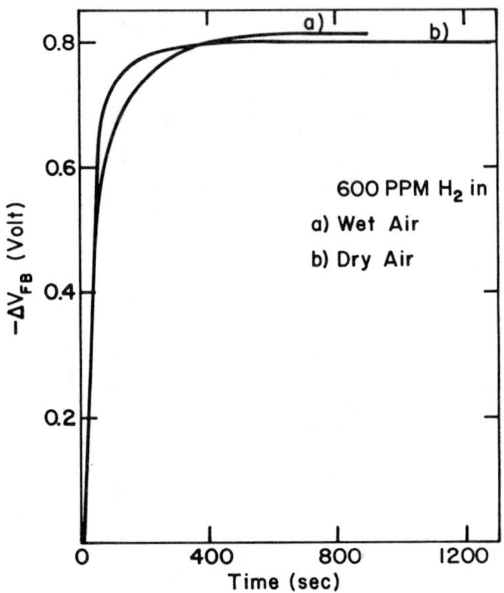

Figure 17.   Effect of water vapor on the device hydrogen response
in air at 160°C. Pd/SiO$_2$ ($\sim$1000 Å)/n-Si capacitor is used. Dry
air and wet air (saturated with water vapor) are initial ambients,
respectively.

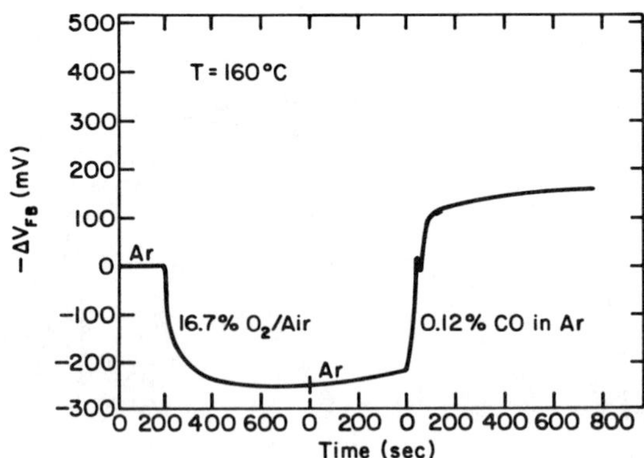

Figure 18.   O$_2$ and CO response at 160°C of a Pd/SiO$_2$ ($\sim$1000 Å)/
n-Si capacitor. The exposure cycle is 100% Ar→16.7% O$_2$ in Ar→
100% Ar→0.12% CO in Ar. Device CO response is mainly due to
removal of oxygen from the device structure.

hydrogen at room temperature. This effect is also seen in Table II.
However, when devices are operated at 160°C the data of Figure 17
show that water vapor has little influence on hydrogen sensitivity.
In general, for T $\geq$ 100°C, water vapor gives little interference to
hydrogen sensitivity due to its volatilizing from the device surface.
We also note that CO is found to provide little interference to
device H_2 sensitivity at room temperatures (15).

Temperature Effects. In general, temperature increases the speed of
response of C-I-S capacitor sensors as seen in Figure 12. Increasing
the temperature of device performance tends to decrease device saturation sensitivity to hydrogen. For devices being used to detect CO,
however, increasing the operating temperature increases the device
sensitivity to CO. Except for CO, increasing the temperature tends to
make interferences to hydrogen detection from other gases less important due to the dominating temperature-caused loss of sensitivity
and/or due to faster desorption of other gas species from the metal
surface at elevated temperature.

Comparison of Diode-type and Capacitor-type Sensors

The first question to ask when comparing various diode and capacitor
sensor structures is how do their sensitivities compare. This question is answered for several hydrogen sensing structures in Table V.
In order to make a valid comparison, the barrier height changes on
exposure to the various hydrogen ppm levels are shown. These are
deduced from $\Delta V/V|_I$ for the diodes and from ΔV_{FB} for the capacitors
(13). As may be seen from the data, the $Pd/TiO_x/n$-Si C-I-S diode
structure and the $Pd/SiO_2/Si$ capacitor structure are extremely sensitive to very low ppm H_2 levels even in 100% O_2. At higher H_2 ppm
levels the $Pd/SiO_2/Si$ capacitor structure becomes more sensitive to
hydrogen in O_2. This is probably the result of the capacitor   responding to H_2 due to some mixture of the five capacitor response
mechanisms listed in the Principles of Operation Section, including
those that give rise to the HID effect. It can also be due to the
difference in metal film morphology resulting from different substrates used for metallization (i.e., thin SiO_x and thick SiO_2).
     We note from Table V that the diode structures, except for the
TiO_x-based I-layer diode at very low H_2 ppm levels, tend to have
lower sensitivities than the capacitors when looked at in terms of
$\Delta V/V|_I$. However, small changes in ΔV_{FB} (or ΔV_{TH} in the MOSFET) must
be electronically amplified to increase sensitivity. If the diode
structures are operated in the $\Delta I/I$ $_V$ sensitivity mode, electronic
amplification is generally not needed since ΔI is related to $\Delta \phi_B$
through Equations 1 and 2. That is, when operated in the $\Delta I/I|_V$ mode,
the device physics of the diode gives built-in amplification of small
barrier changes as seen in Table II.
     As may be seen from Table I, the response times in hydrogen
exposures of capacitor structures tend to be comparable to those of
diode structures; however, the capacitor structures can be susceptible to the HID phenomenon (16) especially at elevated temperatures.
In general, the presence of water vapor or oxygen reduces the response and recovery times of both device classes. There are differences in gas sensing ability between the two structures. For example, the $Pd/TiO_x/Si$ and $Pd/SiO_x/Si$ diodes do not respond to CO in

Table V.   Comparison of the Barrier Height or Flatband Voltage
Change for Different Sensor Structures

Sensor Structure H_2 conc. in O_2 Ambient	$Pd/TiO_x/n$-Si C-I-S Diode $\Delta\phi_B$ (volt)	$Pd/SiO_x/n$-Si C-I-S Diode $\Delta\phi_B$ (volt)	$Pd/SiO_2/n$-Si C-I-S Capacitor ΔV_{FB} (volt)
4 ppm	-0.10	∿0.00	-0.06
40 ppm	-0.22	∿0.00	-0.36
80 ppm	-0.23	∿0.00	-0.43
600 ppm	-0.26	-0.02	-0.63
0.4%	-0.30	-0.10	-0.79
1.0%	-0.31	-0.20	-0.80

O_2 but the $Pd/SiO_2/Si$ capacitor does respond at elevated temperatures. This difference is believed to be due to the role of O_2 in
the I-layer of the capacitor structures (15).

As Table III notes, the interferences from O_2 to hydrogen detection increases for $Pd/SiO_x/Si$ diode structures with increasing concentration of O_2 relative to that in air. The interference from O_2
to hydrogen detection for $Pd/SiO_2/Si$ capacitor structures, on the
other hand, is much less significant although, as we mentioned
before, high concentrations of oxygen compared to that in air
slightly enhance the hydrogen sensitivity at room temperature.

The effect of water vapor on the device hydrogen sensitivity
however, is the same for both C-I-S diode and C-I-S capacitor structures; that is, high concentrations of water vapor speed up the response but significantly reduce the sensitivities.

<u>Summary</u>

Solid state C-S and C-I-S didoe and C-I-S capacitor structures are
sensitive detectors of gaseous species. They have the advantage of
being very compatible with microelectronics since they use the same
materials, devices, and operating voltages and currents. The overwhelming research and development effort of these devices has concentrated on structures for sensing hydrogen or hydrogen-bearing
gases. However, with selective membranes and properly chosen C-
layers and I-layers, other gases can be detected. As an example,
Dobos and Zimmer, using a split-gate MOSFET structure with a Pd or
Pt gate that has a gap (of the order of the channel length in dimension) directly above the channel, have recently reported PPM level CO
sensitivity in air at elevated temperatures (17).

In summarizing the response of the hydrogen diode sensors, we
note that these devices are faster at elevated temperatures than at
room temperature. At room temperature the SiO_x-based I-layer diodes
are faster in responding at higher oxygen concentrations and at
higher concentrations of water vapor; however, the presence of CO
slows down the response of these devices to hydrogen. At higher
temperatures interfering gases in the environment (e.g., O_2, H_2O,
etc.) have less effect on response times; i.e., temperature effects
eventually dominate. Sensitivity is affected by increased temper-

ature in the following way: the lower detection limit for hydrogen is generally extended to lower concentrations, the upper detection limit is generally reduced due to saturation at lower concentrations at higher temperatures, and sensitivity at saturation is reduced. Various gases interfere differently with the hydrogen sensitivity of diode structures. This interference depends on the materials used in the sensor. In general, the SiO_x-based hydrogen detecting diodes, CO has no effect on sensitivity. On the other hand, high concentrations of oxygen in the ambient decrease hydrogen sensitivity. Water vapor has the same effect but only for $T \leq 100°C$.

The response of the capacitor hydrogen sensors is somewhat different than the diode devices due to the richer interplay of detection mechanisms in the capacitor structures. In general, response times are more of a function of device materials and structure in the capacitor sensors. Water vapor is observed, as with diode structures, to speed up device response to hydrogen at room temperature as is oxygen. Again in agreement with the diode structures, CO slows down room temperature response to hydrogen. In general, increased operating temperature leads to decreased response times. Temperature affects the sensitivity to hydrogen in capacitor structures in the same manner as it does for diode structures; i.e., the lower detection limit is better but the upper detection limit, and sensitivity, are reduced by saturation at lower concentration. The interference to hydrogen detection is found to be complex: hydrogen sensitivity is the most pronounced in the total absence of oxygen; however, hydrogen sensitivity in 100% O_2 is somewhat better than that in dry air. In the case of water vapor the results are similar to those observed for diode H_2 sensor structures based on SiO_x; i.e., H_2O vapor reduces the sensitivity at room temperature but is unimportant to $T \geq$ 100°C. We also note that CO, at room temperature, does not interfere with H_2 detection in SiO_2-based structures; however, CO itself when introduced into O_2 at elevated temperatures will produce shifts in V_{FB} (or V_{TH} in a transistor) which are similar to those observed for hydrogen injected into oxygen.

Although C-S and C-I-S diode devices and C-I-S capacitor structures are relatively stable, elevated operating temperatures can give rise to stability problems for these structures. Stability problems may appear due to a number of reasons. For example, several detection mechanisms (from the list in the operation section), with different time constants, may compete as the dominant detection mechanism. The result can be the so-called HID phenomenon-hydrogen induced drift (16). Another problem can arise from changes in the structure and composition of various device layers due to interactions at elevated operating temperatures. Blisters formed on the Pd thin film used as the C-layer, for example, have been found to significantly reduce, or even destroy hydrogen sensitivity of C-I-S sensor structures (18,19). The cause is believed to be due to a reaction of hydrogen and oxygen in the C-layer over a certain range of H_2/O_2 mixture at elevated temperatures (20). Considerable effort has been directed toward solving those instability problems in recent years. Among the approaches that have been used are: (1) use of double C-layer, with the top C-layer acting as a coating to prevent chemical reaction from taking place in the more critical bottom layer (which has partially solved the blister problem) (19) and (2) use of double I-layers, with the top I-layer acting to prevent absorbed gas species from diffusing into the more critical bottom I-layer and/or the S-layer (which has partially solved the HID problem) (16). More

research, however, is needed in order to completely understand and
solve device instability problems.

It is interesting to note that, although the discussions in
this review have been restricted to the C-S and C-I-S type of sensor
structures, newly evolving, related sensor structures may prove to be
even more effective devices in some particular circumstances. For
example, a $Pd/TiO_x/Ti$ (C-I-C) hydrogen sensor with the I-layer serving
as the dominating sensing element has been reported to be extremely
sensitive to hydrogen in an oxygen-rich ambient at room temperature
(21). This device can be fabricated in any configuration and even on
plastics rendering it extremely versatile.

Literature Cited

1.   Metal-Semiconductor Schottky Barrier Junctions and Their Appli-
     cations, B. L. Sharma, Ed., Plenum Press, NY, NY (1984).
2.   S. J. Fonash, H. Huston, and S. Ashok, Sensors and Actuators, 2,
     363 (1982).
3.   M. C. Steele, J. W. Hile , and B. A. MacIver, J. Appl. Phys., 47,
     2537 (1976).
4.   K. I. Lundstrom, M. S. Shivaraman, and C. N. Svensson, J. Appl.
     Phys., 46, 3876 (1975).
5.   N. Yamamoto, S. Tonomura, T. Matsuoka, and H. Tsubomara, Surf.
     Sci., 92, 440 (1980).
6.   B. Keramati and J. N. Zemel, J. Appl. Phys., 53, 1091 (1982).
7.   T. H. Far, I. Lundstrom, and J. N. Zemel, Digest of Technical
     Papers, Transducer 1985, 226-228, IEEE, NY, NY (1985).
8.   S. Sze, Physics of Semiconductor Devices, 2nd Edition, 366-379,
     John Wiley & Sons, NY, NY (1981).
9.   F. Winquist, A. Spetz, M. Armgarth, C. Nylander, and I Lundstrom,
     Appl. Phys. Letts., 43, 839-841 (1983).
10. N. Yamazoe and T. Seiyama, Digest of Technical Papers, Trans-
     ducer 1985, 376-379, IEEE, NY, NY (1985).
11. Masami Ogita, Dong-bai Ye, Kazuhiko Kawamura, and Tatsuo
     Yamamoto, Digest of Technical Papers, Transducer 1985, 229-231
     IEEE, NY, NY (1985).
12. K. Kawamura and T. Yamamoto, IEEE Electron Dev. Lett., EDL-4,
     88 (1983).
13. S. J. Fonash, Zheng Li, and M. J. O'Leary, accepted for publi-
     cation in J. Appl. Phys., November, 1985.
14. "Development of the Solid State Hydrogen Sensor," Z. Li,
     D. Restaino, and S. J. Fonash, Final report for Sandia Contract
     No. 52-4585 (1984).
15. Zheng Li and S. J. Fonash, submitted for publication.
16. K. Bobos, M. Armgarth, G. Zimmer, and I. Lundstrom, IEEE Transi-
     tions of Elec. Devices., ED-31, 508 (1984).
17. K. Dobos and G. Zimmer, Digest of Technical Papers, Transducer
     1985, 242-244, IEEE, NY, NY (1985).
18. M. Armgarth and C. Nylander, IEEE Electron Device Letters, Vol.
     EDL-3, No. 12, 384 (1982).
19. S.-Y. Choi, K. Takahashi, and T. Matsuo, IEEE Electron Device
     Letters, Vol. EDL-5, No. 1, 14 (1984).
20. Zheng Li and S. J. Fonash, to be published.
21. Zheng Li and S. J. Fonash, accepted for presentation at IEEE
     International Device Meeting, Washington, DC, December 1985.

RECEIVED November 18, 1985

Amperometric Proton-Conductor Sensor for Detecting Hydrogen and Carbon Monoxide at Room Temperature

Norio Miura, Hiroshi Kato, Noboru Yamazoe, and Tetsuro Seiyama

Department of Materials Science and Technology, Graduate School of Engineering Sciences, Kyushu University, Kasuga-shi, Fukuoka 816, Japan

A new type of amperometric gas sensor using a proton conductor (antimonic acid) and its sensing mechanism are proposed for detecting small amounts of H_2 (or CO) in air at room temperature. The sensing element is composed of the following electrochemical cell : (counter electrode) air, Pt | proton conductor | Pt, sample gas (sensing electrode). The short circuit current of the cell is found to be in direct proportion to the concentration of H_2 (or CO). It is also shown that the sensor can be modified into a simpler construction which eliminates the reference gas (air). This modified sensor is found to exhibit performances as good as that of the original one, with satisfactory stability for about two months.

In recent years, gas sensors operating at room temperature are becoming increasingly more important in many fields. These sensors can be used as so called "cordless sensors", because they need no external electric sources to heat the sensor elements. Although electrochemical gas sensors which utilize liquid electrolytes are available to detect inorganic gases, e.g., O_2, CO, Cl_2, H_2S, etc. at room temperature (<u>1-3</u>), they often have time-related problems such as leakage and corrosion. The problems are minimized if solid electrolytes are used in place of liquid electrolytes.

It has been reported (<u>4,5</u>) that solid electrolyte sensors using stabilized zirconia can detect reducible gases in ambient atmosphere by making use of an anomalous EMF which is unusually larger than is expected from the Nernst equation. However, these sensors should be operated in a temperature range above ca. 300°C mainly because the ionic conductivity of stabilized zirconia is too small at lower temperatures. On the other hand, solid state proton conductors such as antimonic acid (<u>6,7</u>), zirconium phosphate (<u>8</u>), and dodecamolybdophosphoric acid (<u>9</u>) are known to exhibit relatively high protonic conductivities at room temperature. We recently found that the electrochemical cell using these proton conductors could detect

0097–6156/86/0309–0203$06.00/0

small amounts of H_2 or CO contained in air (<u>10</u>-<u>12</u>). This solid-state sensor can be operated even at room temperature and belongs to a potentiometric class which utilizes electromotive force (EMF) as a sensor signal.

The sensing performances of the potentiometric proton conductor sensor are shown in Figure 1. It is seen that the 90 % response time to 2000 ppm H_2 in air at room temerature is about 10 seconds and the EMF value of the sensor cell changes almost in proportion to a logarithm of H_2 concentration. It has been shown that the sensing electrode is at a potential determined by electrochemical oxidation of H_2 (or CO) and electrochemical reduction of O_2 (<u>11</u>). Although such a potentiometric sensor is convenient for detecting a broad range of gas concentration, the accuracy of gas detection is inferior to an amperometric sensor in which the sensor signal is in direct proportion to the sample gas concentration. Thus we tried to develop an amperometric sensor using proton conductors and found that the short circuit current of the proton conductor sensor is proportional to the H_2 (or CO) concentration in air (<u>13</u>). We describe here the fabrication, performances, and the sensing mechanism of the new type of amperometric proton conductor gas sensor.

<u>Experimental</u>

The first sensor element examined is represented as follows:

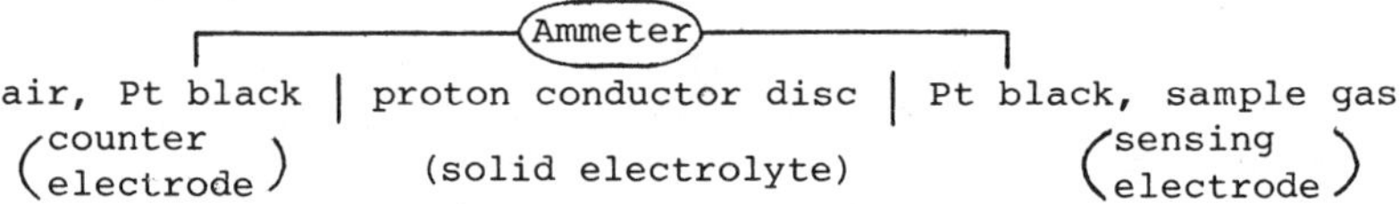

air, Pt black | proton conductor disc | Pt black, sample gas
(counter electrode)   (solid electrolyte)   (sensing electrode)

Figure 2 shows the structure of this sensor which is similar to that of the potentiometric sensor reported earlier (<u>10</u>). The only difference is that in this sensor a short circuit current between the sensing electrode and the counter electrode is measured with an ammeter. The proton conductor, antimonic acid ($Sb_2O_5 \cdot 2H_2O$), was prepared from antimony trioxide and hydrogen peroxide according to a method described elsewhere (<u>7,14</u>). The sample powder was mixed with 20 wt% Teflon powder binder (Lubron L-2, Daikin Ind. Co. Ltd.) and then cold-pressed at 4200 kgf/cm^2 into a compact disc 10 mm in diameter and 1 mm in thickness. Platinum black powder was applied on both ends of the thin disc to form a sensing- and a counter-electrode with a geometric area of ca. 0.4 cm^2. The disc was then fixed to an end of a glass tube by means of Epoxy resin. Electrical contacts between each electrode and Pt leads were made by carbon paste. The sample gas prepared by mixing small amounts of H_2 with air (or N_2 in some cases) was passed over the sensing electrode at 90 cm^3/min, while only air was fed to the counter electrode at the same gas flow rate. Commercial gases without purification were used for all the experiments. The gases were humidified by passing them through water. This is necessary to prevent the proton conductor from drying. The sensor signal, a short circuit current of the cell, was measured at room temperature by means of an ammeter (Hokuto Denko Co, Ltd., Zero Shunt Ammeter HM-101).

Results and Discussion

Sensing performance for H_2. Sensing performance of the amperometric sensor was examined for the detection of H_2 in air. Figure 3 shows the response curve for 2000 ppm H_2 in air at room temperature. The response was studied by changing the atmosphere of the sensing electrode from an air flow to the sample gas flow. With air the short circuit current between two electrodes was zero. On contact with the sample gas flow, the current increased rapidly. The 90% response time was about 10 seconds and the stationary current value was 10 μA. When the air flow was resumed, the current returned to zero within about 20 seconds.

Figure 4 shows how the short circuit current depends on the concentration of H_2 which is diluted with air or N_2. It is noteworthy that for H_2 in air the short circuit current is approximately in direct proportion to the H_2 concentration. As mentioned before, this fact suggests that for practical purpose the amperometric sensor is more accurate than a potentiometric sensor. When H_2 was diluted with N_2, the sensor exhibited a very different behavior with far greater current values and a nonlinear dependence on H_2 concentration. In this case, the cell is actually an H_2-O_2 fuel cell which accounts for the greater current values.

Sensing mechanism. As shown previously (<u>11</u>), when the circuit of the cell is open, the potential of the sensing electrode is determined by the following reactions (1) and (2).

$$H_2 \longrightarrow 2H^+ + 2e \tag{1}$$
$$1/2O_2 + 2H^+ + 2e \longrightarrow H_2O \tag{2}$$

In an atmosphere of H_2 and air, these reactions proceed simultaneously to form a local cell on the sensing electrode. In this situation the potential of the sensing electrode is a mixed potential (E_M) where the anodic current $i_{(1)}$ is equal to the cathodic one $i_{(2)}$. The mixed potential is determined as an intersection of both anodic and cathodic polarization curves as shown in Figure 5. It is important that the anodic polarization curve have a limiting current region as mentioned before (<u>11</u>). On the counter electrode where air is passed over, only reaction (2) takes place endowing the counter electrode with a potential near E_C. Thus, the potentiometric sensor observes the potential difference between the sensing electrode and the counter electrode E_M-E_C. On the other hand, when the two electrodes are electrically connected with a lead (short circuit), both the sensing and the counter electrodes are forced to be at the same potential (E_{SC}) as shown in the same figure. This means that the potential of the sensing electrode shifts to the direction where reaction (2) is unfavorable, while that of the counter electrode shifts favorably for reaction (2).

In order to confirm such potential shift we observed the actual behavior of the sensing and counter electrode potential under both open and short circuit conditions. Each potential was measured against a silver reference electrode which was attached to the sensor element as shown in Figure 6. Figure 7 depicts the response curves against 500 ppm H_2 in air under both conditions. When the circuit is open, the change in potential occurs only at the sensing

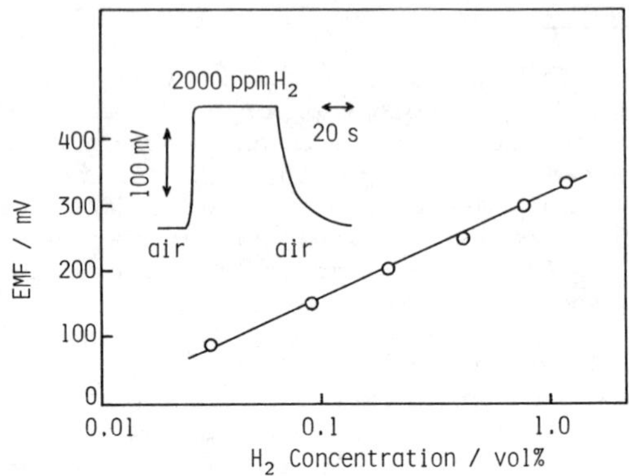

Figure 1. Response curve of the potentiometric sensor to
2000 ppm H_2 and dependence of EMF of the sensor on H_2 concentration in air.

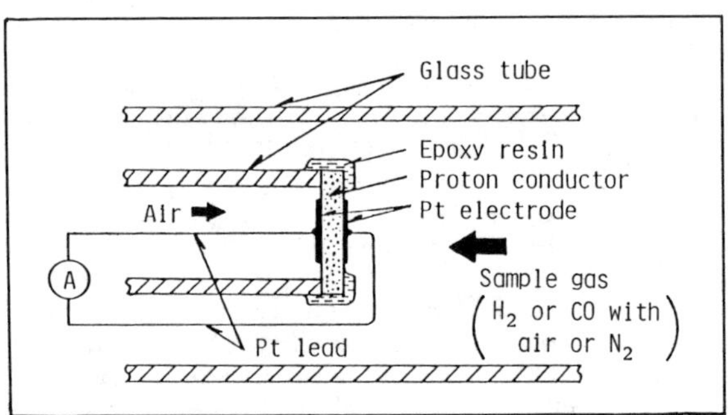

Figure 2. Structure of the amperometric sensor.
"Reproduced with permission from Ref. 13. Copyright 1984, 'The
Chemical Society of Japan'."

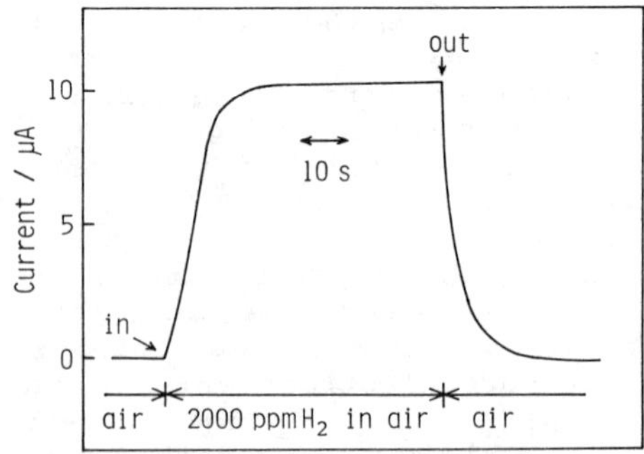

Figure 3. Response curve of the amperometric sensor to 2000 ppm
H_2 in air at room temperature.

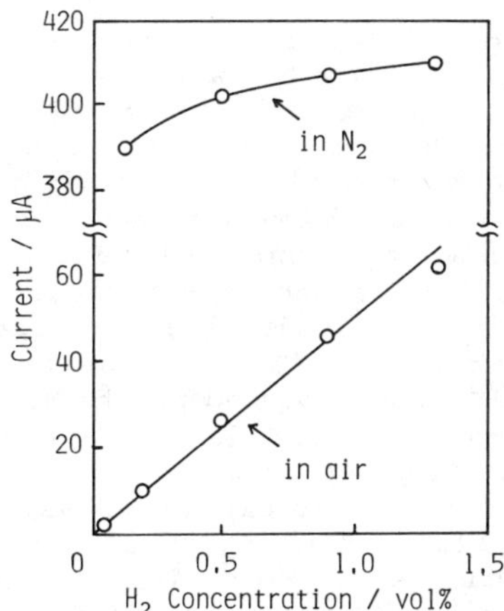

Figure 4. Short circuit current of the amperometric sensor vs.
H_2 concentration in air or N_2.
"Reproduced with permission from Ref. 13. Copyright 1984, 'The
Chemical Society of Japan'."

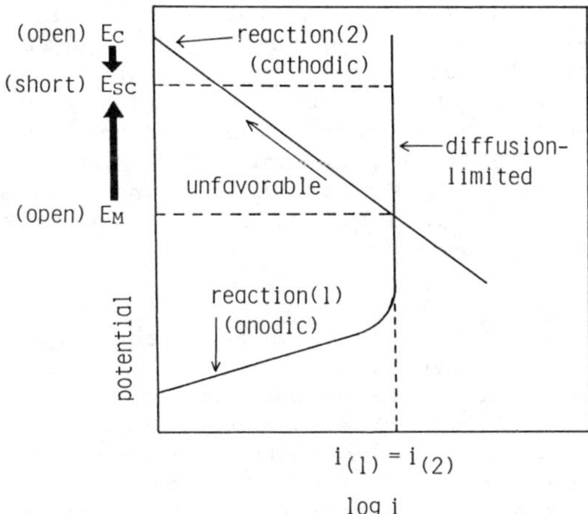

Figure 5. Schematic polarization curves for reactions (1) and (2).
"Reproduced with permission from Ref. 13. Copyright 1984, 'The
Chemical Society of Japan'."

electrode. Under the short circuit condition the sensing and the counter electrode are forced to shift to a potential in between as mentioned above. Moreover, it is noticed that the potential shift is larger for the sensing electrode.

Thus, we illustrate schematically the sensing mechanism of the amperometric sensor in Figure 8. When the circuit is open, reactions (1) and (2) are balanced at the sensing electrode; the proton produced by the electrochemical oxidation of H_2 will be consumed at the sensing electrode by the electrochemical reduction (2). It is important that the anodic reaction has been shown to be a diffusion-limited process. Under the short circuit condition, these two reactions are not balanced at the sensing electrode. The cathodic reaction at the sensing electrode becomes unfavorable, so that the consumption of H^+ by reaction (2) at the sensing electrode is suppressed. The excess H^+ thus produced on the sensing electrodes migrates toward the counter electrode through the proton conductor membrane to be consumed by the reaction (2). This process is accompanied by a flow of equivalent electrons as an external current. The anodic oxidation reaction of H_2 remains as a diffusion-limited process under the short circuit conditions, so that the amount of H^+ produced by the anodic reaction is proportional to the H_2 concentration. This eventually gives rise to an external current roughly proportional to the H_2 concentration in the gas phase.

<u>Modification of the sensor structure.</u> The above amperometric sensor has a rather complicated construction, because the sample gas (H_2 + air) is separated from the reference air. So, we tried to simplify the sensor structure as shown in Figure 9. As proton conductor we used a thin antimonic acid membrane (mixed with Teflon powder) of 0.2 mm thickness. This membrane is thin and porous enough to allow a part of the sample gas to permeate. On the other hand, the counter Pt electrode was covered with Teflon and Epoxy resin in order to avoid a direct contact with the sample gas.

The modified sensor was found to exhibit the excellent response to small amounts of H_2 in air as shown in Figure 10. The 90 % response time for 2000 ppm H_2 was about 10 seconds and the short circuit current depends linearly on the H_2 concentration in the same manner as observed in the original sensor.

<u>Sensing mechanism of the modified sensor.</u> The sensing mechanism in this modified sensor should be essentially the same as that of the unmodified one. It is noteworthy that a stationary short circuit current was obtained in spite of such sensor construction that the counter electrode was covered with Epoxy resin. Since the sensing electrode is placed in the same situation as the unmodified sensor, this fact indicates that the cathodic reaction is allowed to take place stationarily at the counter electrode. The proton conductor membrane is as thin as 0.2 mm, so that the reactant O_2 and the produced H_2O will permeate the membrane as shown in Figure 11. A part of H_2 will naturally also permeate through the membrane, but the transfered H_2 will be consumed by the reaction with O_2 electrochemically or catalytically at the counter electrode.

Furthermore, the rate of H_2 supply to the counter electrode through the membrane is rather small as compared with that to the

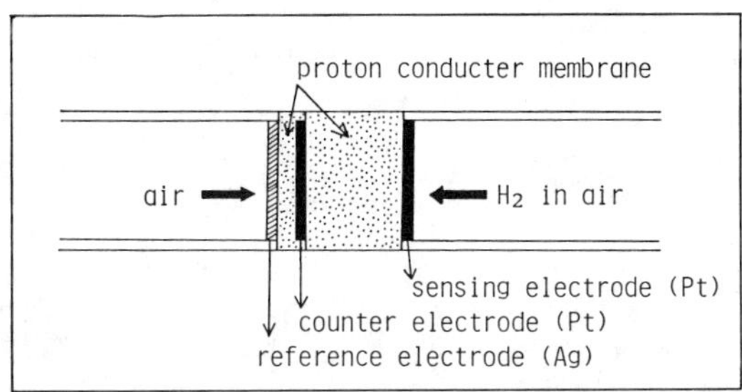

Figure 6.   Configuration of the sensor attached with reference Ag electrode.

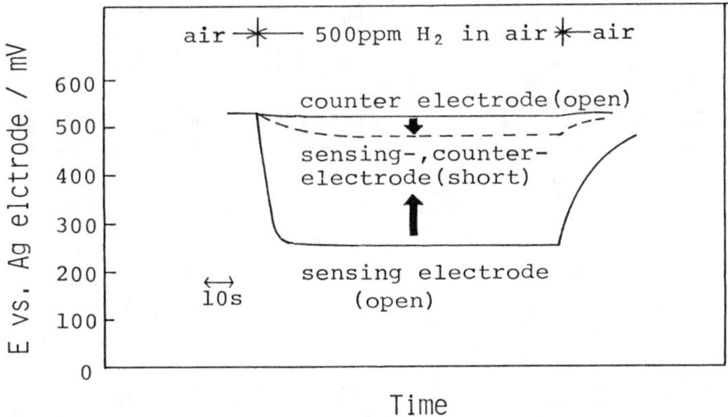

Figure 7.   Behaviors of respective electrode potential under open circuit and short circuit conditions.

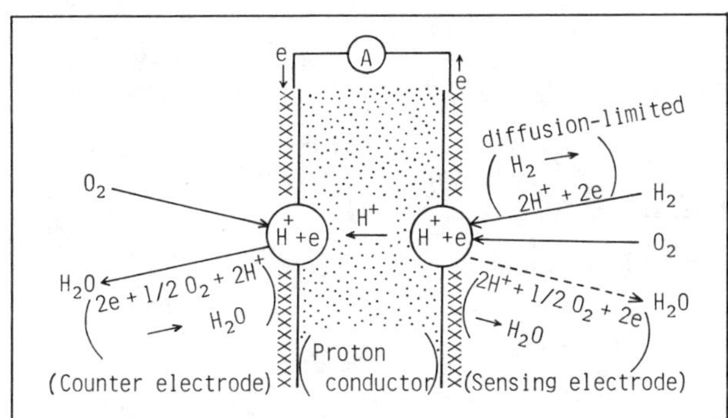

Figure 8.   Sensing mechanism of the amperometric sensor. "Reproduced with permission from Ref. 13. Copyright 1984, 'The Chemical Society of Japan'."

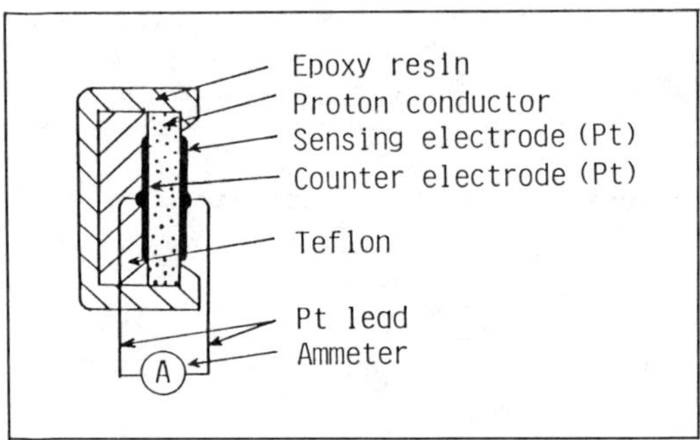

Figure 9. Structure of the modified amperometric sensor.
"Reproduced with permission from Ref. 13. Copyright 1984, 'The
Chemical Society of Japan'."

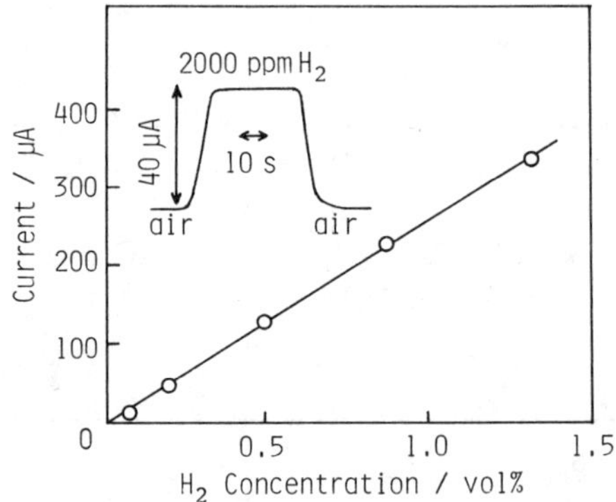

Figure 10. Response curve of the modified amperometric sensor to
2000 ppm H_2 and dependence of short circuit current of the sensor
on H_2 concentration in air.
"Reproduced with permission from Ref. 13. Copyright 1984, 'The
Chemical Society of Japan'."

sensing electrode from the gaseous bulk. Therefore, the H_2 concentration around the counter electrode is considered to be lower than that around the sensing electrode. This difference in H_2 pressure between the two electrodes can account for the stationary flow of short circuit current.

Performances for CO. This modified sensor was found to be also sensitive to small amounts of CO in air. Figure 12 shows response curves to CO in air at room temperature. Although the value of short circuit current is rather small as compared with that for H_2 detection, the response is still rapid enough. It is noteworthy that the current value is also in direct proportion to CO concentration as shown in Figure 13. The sensing mechanism is considered to be almost the same as in the case of H_2 detection. The anodic reaction for CO can be expressed by the following reaction.

$$CO + H_2O \longrightarrow CO_2 + 2H^+ + 2e \qquad (3)$$

It was confirmed that this sensor was insensitive to methane (15000 ppm) and propane (7000 ppm) in air.

Long-term stability. As for a practical use, the long-term stability is one of the important factors. Figure 14 shows the results of a long-term stability test for the modified sensor at room temperature. Except for the beginning of the test period, the short circuit current to 1.3 vol% H_2 in air was stable for about two months. The anomalously large current at the beginning has not been understood well yet.

Conclusions

The results of our work may be summarized as follows:
1) A new type of amperometric sensor using a proton conductor could detect small amounts of H_2 or CO in air at room temperature.
2) The short circuit current of the sensor cell was in direct proportion to the sample gas concentration.
3) The possible sensing mechanism of this sensor was proposed.
4) The sensor could be modified into a simpler construction where the reference gas (air) was no longer necessary.
5) The modified sensor was stable for about two months.

Acknowledgments

This work was partially supported by the Grant-in-Aid for Developmental Scientific Research (No. 60850151) from the Ministry of Education, Science and Culture, Japan.

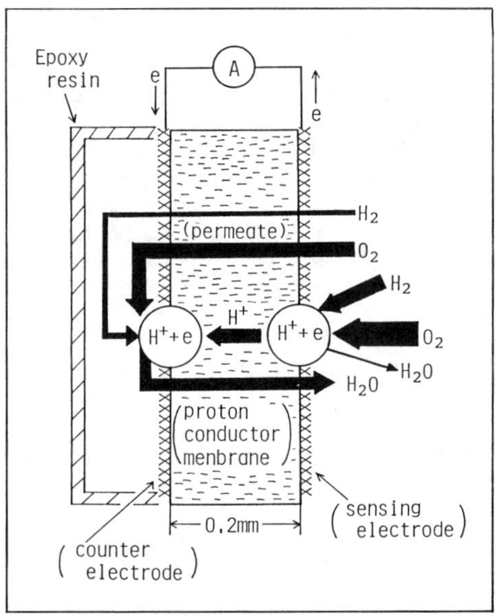

Figure 11.   Sensing mechanism of the modified amperometric sensor.

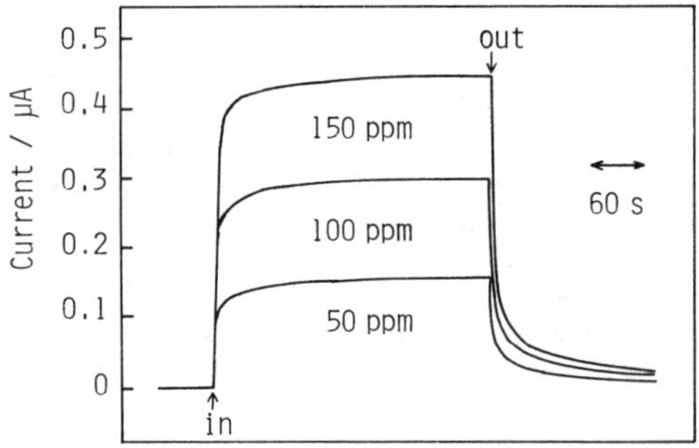

Figure 12.   Response curves of the modified amperometric sensor to CO.

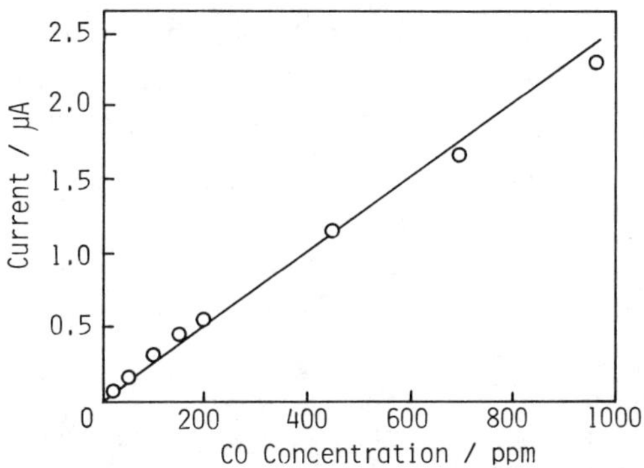

Figure 13.   Short circuit current of the modified amperometric sensor vs. CO concentration in air.

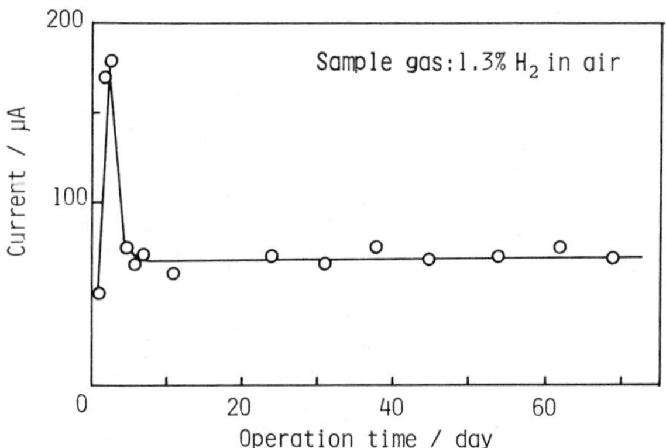

Figure 14.   Long-term stability of the modified amperometric sensor at room temperature.

<u>Literature Cited</u>

1. Dietz, H. ; Haecker, W. ; Jahnke, H. "Advances in Electro-
    chemistry and Electrochemical Engineering " Gerischer H. ;
    Tobias, C., Ed. ; John Wiley & Sons: New York, 1977, Vol. X,
    p.1.
2. Belanger, G. <u>Anal</u>. <u>Chem</u>. 1974, 46, 1576.
3. Isobe, M. <u>Keisoku Gijustu</u> 1976, 4, 43.
4. Shimizu, F. ; Yamazoe, N. ; Seiyama, T. <u>Chem</u>. <u>Lett</u>. 1978, 229.
5. Okamoto, H. ; Obayashi, H. ; Kudo, T. <u>Solid State Ionics</u> 1980,
    1, 319.
6. England, W. A. ; Cross, M. G. ; Hammett, A. ; Wiseman, P. J. ;
    Goodenough, J. B. <u>Solid State Ionics</u> 1980, 1, 231.
7. Ozawa, Y. ; Miura, N. ; Yamazoe, N. ; Seiyama, T. <u>Nippon
    Kagaku Kaishi</u> 1983, 488.
8. Jerus, P. ; Clearfield, A. <u>Solid State Ionics</u> 1982, 6, 79.
9. Nakamura, O. ; Kodama, T. ; Ogino, I. ; Miyake, Y. <u>Chem</u>. <u>Lett</u>.
    1979, 17.
10. Miura, N. ; Kato, H. ; Yamazoe, N. ; Seiyama, T. <u>Denki Kagaku</u>
    1982, 50, 858.
11. Miura, N. ; Kato, H. ; Yamazoe, N. ; Seiyama, T. " Proc. of
    the Int. Meeting on Chemcal Sensors " Seiyama, T. ; Fueki, K.;
    Shiokawa, J. ; Suzuki, S., Ed.; Kodansha/Elsevier, Tokyo, 1983;
    p.233.
12. Miura, N. ; Kato, H. ; Yamazoe, N. ; Seiyama, T. <u>Chem</u>. <u>Lett</u>.
    1983, 1573.
13. Miura, N. ; Kato, H. ; Ozawa, Y. ; Yamazoe, N. ; Seiyama, T.
    <u>Chem</u>. <u>Lett</u>. 1984, 1905.
14. Miura, N. ; Tokunaga, K. ; Harada, T. ; Yamazoe, N. ; Seiyama,
    T. <u>Denki Kagaku</u> 1978, 46, 113.

RECEIVED October 31, 1985

SENSORS FOR LIQUIDS AND SOLIDS

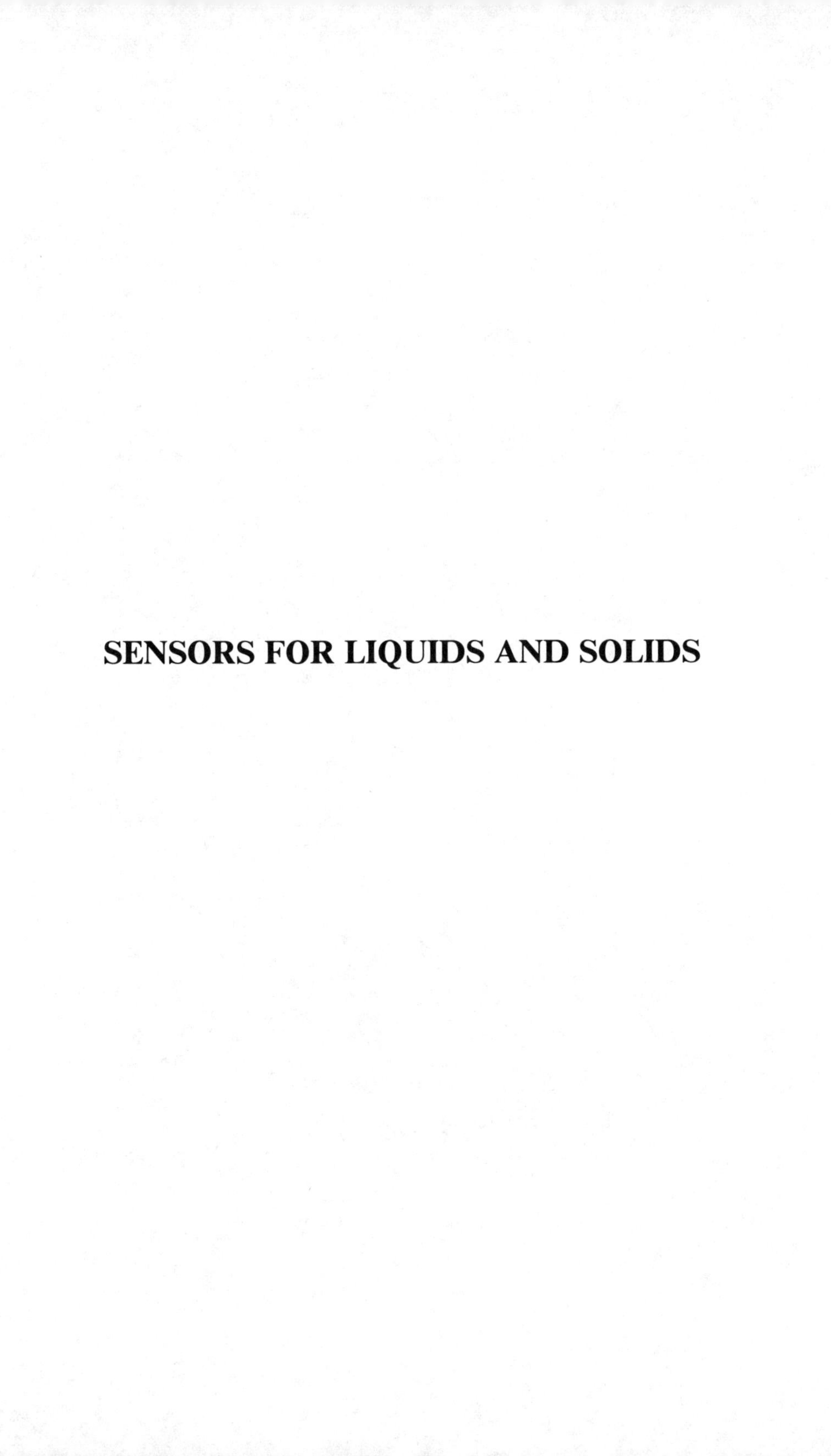

Advances in Atmospheric Gas Composition Determinations Using IR Spectroscopy
A Fast-Response, High-Resolution, In Situ, CO_2 Sensor and a Rocket-Borne, Emission-Measurement System

Gail E. Bingham[1] and Allan J. Steed[2]

[1]Soil Science and Biometeorology Department, Utah State University, Logan, UT 84322
[2]Space Dynamics Laboratories, Utah State University, Logan, UT 84322

Studies of atmospheric properties using IR spectroscopy techniques have been reported in the literature for nearly 100 years. This paper presents a brief historical review of the development of this area of science and discusses the common features of spectrographic instruments. Two state of the art instruments on opposite ends of the measurement spectrum are described. The first is a fast response in situ sensor for the measurement of the exchange of CO_2 between the atmosphere and the earth's surface. The second is a rocketborne field-widened spectrometer for upper atmosphere composition studies. The thesis is presented that most improvements in current measurement systems are due to painstakingly small performance enhancements of well understood system components. The source, optical and thermal control components that allow these sensors to expand the state of the art are detailed. Examples of their application to remote canopy photosynthesis measurement and upper atmosphere emission studies are presented.

Electromagnetic radiation from atmospheric gases is rich with information on species concentrations, temperatures, chemical reaction processes, and other parameters. Measurement of many of the properties of gases using infrared techniques, i.e., by measuring the absorption and emission characteristics of the gases is now common. Most atmospheric visible and UV absorption and emission involves energy transitions of the outer electron shell of the atoms and molecules involved. The infrared spectrum of radiation from these atmospheric constituents is dominated by energy mechanisms associated with the vibration of molecules. The mid-infrared region is rich with molecular fundamental vibration-rotation bands. Many of the overtones of these bands occur in the near infrared. Pure rotation spectra are more often seen in the far infrared. Most polyatomic species found in the atmosphere exhibit strong vibration-rotation bands in the $1 - 25$ μm region of the spectrum, which is the region of interest in this paper. The richness of the region for gas analysis studies can be seen in Figure 1, which shows the spectral location of some of the atmospheric infrared absorption and emission bands located in the 1 - 5 μm region. A detailed treatise of molecular polyatomic radiation characteristics has been presented by Herzberg (1). A band model which will represent low resolution gas transmittance is often required to interpret measurements made using absorption techniques. Two models which have found wide acceptance are the Elsasser model (2), and the Goody random band model (3). Other models are continuing to be developed and updated such as the one based on an equivalent absorption coefficient (4).

In situ measurements of the emission and absorption characteristics of the atmosphere always lag behind theoretical developments and laboratory studies. This is primarily attributable to equipment limitations. The laboratory environment is basically friendly, and there, experimenters are not usually faced with limitations of equipment weight, size, and power, and there is no necessity to design to meet adverse environmental conditions. This is not the case when field measurements are undertaken. In the field the elements mentioned above must be considered and solutions provided in order to conduct successful measurement programs. This paper provides a brief synopsis of developments in IR spectroscopy, compares basic system components, and discusses some of our recent efforts to extend measurements techniques, which are now common under controlled laboratory conditions, to the more difficult situation of actual atmospheric measurements. We have not presented a detailed study of a specific single example. Rather, we chose to discuss two typical field instruments and highlight the development of the components of these instruments that ultimately allowed successful system deployment.

History

The late 1800's saw the first application of infrared spectroscopy to the measurement of the composition of the atmosphere (5-9). The pioneers of this science used rotating prisms as dispersing elements and photographic plates as detectors. As scientists designed and developed more sophisticated lenses and gratings, and began to use thermocouples and thermopiles as detectors (10-16), they were able to identify and quantify specific major absorption bands such as those from CO_2 (17-19) and water vapor (20-22). The earliest systems used the sun or a stellar object as a source of illumination to backlight the sample of interest (5,7). (Early measurements that used this technique have recently been re-examined to determine the average atmospheric CO_2 concentrations of the late 1880's. The measurements were found to be of high quality and the information available from them is being used to improve our understanding of the "Greenhouse Effect" (23)).

Studies of the effects of the environment upon absorption spectra proceeded as rapidly as available instrumentation would allow, and included such parameters as thickness (24), pressure (20,25,26), temperature (27-29), and background gas composition (30).

The different kinds of spectra were understood early (absorption (31), emission (32,33), and Raman (19)), but application of this knowledge to atmospheric problems was again delayed due to a lack of suitable equipment.

Instrumentation developments in the 1920's and 30's led to a rapid expansion of spectroscopic methods in the laboratory (28, 34-39). These included further penetration into the infrared regime and some applications to infrared transmission in the atmosphere. Additional equipment was developed during World War II as a result of military requirements. This period was a fruitful one for the science of spectroscopy, and saw the first applications of infrared equipment as gas measurement tools (40-41) and as "routine" process controllers (42).

The tremendous developments in electronics and optical coating methods and the increased need for data pertaining to atmospheric composition, essentially led to an exponential increase in the use of spectroscopic analysis in the years after World War II.   In recent years investigators have carefully studied atmospheric windows using remote sources and extremely long absorption paths (43,44).   During this same period other experimenters were attempting to compress long absorption paths into small instruments.   These efforts have been very successful (45-47).   The more recent development of lasers, solid state detectors and electronic miniaturization currently brings this field to a high degree of activity and sophistication.

The Basic System

The basic components of an optical sensor remain essentially similar even though the wavelength regime and type of measurement desired may change.   Figure 2 illustrates the basic components that are common to all such detection systems.   The common components consist of collection optics, a method of wavelength control and selection, a beam modulation system, a detector, and a signal processor.   These components are essential for either absorption or emission measurements.   To these common components, various "front end" arrangements are added to tailor the system to a specific application.

Absorption-based measurements, which encompass nearly all laboratory infrared gas analysis studies, require a source of "within band" radiation arranged to backlight the sample of interest.   This can be a spectral or broad band (blackbody) source located within the instrument, or one located at a great distance from the detector. Early solar measurements were of this second type.   High-flying and space systems often use a reverse design and probe the atmosphere by using the earth as the source.   Where gas temperatures are high compared to the background, emission from the gas itself can be used as the source of radiation.   The weak nature of these emissions usually requires a much more sensitive detector system than that required for absorption measurements. The history of the development of infrared gas measurement systems is one of small improvements in these same basic components to allow increased system capacity.

Absorption-Based Spectroscopy Systems

One of our recently-developed systems is a good example of adapting laboratory methods to in-situ field measurements.   We developed this system to measure the exchange of CO_2 between the atmosphere and the surface of the earth.   The transfer of matter and energy from a

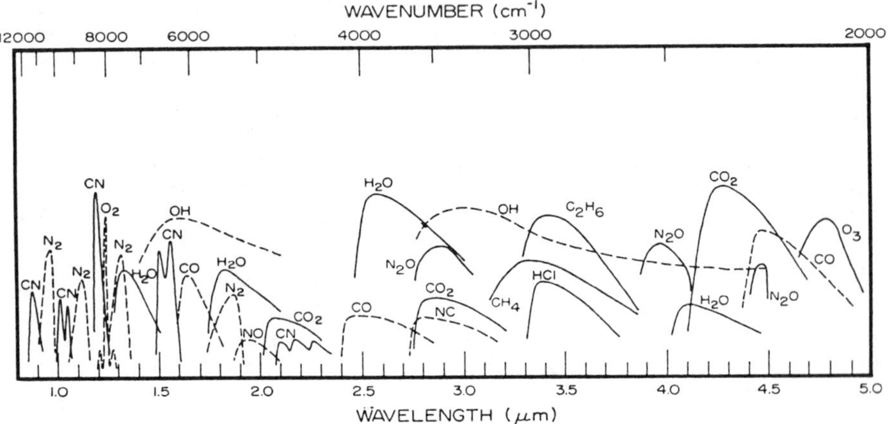

Figure 1. The spectral absorption/emission of selected atmospheric gases in the middle infrared regime. The illustration shows the wavelength occurrance without specific regard to relative intensities.

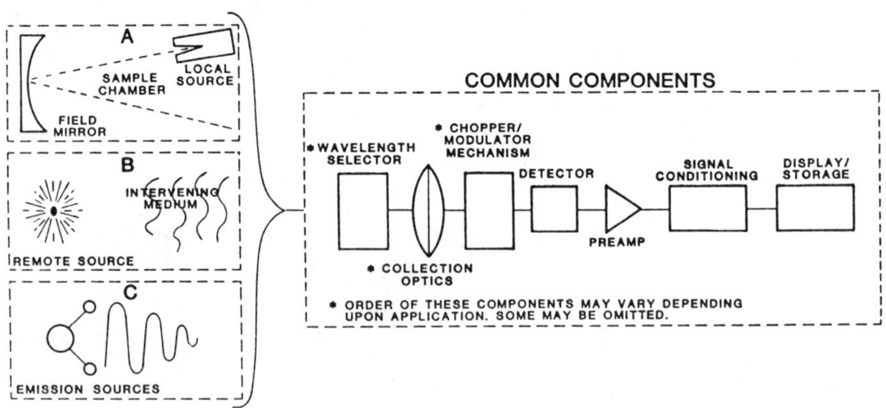

Figure 2. Common components of basic infrared detection systems. The illustration shows examples of three typical infrared energy inputs.

surface to the atmosphere is driven by the temperature and concentration differences that exist between the surface and the air above it. In the boundary layer close to the earth's surface, mixing and the transport of atmospheric properties is accomplished essentially by turbulent transfer. For many properties, no sources or sinks exist in the boundary layer, and the fluxes of these properties (mass per unit transfer area per unit time) are approximately constant with height. An exchange measurement made within this "constant flux" region provides an accurate estimate of the exchange at the surface.

One of the most accurate methods for determining gas exchange in this region of the boundary layer is called the eddy covariance or eddy correlation method. To apply the method, scientists measure the individual upward and downward components of the turbulent exchange and calculate a net difference (48,49). Stated mathematically, the relationship becomes:

$$F_s = \overline{\rho w's'} \tag{1}$$

where; ρ is the mean air density, w' is the vertical velocity, and s' is the fluctuating portion of the scalar property being measured. The bar represents an average of the products of instantaneous measurements over the time interval of interest.

One of the most difficult measurements to accomplish is that of CO_2. Applied to CO_2, determined by open path infrared absorption, the relationship described by Equation 1 becomes:

$$F_c = \overline{\rho w'c'} \tag{2}$$

where; c is the CO_2 mixing ratio.

<u>Fast Response CO_2 Sensor.</u>   The sensor requirements for eddy covariance measurements are extreme. To be used within a few meters of a plant canopy, the sensor must have a frequency response in excess of 20 Hz. Additionally, because the large mean density of CO_2 in the atmosphere (about 560 mg m^{-3}) and the deviations around the mean associated with turbulent transfer are small ($\geqslant$10 mg m^{-3}), the sensor must have a signal to noise ratio in excess of 3500:1. The sensor must maintain these specifications for long durations, while mounted on a tower above the canopy, where it is exposed to constant changes in temperature, solar irradiation, and background gas concentrations. The instrument must unobtrusively sense the natural turbulant fluctuations of the atmosphere. To effectively accomplish this it must be small and streamlined.

Figure 3 is a schematic diagram of the electro-optical system in our fast response CO_2 sensor. As we noted in the discussion of Figure 2, the sensor contains many components common to other electro-optical systems. The novel features of this instrument are the arrangement and type of components used. The instrument operates as a dual wavelength spectrometer, using the strong CO_2 absorption band at 4.27 μm for detection and a reference (nonabsorbed) bandpass centered at 3.85 μm. During operation the instrument couples a diverging beam of energy from the conical cavity radiator into a folded optical system capable of 2.5 m of absorption path. After the beam interacts with the gas in the cell, an optical system collects

the beam and passes it through a combined chopper/filter wheel.   The
modulated, filtered beam then impinges on a detector which develops a
pulsed electronic signal (as the result of chopping) with a pulse
height equal to beam intensity.   The instrument separates the pulses
into two channels, representing sample and reference beam intensity,
and passes them to matched peak-detect and sample-hold circuits.
Processing of the dc output of the hold circuits removes instrument
and sample cell effects, resulting in a fast response dc signal
directly proportional to the optical density of CO_2 in the sample
cell.

     Although this is not the first application of the basic
techniques used in this instrument, major improvements of several
components make this sensor a significant improvement over previous
ones.   Some of the more significant improvements are detailed below.

<u>Temperature Control</u>.   A prime factor in the superior performance of
this sensor is the temperature control system in the electro-optical
compartment.   The source, operating at 800 °C, and the detector, PbSe
at -35 °C, are located less than 50 mm apart.   Both are precision
temperature-controlled devices and required careful thermal
engineering to assure a 3500:1 sensor signal to noise ratio.   The
sensor uses a three-level temperature control system.   Overall heat
removal is accomplished by a recirculating water heat transfer system
connected to a small radiator in front of the control chassis cooling
fan.   The water cools the power transistors and the source
containment jacket and removes (or supplies) waste heat to the second
level system.   The chopper/filter motor, filter assembly, preamp,
voltage reference, and detector cooler are located in a temperature
stabilized housing controlled to 20 ±0.05 °C.   A Peltier stack
cooling system accomplishes this task.   The detector cooler, a dual
Peltier stack located within the hermetically sealed TO-8 detector
can, pushes down from this temperature stabilized base to hold the
detector at -35 ±0.001 °C.

<u>IR Source</u>.   Sources with discrete spectral emittance in the middle
and far infrared are rare.   Some laser diodes exist that radiate at
wavelengths shorter than 2 μm, but laser diodes that radiate at
longer wavelengths require cryogenic cooling and as yet are not
commonly used in portable <u>in situ</u> sensors. A major limitation of the
blackbody type source in a miniature sensor is the high input power
requirement and the resulting waste heat which must be dissipated.
Also of concern in extremely sensitive systems is the need to
efficiently eliminate out-of-band radiation so that excessive heating
of coatings and optical surfaces by this unused energy does not
affect instrument performance.   The close proximity of the source and
detector in our sensor required the development of a small, reliable,
highly efficient source that produced very little waste heat.

     In a thermal source, the energy emitted is dependent upon source
area and temperature; i.e., the larger the source area and the higher
the source temperature, the more energy emitted.   Although the ir-
radiance is proportional to the emitting area, waste heat is a function of the overall surface area of the heated piece.   To achieve our
high signal to noise ratio, source temperature control was critical.
Since we also required high source reliability, we chose a classical,
wound slug source that could be heated and regulated with only two,

small-diameter wires. The increased mass of the slug-type source adds stability, but also adds the capability for heat transfer to its surroundings. Our challenge was to design a containment system that would allow a high output source to be operated under tight control with low power consumption. Figure 4 shows our source containment system. It consists of a precision, commercial, conical-cavity radiator (Electro-optical Industries, Colita, CA, Model 19307) in an inverted, quartz, vacuum-Dewar flask. The design was prompted by the need for exceptionally good insulation at the front of the source. To couple high output from as small a source as possible, one must collect at a low f#. This places the slug close to the front wall of the source, leading to high input power requirements, overheating of the collection optics, and premature source failure. The combination of a ceramic wafer support and the quartz flask reduced operating source power input (waste heat output) by 60 percent. It reduced the amount of heat radiated to the collection optics by more than a factor of 10 when compared to some conventional source designs. The resulting 5.0 cm diameter source package, operating at 800 °C, maintains a cool enough exterior that it can be held in the fingers. During reliability trials, this design has operated at 1000 °C for over two years without a failure. When installed in an instrument the flask is encased in a thin aluminum jacket to control the flow of the small amount of waste heat that finally escapes from the source.

<u>Sample Cell</u>. Energy from the source is collected and coupled to an extremely large aperture (f/3) folded optical absorption cell that has a source to detector throughput of 40 percent. Lavery (<u>46</u>) provides an excellent tutorial on the design and operation of folded optical systems in environmental instrumentation. An important consideration in the design of <u>in situ</u> sensors is the rate at which a sample can be updated in the absorption cell, and the power required to exchange the sample. When flushing an enclosed but well stirred cell, Lavery shows that 5 cell volumes of gas are needed to reach 99 percent exchange. Since our application required that the sensor follow the rapid atmospheric concentration changes occurring near an actively exchanging surface, the sample smearing and high pump power required to exchange samples in an enclosed cell were unacceptable. We approached this problem by removing the cell walls and allowing natural atmospheric motion to exchange the whole sample, plug fashion. This design requires special fabrication considerations, but has been extremely successful. Figure 5 shows our sensor head on a tower over a soybean field, with the other instrumentation necessary to make eddy covariance measurements of canopy net primary productivity.

The cell has an optical capability of 2.5 m, using 12 transversals. When applied to CO_2, however, the absorption coefficient (a_c) is of sufficient size to allow measurements using an 0.5 m (4 transversals) path. Due to the magnitude of a_c, a large difference in intensity develops as the beam passes through the cell. Maximum detection and amplification efficiency can only be achieved when the intensity levels of the sample and reference beams are approximately the same. To help balance the intensity levels, the mirror coatings (multi-layer silicon/silicon dioxide) have been optimized to pass the sample beam (99 percent reflectance at 4.3 μm) and partially attenuate the reference wavelength (95 percent

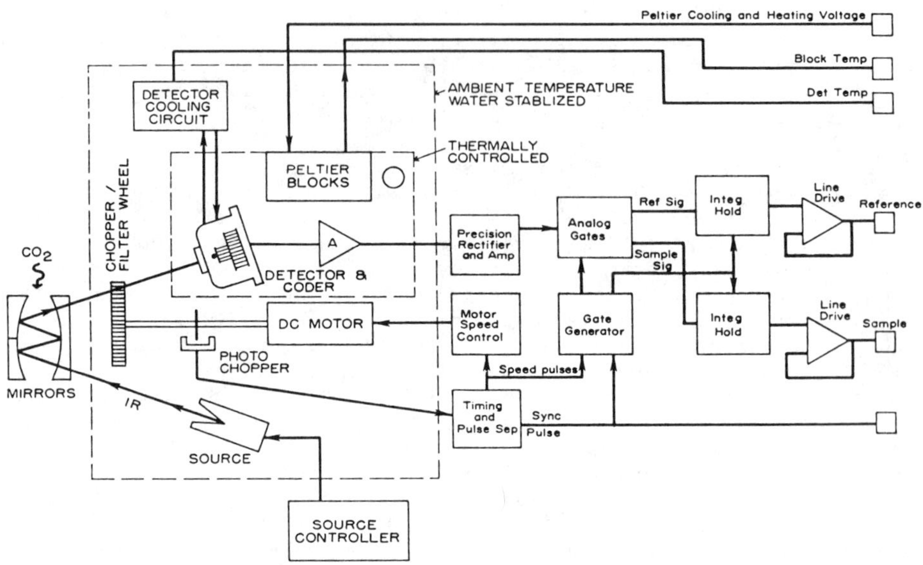

Figure 3.   A block   diagram showing the electro-optical
components of the fast response <u>in-situ</u> CO_2 sensor.   The
illustration shows a specific arrangement of the components
identified in Figure 2.   Temperature controlled regions are shown
within dashed lines.

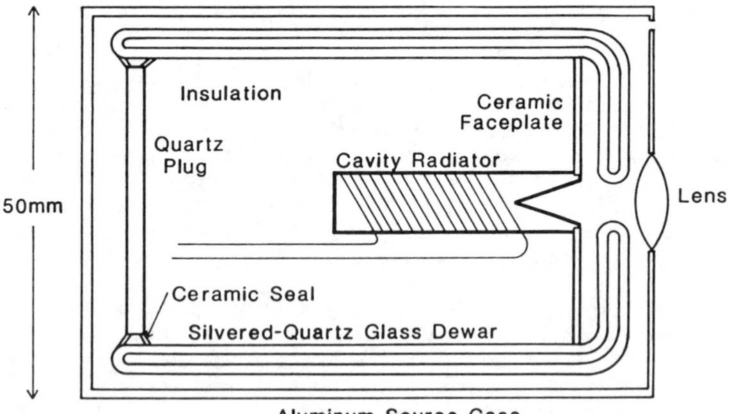

Figure 4. The source containment system used in the CO_2 sensor.
The quartz glass dewar allows good thermal insulation of the
forward portion of the source plug, thereby minimizing heat
transfer from the source to the optical system.

reflectance at 3.85 μm). The mirrors are of marriaging steel, which can be worked and polished during fabrication, and then hardened by heat treatment. After treatment, the mirrors have a low temperature response and readily accept optical coatings. Cell standoffs are of Invar, which minimizes cell distortion under varying radiation loads.

Filter/Chopper. As the beam exits the cell the instrument collimates it and passes it to the wavelength discrimination and modulation section. Our design uses a single chopper/filter wheel to provide precise filter changing and beam chopping. The wheel, which rotates at 80 rps, contains 8 filters (4 sample and 4 reference) and 8 beam interrupters that modulate the beam at 640 Hz. The filters and beam interrupters are coated on a 3.2 cm diameter Sapphire substrate, 0.5 mm thick (see Figure 6). An important problem that must be faced when using narrow bandpass filters in in situ instruments arises from temperature effects on the filters. The general effect of temperature on filter performance is shown in Figure 7 for a set of commercial filters of the type used in our early instruments (50). To minimize differential temperature effects, both sets of filters use a similar design and are coated on a single substrate. Temperature still affects the filters, but the differential effect is small and can be largely eliminated in the signal processor.

While our thermal control scheme does much to stabilize filter temperature, early versions of the sensor still showed an annoying warmup drift. This was eventually tracked to heating within the coated layers of the filters and was due to excessive out-of-band radiation. Even with filters of similar design operating in a temperature-stabilized cavity, the sensor had sufficient resolution to detect the differential response of the sample and reference filters. Since this was the result of a very small difference, it took a long period to stabilize. This problem was solved by blackening the inside of the temperature controlled filter housing to stimulate radiative equilibration of the coatings when not exposed to the beam, and by coating a broad band, thermally conductive filter on the sample cell entrance lens/window to eliminate much of the radiation with wavelengths shorter than the reference and longer than the measurement band.

Common practice would have the reference bandpass as close to the measurement region as possible (typically at 4.0 μm). In this application, however, interference considerations dictate the chosen reference location. The middle infrared regime is very rich in absorption regions; and hence, it is nearly impossible to make an accurate determination of gas concentration without some concern for the cancellation of interference effects. These effects arise from absorption of other gases in the band or bands where the measurements are being made (see Equation 4). One of the simplest ways to eliminate interference from another gas, which is known to be fluctuating in the sample volume, is to choose the reference wavelength in a region where this gas has an absorption coefficient of similar magnitude as at the measurement wavelength.

In our sensor, the interfering gas is water vapor. Water vapor does not have strong absorption bands in the 4.27 μm region, but it does have a relatively strong continuum absorption and a few scattered lines. The variations in water vapor in the open

Figure 5. A photograph of the fast response _in-situ_ CO_2 sensor monitoring carbon uptake by a soybean canopy. The open, folded optical absorption cell allows for fast sample exchange. The surrounding instruments measure air flow and humidity variations.

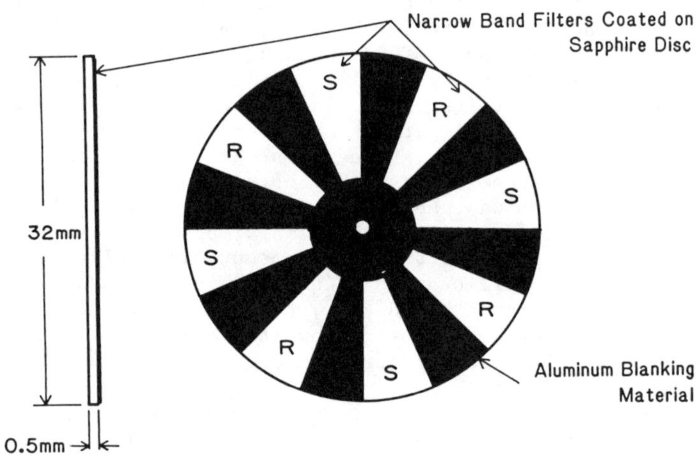

Figure 6. A diagram of the chopper/filter wheel used in the fast response, _in-situ_ CO_2 sensor. Combining both functions on a single wheel provides the precise timing necessary to minimize electrical filter interference and significantly reduces instrument weight and volume. Coating sample and reference filters of similar design on a single substrate minimizes the effects of ambient temperature changes.

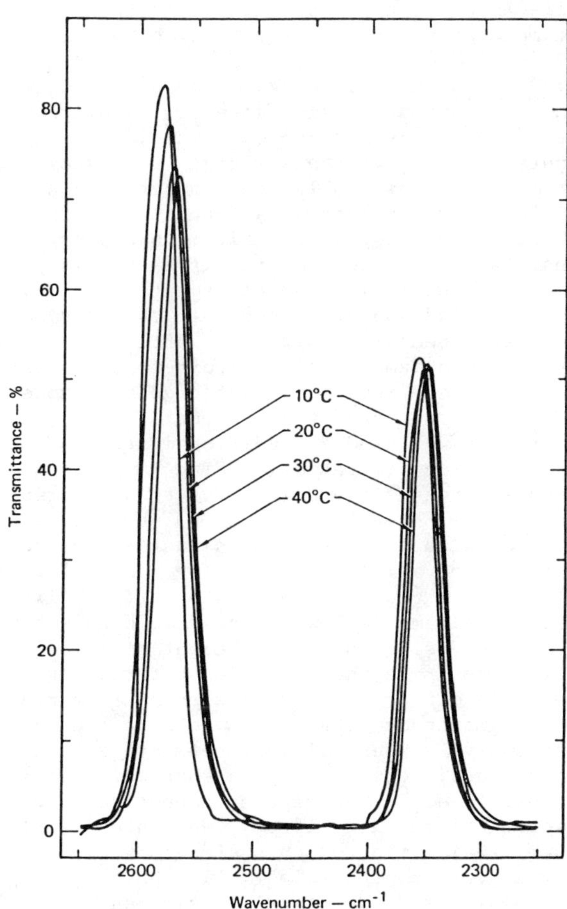

Figure 7. The effect of ambient temperature changes on the transmission of narrow bandpass optical filters of different design on separate substrates.

atmosphere are so large (up to 35 g m^{-3}) that effects are sufficient to cause a significant error in an uncorrected sensor. After a review of the available literature (51-53), we decided to depart from common practice and place the reference at a shorter wavelength. The review indicated a broad peak in H_2O absorption near 4.0 μm, tapering back to a similar level near 3.85 μm. Extensive tests of the instrument showed that this location for the reference exactly cancels the effects of water vapor at temperatures below 25 °C, but slightly overcompensates for its effects at higher air temperatures.

<u>Preamplifier and Signal Conditioning</u>. The sensor uses an autozeroed preamp that has an automatic gain feature. The preamp maintains a constant reference pulse height and contains just enough filtering to shape the pulse and eliminate high frequency noise. The chopper/filter motor is controlled by an active controller to 0.01 percent to eliminate random signal attenuation by this filtering as motor speed slips in and out of the filter bandpass. The output of the preamp consists of four sets of multiplexed sample and reference pulses for each chopper/filter wheel rotation. This multiplexed pulse train is rectified and separated into channels representing the sample and reference beam intensities. After peak detecting, each channel is applied to a sample-hold circuit that converts the pulses to dc levels. These dc levels are buffered and passed to a signal processing chassis at the base of the tower.

In the rectifier we used an additional step to achieve maximum resolution and linear amplification. Since our information is coded in the difference between the height of the reference and sample pulses, the absolute height of the pulses contains no useful information. Most signal conditioners would pass the full pulse, which would limit the ability to amplify the difference, thereby reducing resolution. We used a precision offset circuit to reduce pulse height to a minimum, allowing the full bandpass of the system to be applied to processing the pulse height difference.

Slight differences in the diffusion path during filter coating results in slightly different transmission characteristics for each of the 8 filter segments on the wheel, despite each set having been coated simultaneously. These differences are evident in the sensor output as a continuous, repeated ripple that significantly increases the sensor output noise. This ripple cannot be filtered adequately using conventional electronic filter techniques while maintaining sensor response. We eliminated this effect by using a technique we refer to as true average processing. In the signal processing, we demodulate the sample-held dc signal again, and apply an amplitude matching circuit to each pulse to correct for these differences. After amplitude matching, the pulses are recombined through a summing circuit to provide a four-filter average with a frequency response of 80 Hz.

<u>Sensor Performance</u>. The intensity of electromagnetic radiation transmitted through an optical path of length L of an atmosphere that is not optically thick can be described by:

$$I = I_o \exp(-a_j L) \tag{3}$$

where; I_o is the source intensity, and a_j is the integrated extinction coefficient of a specific spectral bandpass j. In our application, the extinction coefficient can be considered to consist of two components:

$$a_j = a_{cj} + a_{ij}$$ (4)

where; a_{cj} consists of a change in transmission in our chosen bandwidth due to absorption by CO_2, and a_{ij} is a transmission change due to additional absorption by interfering gases. As can be seen by (Equation 4), the sensitivity of our sensor will be proportional to the absorption path length and the number of absorbing molecules in the beam, or in other words, the sample optical density. Since a_{cr} is small, and the reference wavelength was chosen such that $a_{ir} = a_{is}$, the sensor detects the number of CO_2 molecules in the cell.

The density of CO_2 in the absorption cell, however, is a function of both concentration and bulk air density. In normal process analyzers, where temperature and pressure within the absorption cell are controlled, measurements can be easily referred to gas density by a simple calibration curve. In an open path system, changes in bulk air density must be measured. Indeed, one of the major problems faced in testing the sensor was the development of test facilities where we could control the temperature, pressure and CO_2 more accurately than the sensor could measure. Even the small changes in building pressure associated with ventilation system fluctuations resulted in output signal changes three to four times the sensor signal to noise level. In operation, pressure and temperature near the open cell are measured and used to calculate gas density.

The system has a typical corrected output sensitivity of 0.1 mg CO_2 m^{-3} optical density change in a single range. To allow a flexible operation under a wide range of conditions, matched electronic offset circuits are used to control the dc level of each signal during processing such that the output is near midscale for any given atmospheric mean concentration. The high resolution of the sensor limits its overall range to about 300 mg CO_2 m^{-3}.

Figure 8 shows the high frequency nature of the vertical velocity, water vapor, and CO_2 densities at 2 m above a soybean canopy during a three minute period. The instantaneous fluxes at the location of the sensors are also shown. Note the high degree of correlation between increases in CO_2 density, decreases in water vapor density, and the bursts of downward moving air. The opposite response occurs for nearly all updrafts, indicating that the surface is a source of water vapor and is actively removing CO_2. To calculate the average exchange of these vital gases for the surface in question, long period averages of the instantaneous values are calculated. Average fluxes for the 15 minute period from which the data in Figure 8 were taken are also shown in the Figure. An example of the application of this sensor to the determination of crop climate response can be seen in the work of Anderson <u>et al.</u> (<u>54</u>).

<u>Emission Based Spectroscopy Systems</u>

In the preceding sections we presented descriptions of absorption spectroscopy systems that use folded optics and small sample chambers to achieve the desired measurements. Other experimenters have

studied atmospheric gases using long-pathlength cells, which are
basically scaled-up variations of the type of cells described in the
preceding section, mated to spectrometers with large free spectral
ranges (55). In many instances, however, it is either not possible
or not convenient to utilize a sample chamber. To make measurements
with longer effective path lengths than are permitted by a sample
chamber, a remote infrared source, preferably at a temperature
substantially above that of the atmospheric gases to be measured, can
be used with the basic system described earlier. Large infrared
sources have commonly been used by experimenters to measure
integrated atmospheric properties over long pathlengths and to verify
and update atmospheric transmission models (56). Experimenters use
this technique to make measurements of integrated atmospheric
pollution. Scientists have successfully used the sun as a celestial
source to discover and study trace gases in the atmosphere.
Recently, such an instrument flew on the space shuttle in an effort
to determine global distributions of atmospheric species. The
obvious drawback in using a source such as the sun, is that it
imposes time and position constraints on the measurements, and hence,
limits them in these respects.

Emission measurement techniques have in many applications proven
very useful in providing an alternative to the absorption method.
Emission measurements free the experimenter from the time and
position restraints imposed by a celestial source and remove the
complications imposed by the necessity to position a remote source in
line with the gas sample of interest. One example of the application
of emission measurements and their effectiveness, is their use to
measure the effluents from sources such as smoke stacks (57). In
this application there is usually a temperature differential which
allows discrimination between the target and the ambient atmosphere.
This type of measurement is most effective in monitoring target gases
when they are in close proximity to the source since the target gas
temperature soon becomes the same as the ambient atmosphere and their
measurement becomes much more difficult if not impossible.

Another successful application of emission spectroscopy is in
making measurements of gases in the upper atmosphere. In this case,
the gas density is too thin to utilize absorption cell techniques,
and in situ measurements are difficult because of the inaccessibility
of the region. In some selected spectral regions, especially in the
near infrared, ground-based sensors can provide measurements of upper
atmospheric chemical processes (58). Water and CO_2 absorption mask
most of the upper atmospheric infrared emissions from ground based
observers, but as shown in Figure 9 window regions do exist and
useful measurements can be made in those window regions. Balloon and
aircraft-borne platforms can carry instruments above much of the
atmospheric masking and are effective for accomplishing measurements
in some IR spectral regions (59). Throughout much of the spectral
region between 2.5 and 25 μm, however, sensors operating from a
rocketborne platform, or using limbscanning and inversion techniques
from a satellite platform provide the best method for determining the
IR properties of the upper atmosphere. In these cases, emission
spectroscopy offers some distinct advantages.

Emission instruments in general, utilize the basic components
outlined in Figure 2. The source/sample chamber is replaced by the
target gas itself, which acts as the source. In most atmospheric

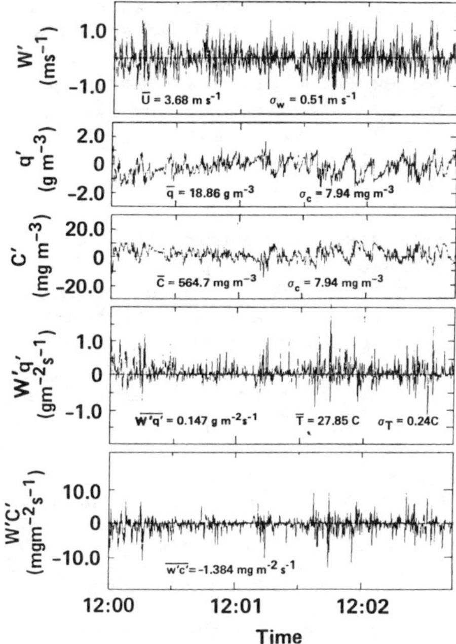

Figure 8. The high frequency nature of the vertical velocity (W'), water vapor (q'), and CO_2 densities (C') at 2 meters above a soybean canopy during a 3 minute period. The illustration also shows instantaneous water vapor (W'q') and carbon dioxide (W'C') fluxes and the mean quantities for the 15 minute period from which these traces were taken. Data courtesy of Center for Agricultural Meteorology and Climatology, University of Nebraska, Lincoln, Nebraska, and Environmental Sciences Division, Lawrence Livermore National Laboratory, Livermore, California.

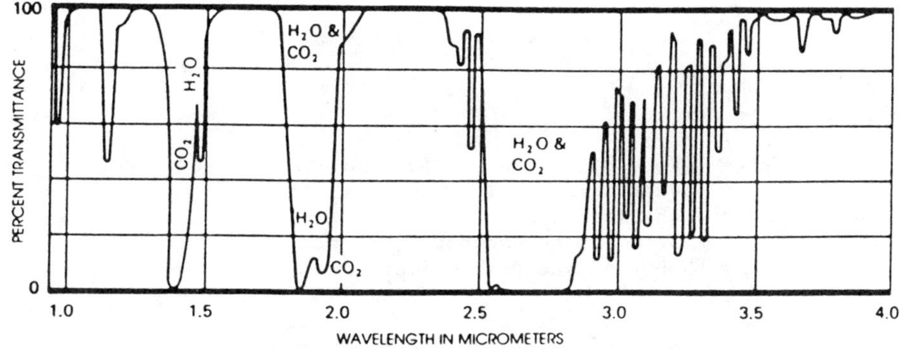

Figure 9. Spectral location of atmospheric absorption bands for a 1000 ft. horizontal air path at sea level with 5.7 mm precipitable water at 26°C.

cases the signal strength of the source to be measured is of a much lower level than for the absorption case, thereby creating the requirement for more sensitive instruments. For applications such as upper atmospheric research, where gas densities are low, emission measuring instruments are required to be orders of magnitude more sensitive than for the absorption case. Often, cryogenic cooling of the entire instrument is required to achieve background limited detector noise characteristics and a level of sensitivity sufficient to make the measurement.

Early infrared atmospheric absorption measurements required a broad free spectral range and long path length, and were made using prism or grating spectrometers with solar energy as the source. In later years the Michelson interferometer-spectrometer essentially replaced the prism and grating instruments for broad spectral range measurements. This results primarily from the throughput and multiplex advantages of the interferometer system (60) and the tremendous advances in digital computers which allow fast Fourier analysis of the interferometer data. Another advantage of the Michelson interferometer is referred to as Connes' advantage. This allows enhanced accuracy of the measured frequency of the spectra and is made possible by the inclusion of a monochromatic, drive mirror position reference system. A He-Ne laser system is incorporated as the monochromatic reference source in most present interferometer instruments.

Equations 5 and 6 demonstrate the sensitivity advantage obtained by an interferometer-spectrometer system over a grating-type instrument. Equation 5 gives the noise equivalent spectral radiance (NESR) for a grating or prism type instrument. Equation 6 gives the same information for an interferometer-spectrometer. By a comparison of the two equations, we can see that, especially for broad-band measurements (large η), the Fourier transform system has significant advantages in sensitivity. Also, the $A\Omega$ of the Michelson interferometer can be much larger, thus improving the NESR. In the upper atmosphere, where emissions are weak, this improved sensitivity makes many heretofore impossible measurements possible. Baker et al. (61) presents a summary of the development of infrared interferometry for upper atmospheric emission studies.

$$\text{NESR (Grating Spectrometer)} = \frac{\text{NEP } \sqrt{\eta}}{A\Omega\Delta\sigma\epsilon_s \sqrt{T}} \qquad (5)$$

Where;

$$
\begin{aligned}
\text{NEP} &= \text{Detector noise equivalent power, 1 Hz Band (watts),}\\
\Delta\sigma &= \text{Spectral resolution } (cm^{-1}),\\
T &= \text{Scan time (sec),}\\
A\Omega &= \text{Instrument throughput } (cm^2 \text{ sr}), \text{ and}\\
\eta &= \text{Number of spectral elements scanned.}
\end{aligned}
$$

$$\text{NESR (Interferometer)} = \frac{\text{NEP}}{A\Omega\Delta\sigma\epsilon_s \sqrt{T}} \qquad (6)$$

Where;

ε_s = Interferometer transmission and modulation efficiency.

A key point in the case of emission measurements of gases in the atmosphere, is that the dim sources are frequently large in spatial extent, and instruments with large solid angle collecting capabilities can be used to great advantage. As illustrated in Equation 6, the NESR of an interferometer system is inversely proportional to the solid viewing angle Ω, and the collecting area A. The collecting area is, of course, limited only by the size of the instrument optics (practical size and cost restraints). However, Ω is another matter, and theoretical limitations come into play. In a standard Michelson configuration, the maximum solid viewing angle is $2_\pi/R$, where R is the resolving power of the interferometer-spectrometer. Researchers have devised many schemes to widen the field of view of a Michelson system, and Baker (62), offers a review of a number of these. We illustrate one such technique in the following pages.

Emission measuring sensors also need an internal calibration source. In the case of absorption sensors, the external source most often served as the standard. In the case of emission, a reference source is usually incorporated in the instrument to certify instrument performance and provide an absolute reference.

Rocketborne Field-Widened Interferometer-Spectrometer. The rocketborne field-widened interferometer-spectrometer (RBFWI) is an example of an instrument that uses emission techniques to make broad-band spectral measurements in the 2 - 7.5 μm spectral region in the upper atmosphere. Figure 10 illustrates the concept of this cryogenically-cooled instrument developed at Utah State University under an Air Force Geophysics Laboratory contract.

Optical System. As shown in Figure 11, the optical section of the instrument consists of a beamsplitter, two optical wedge mirrors and a detector section with associated detector collection mirror. This is basically the Michelson interferometer technique except that the end mirrors have been replaced by optical wedges mirrored on the back side. The two windows are necessary only to maintain an ambient pressure in the interferometer section, and a vacuum in the detector section. The window on the detector can be replaced with an optical filter if only a selected spectral region is to be investigated.

The described system has a field of view of 12 degrees full angle, and therefore, one of the most difficult problems to overcome is that of vignetting of the rapidly expanding bundle of rays as it passes through the system. Much emphasis has been placed on a compact, simple design to allow maximum energy transfer to the detector. In most interferometer spectrometers a sandwich-type beamsplitter is used. The beamsplitter coating is placed on an appropriate optically flat substrate, which transmits over the spectral region of interest. An additional compensating flat substrate is placed adjacent to the beamsplitter surface to insure that the transmitted beam and the reflected beam experience the same optical media. Since the beamsplitter surface must be optically flat, both the beamsplitter substrate and the compensator must be reasonably thick. This required large size of the sandwich-type

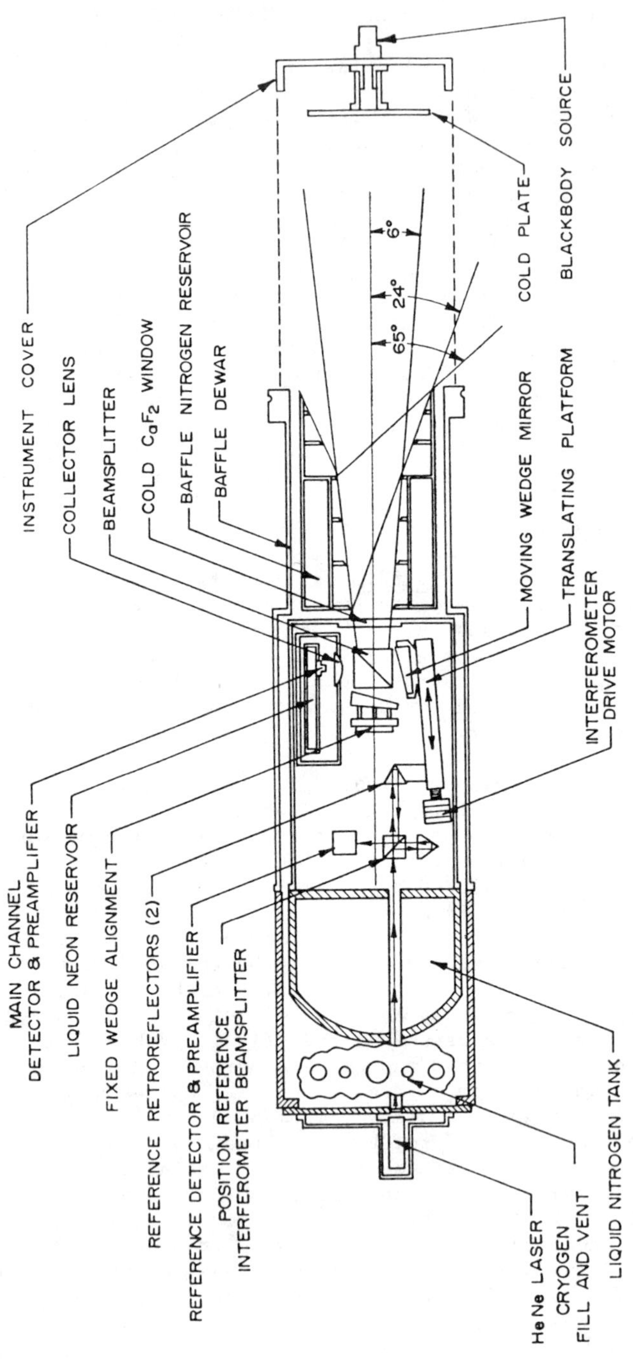

Figure 10. A cutaway diagram showing the electro-optical components of the RBFWI.

beamsplitter is not acceptable in the RBFWI. Instead, we chose a cube-type beamsplitter as it allows the end wedges to be placed closer to the beamsplitter and greatly reduces vignetting by providing a longer ray path within the beamsplitter.

One problem that is peculiar to the cube beamsplitter, however, is that if the two beamsplitter sections are assembled with an air gap spacing between them, total internal reflection can result. This is unacceptable and results in the need for a bonding agent which matches the index of refraction of the two beamsplitter sections. In the case of this instrument, it proved impossible to find a bonding agent which matched the index of refraction, had good IR transmission over the spectral region of interest, and remained useable at cryogenic temperatures. The best solution proved to be a halogenated chlorofluorocarbon oil pressed between the two beamsplitter sections which were held together with spring-loaded mounts. The optics are securely mounted to survive the launch environment of a sounding rocket and maintain interferometer alignment. The system has a remote alignment system that allows the instrument to be optimized at cryogenic temperatures, but there is not sufficient time for an automatic alignment during the flight of the instrument, therefore the prelaunch alignment must be maintained.

<u>Detector</u>. In the case of a field widened system where ultimate sensitivity is the goal, it is imperative that an optimum detector system be employed. We used a parabolic collector with the detector system positioned at the focal point of the parabola. The detector is actually four, indium doped silicon detectors, connected in series and positioned on each face of the square cross-section detector post. The indium doped silicon matches the wavelength response of the calcium fluoride optics and yields excellent results. We have not attempted to image the incoming energy, only to collect it as efficiently as possible. The entire detector system is cooled to 27°K with liquid neon. It would have been more convenient to use a liquid nitrogen cooled detector, but the improved sensitivity obtained with the neon cooled system was determined to be worth the added complication of operating a cryogen system inside a cryogen system.

<u>Wedge Translation</u>. We achieve optical retardation by translating one of the wedges parallel to its "apparent" mirror plane as viewed through the wedge, thus inserting more or less wedge material in one of the interferometer's optical paths. This system requires longer drive lengths than a standard Michelson configuration, but makes possible more than a factor of ten increase in system throughput by substantially increasing the acceptance solid angle of the instrument. Figure 12 illustrates the increase in throughput. The Figure shows the theoretical maximum solid angle possible for a standard Michelson system as a function of resolving power. Also shown on the plot is the operating point of this field-widened system. The throughput gain of 10 actually makes possible a measurement time improvement of 100 (required measurement time to achieve a specified signal to noise is proportional to $(NESR)^2$ (see Equation 6)) and allows measurements, not previously possible, of transient events. This wedge method of field widening was first suggested by Bouchareine and Connes in 1963, but practical field instruments have not demonstrated significant

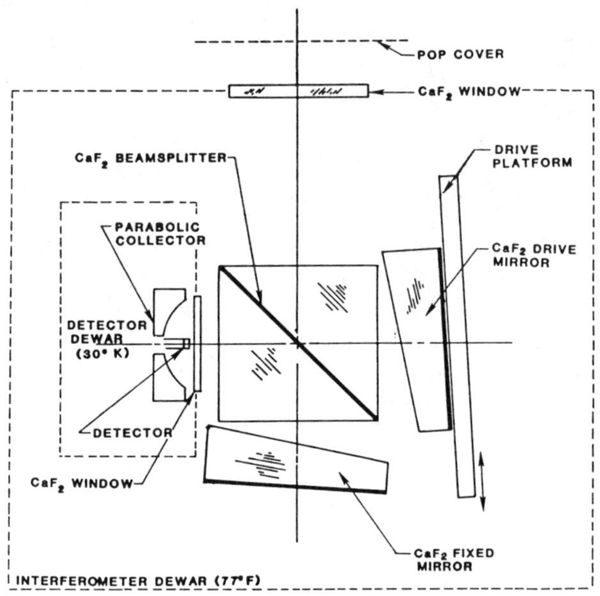

Figure 11. A diagram of the optical section of the RBFWI.

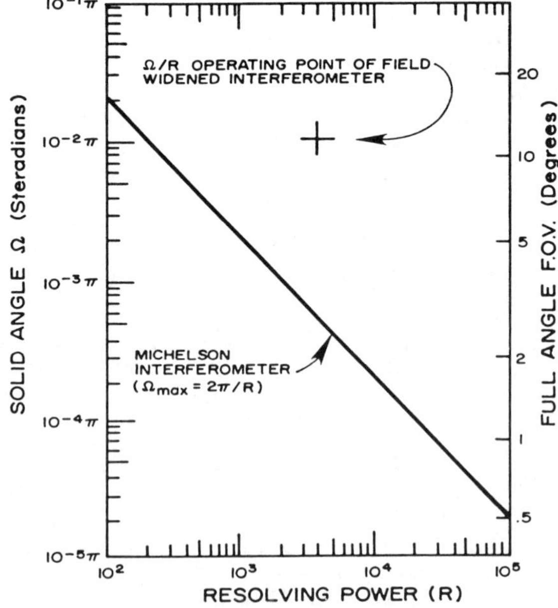

Figure 12. Theoretical maximum solid angle for a standard
Michelson system as a function of resolving power compared to
field-widened interferometer operating point.

measurement improvements over standard Michelson techniques until this generation of field-widened instruments (62).

One of the major advances associated with this instrument is the use of a long pathlength, gas lubricated translation platform capable of supporting the large wedge optics required by this field widening method. The gas bearing developed for this instrument, has ways that are lapped to an optical flatness of one wavelength and is constructed of a material designed to match the thermal expansion of the CaF2 optics. The gas bearing is supported by boil-off gas from the liquid nitrogen tank and the interferometer enclosure is kept at ambient pressure throughout the flight. A more thorough description of the instrument technique can be found elsewhere (63). A summary of instrument specifications is shown in Table 1.

Table I. Rocketborne Field-Widened Interferometer-Spectrometer
Specifications

Function	Specification
Free Spectral Range	2.0 to 7.5 μm (5000 to 1250 cm^{-1})
Resolution	1.2 cm^{-1}
Scanning Speed	1.5 seconds/scan
Field of View	12° full angle
Optics	CaF$_2$ beamsplitter, compensators and external window
Cryogen	Liquid neon (detector) Liquid nitrogen (instrument)
Reference	He-Ne laser (6328Å)
Detector	InSi
Calibration Sources	Inflight, 1000° K blackbody, 3.2 μm diode laser
AΩ	0.67 cm^2 sr
Weight	294 lb. dewar and electronics (without cryogen or payload skin)
Size	44.4 cm diameter, 165 cm long (dewar and electronics)
Instrument Sensitivity	2 x 10^{-13} w cm^{-2} sr^{-1} cm at 5 μm in 1.5 sec scan

Instrument Application. This particular system is operated from a rocketborne platform. A photo of the instrument as it is mounted in the payload configuration is shown in Figure 13. Measurements begin at about 80 kilometers on ascent and continue in a Zenith looking configuration through apogee and descent. Upper atmospheric gas emissions are measured and transmitted to a ground station by telemetry until the instrument descends to a point where the parachute deploys. An example of the spectrum obtained from this instrument when flown on April 13, 1983, is shown in Figure 14. The rocket flight was from Poker Flat, Alaska (near Fairbanks) at 0906:23 (GMT). The rocket had a set apogee of 140 km to maximize measurement time in the 80 to 120 km altitude region. We took the plot in Figure 14 from data at 88 kilometers on ascent. It identifies the emission from a number of species of gases that are primarily excited as a

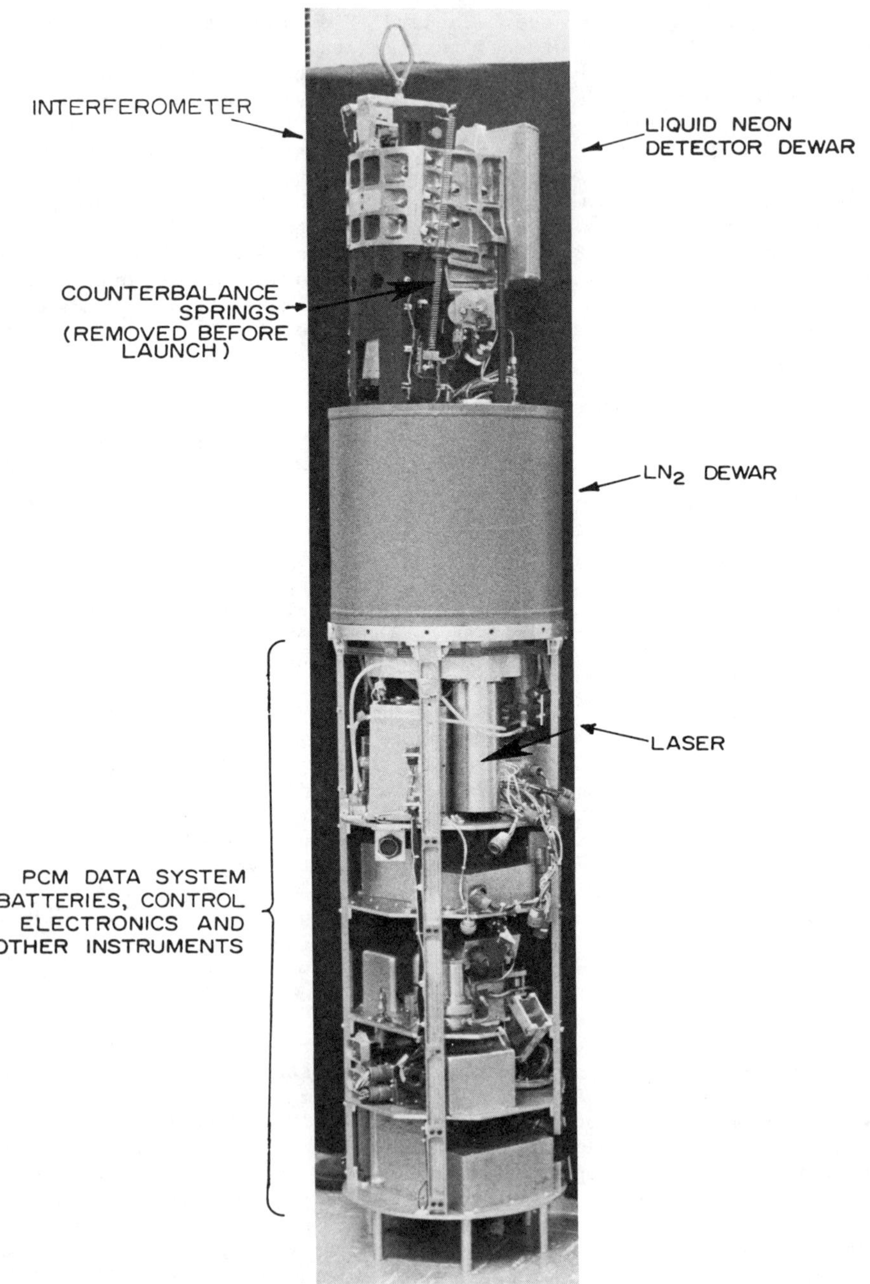

Figure 13. A photograph of the RBFWI integrated into the payload of an atmospheric sounding rocket.

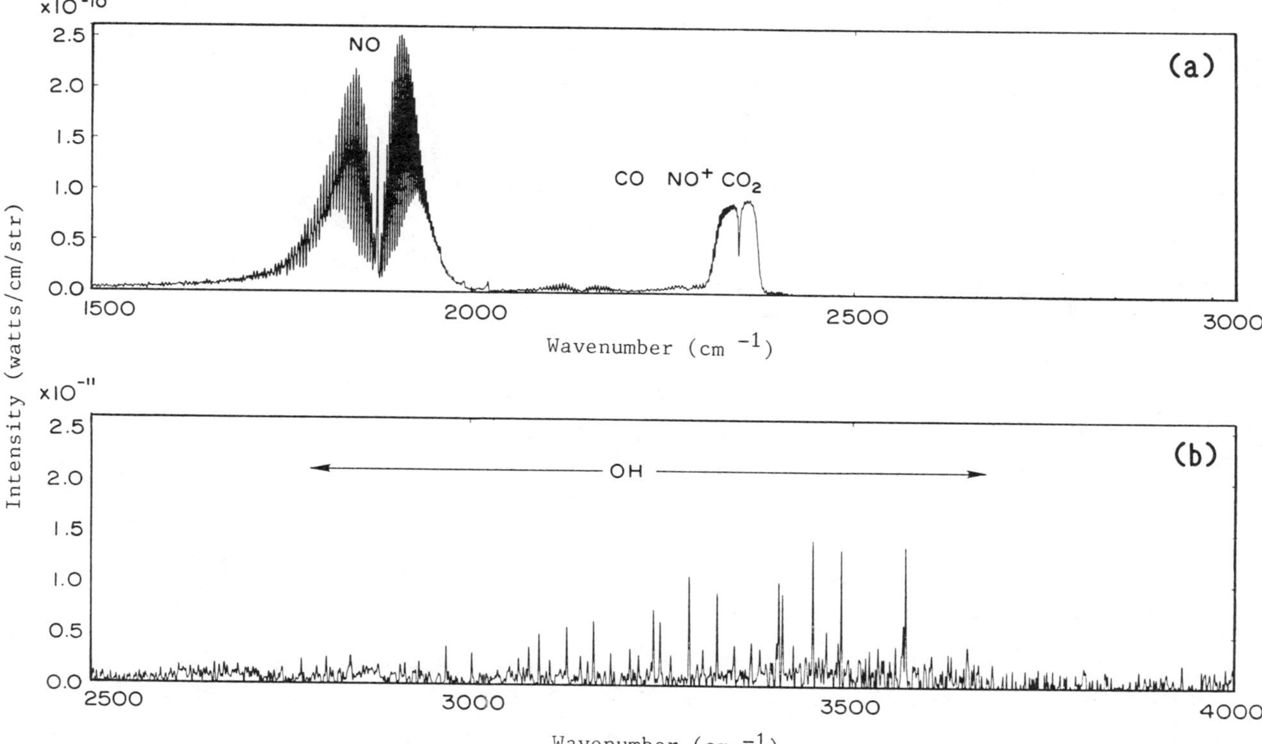

Figure 14. Spectra   obtained from data accumulated by the RBFWI during a recent flight of the instrument.   Atmospheric radiators are identified.

result of thermal radiation absorbed from earthshine. Species identified are NO, CO_2, CO, and OH. In addition to the normally existing airglow thermal emissions, there are other emissions from aurorally enhanced NO+, CO_2, and NO. By recording data throughout the flight, we can derive altitude profiles of species concentrations as well as temperature of gases.

With the exception of CO_2 all of the gases shown are optically thin which means that good altitude profiles can be obtained by basically differentiating the vertical profile data. The results therefore, are more accurate than limbscan or inversion measurement techniques since the measurements are made against a cold, well defined space background and rocket altitude is well documented throughout the flight. Also, rocket probes are especially effective when a specific atmospheric event such as an auroral form is to be investigated. The measurement can be activated when the event occurs.

Conclusions

Measurements of the physical phenomenon present in the world we live in often demonstrate far more complexity and variety than our preconceived models allow. When we first develop a new measurement technique, its use may open a new window on the world. During the first few years of measurements in a new field, discoveries often come rapidly with exciting expansion of our understanding. After a few years, however, the discoveries slow down, and when a field has been open for a century, new information and measurement capability come only at the expense of major effort.

The application of IR spectroscopy to atmospheric studies is nearly 100 years old. The past century has been a rich one, greatly broadening our understanding of the fragile system and expansive universe that forms the abode of man. We have reached the stage in this science where new applications and capabilities often come only at the end of long and difficult struggles with the simplest and best understood components. At the end of such a struggle, and often while enjoying the implications of the newly captured data, we are left to ponder if, had we known the difficulty of those seeming simple tasks from the beginning, would we have had the courage to begin? Thankfully, much of the development effort is now spread over a wide range of disciplines and the whole of the effort is not seen by one investigator or research group. This camouflage prompts us to ask the next question and allows the search to continue. The emergence of better spectral sources, infrared fiber optics, improved cryogenic containment and regular access to high altitude and space platforms should provide a fruitful future for this area of sensor development.

Acknowledgments

The CO_2 sensor development work reported here was perfomed under the auspices of the U.S. Department of Energy during which time the senior author was employed by the Lawrence Livermore National Laboratory, Livermore, California.   Optical and electrical design were provided by C.H. Gillespie and R.B. Kennedy, respectively. Much of the field testing was completed under the direction of Drs. S.B. Verma and D.E. Anderson at the Center for Agricultural Meteorology and Climatology, University of Nebraska, Lincoln, Nebraska.

The Field Widened interferometer spectrometer described herein was developed under a contract from the Air Force Geophysics Laboratory. We acknowledge the major contributions of Ralph Haycock, Alan Thurgood and Val King of Utah State University in the sensor development and of Ron Straka of AFGL for contract support. The authors especially wish to thank Glenn Allred for his role in editing this paper.

Literature Cited

1. Herzberg,   G., *Molecular spectra and molecular structure*, D. Van Nostrand Company, Inc., Princeton, 1950.

2. Elsasser, W., Mean absorption and equivalent absorption coefficient of a band spectrum, *Phys. Rev.*, *34*, 1938.

3. Goody,   R.M., *Atmospheric radiation*, Clarendon Press, Oxford, 1964.

4. Jarem,   J.M., Pierluissi, J.H., and Ng, W.W. , A transmittance model for atmospheric methane, *SPIE Vol 510, Infrared technology X*, 1984.

5. Abney, W. De W., Photography of the least refracted portion of the solar spectrum, *Phil. Trans.*, *171A*, 1880.

6. Abney, W.De W., Photography of the least refracted portion of the solar spectrum, *Phil. Trans.*, *177A*, 1886.

7. Langley, S.P., The distribution of energy in the normal solar spectrum, *Compt. rend.*, *93*, 140-3, 1881.

8. Angstrom, K., Contributions to the knowledge of the absorption of heat rays by the various parts of the atmosphere, *Wied. Ann. 39*, 267-95, 1890.

9. Langley, S.P., and Abbot, C.G., *Annals of the astrophysical observatory of the Smithsonian Institution, 1*, Smithsonian Institution, Washington, D.C. 1900.

10. Rubens, H., The dispersion of infrared radiation, *Wied. Ann., 45*, 238-61, 1892.

11. Julius, W.H., Bolometric research on the infrared spectrum, *Arch. Neerland. Sci.*, *22*, 310-82, 1888.

12. Julius, W.H., Bolometric research on absorption spectra, *Verhandl. Akad. Wetenschappen Amsterdam, 1*, 1-49, 1892.

13. Abbot, G.G., *Annals of the astrophysical observatory of the Smithsonian Institution, 2*, Smithsonian Institution, Washington, D.C., 1908.

14. Abbot, C.G., Radiometer measurements of stellar energy spectra, *Astrophys. J.*, *60*, 87-107, 1924.

15. Brackett, F.S., An examination of infrared spectrum of the sun 8900 to 9900 A, *Astrophys. J.*, *53*, 121-33, 1921.

16. Babcock, H.D., Photography of the infrared solar spectrum, Nature, 121, 830-1, 1928; C.A., 23, 1351.6.

17. Arrhenius, S., Heat absorption of CO_2 and its influence on the temperature of the earth's surface, Ann. Physik, 4, 690-705, 1901.

18. Bailey, C.R., and Lih, Kun-Hon, Infrared emission of carbon dioxide, Nature, 121, 941, 1928; C.A., 22, 2887.5.

19. Bailey, C.R., The Raman and infrared spectra of CO_2, Nature, 123, 410, 1929; C.A., 23, 3158; cf. C.A., 22, 2887.

20. von Bahr, Eva, The influence of pressure on the absorption of very long wave radiation in gases, Verhandl. deut. physik. Ges., 15, 673-7, 1913; C.A., 7, 3902; cf. C.A., 5, 24; C.A., 3, 2526.

21. von Hettner, G., The infrared absorption of water vapor, Ann. Physik, 55, 476-95, 1918.

22. Bailey, C.R., The infrared spectrum of water vapor, Trans. Faraday Soc., 26, 203-11, 1930; C.A., 24, 3437.8.

23. Stokes, G.M., Barnard, J.C., Pearson, E.W., Historical carbon dioxide: abundances derived from the Smithsonian spectrobolograms, TR011, Contract No. DE-AC06-76RL01830, Pacific Northwest Laboratory, Richland, WA., 1984.

24. Angstrom, A.K., The dependence of gaseous absorption, especially of carbon dioxide, upon the thickness, Ann. Physik, 6, 163-73, 1901.

25. von Bahr, Eva, The effect of pressure on the specific absorption of infrared radiation by gases, Ann. Physik, 29, 780-96, 1909; C.A., 3, 2526.

26. von Bahr, Eva, Pressure effect on absorption of infrared radiation in gases, Ann. Physik, 33, 585-97, 1910; C.A., 5, 24.

27. von Bahr, Eva, Influence of temperature on infrared absorption of gases, Ann. Physik, 38, 206-22, 1912; C.A., 6, 2352.9.

28. Pfund, A.H., An infrared spectrometer of large aperture, J. Optical Soc. Am., 14, 337-8, 1927.

29. Beutler, H., Influence of pressure and temperature upon the absorption and fluorescence of spectral lines, Astrophys. J., 89, 294, 1939; C.A., 33, 4518.7.

30. von Bahr, Eva, The effect of foreign gases on the absorption lines, Physik. Z., 12, 1167-9, 1911; C.A., 6, 1242.

31. von Helmholtz, R., Die Licht- und Warmestrahlung verbrennender Gase, Berlin, 1890.

32. Paschen, F., Bolometric research in grating spectra, Wied. Ann., 48, 272-306, 1893.

33. Paschen, F., The emission of gases, Wied. Ann., 51, 1-39, 1894; Wied. Ann., 52, 210-37, 1894.

34. Pfund, A.H., A universal spectrometer for the infrared, J. Optical Soc. Am., 22, 281, 1930.

35. Ballard, James W., Infrared sensitivity of cesium oxide photoelectric cells, J. Optical Soc. Am., 20, 618-23, 1930; C.A., 25, 641.8.

36. Strong, John, Transmission of gases from 20μ to 33μ, Phys. Rev., 37, 1003, 1931; C.A., 25, 4792.9.

37. Strong, John, Apparatus for spectrographic studies in the intermediate infrared, Rev. Sci. Instruments, 3, 810-21, 1932, C.A., 27, 662.8.

38. Barnes, R.B., Suggestions concerning infrared prism spectroscopy, J. Optical Soc. Am., 28, 140-3, 1938.

39. Barnes, R.B., and Bonner, L.G., Filters for the infrared, J. Optical Soc. Am., 26, 428-33, 1936; C.A., 31, 8382.3.

40. Dingle, H., and Pryce, A.W., Estimation of small quantities of CO_2 in air by the absorption of infrared radiation, Proc. Roy. Soc. (London), B129, 468-74, 1940.

41. Brattain, R.R., and Beeck, Otto, Rapid precision gas analysis by infrared absorption, Bull. Am. Phys. Soc., 16, 9, 1941.

42. Brattain, R.R., and Beeck, Otto, A rapid infrared analytical method for hydrocarbon mixtures and a routine spectrophotometer for plant control, J. Applied Phys., 13, 699-705, 1942; C.A., 37, 312.8.

43. Cosden, T.H., Curcio, J.A., Dowling, J.A., Gott, C.O., Garcia, G.H., Hanley, S.T., Hought, K.M., Horton, R.F., and Trusty, G.L., Optical Radiation Branch, Naval Research Laboratory, and Agambar, W.L., Potomac Research Inc., Atmospheric transmission measurement program report for 1 July through 30 September 1976 (TQ1976), NRL Report 3104, Naval Research Laboratory, Washington, D.C., 1 September 1977.

44. Dowling, J.A., Haught, K.M., Horton, R.F., Trusty, G.L., Curcio, J.A., Cosden, T.H1, Hanley, S.T., and Gott, C.O., Optical Radiation Branch, Naval Research Laboratory, and Agambar, W.L., Potomac Research In,., Atmospheric extinction measurements at Nd-YAG and DF laser wavelengths performed in conjunction with the JAN propagation tests, June-September 1975, NRL Report 8058, Naval Research Laboratory, Washington, D.C., 10 February 1978.

45. White, J.U., Long optical paths of large aperture, J. Opt. Soc. Am., 32, 285-88, 1942.

46. Lavery, D.S., Chapter 7, Environmental Science, in Infrared and Raman Spectroscopy, Part 1, Practical Spectroscopy, Marcel Dekker, Inc., New York, 1977.

47. Bingham, G.E., Shinn, J.H., and Cedarwall, R.T., Feasibility of large area CO_2 flux measurements: A progress report on LLNL sensor and methods development efforts, UCID-19057, Lawrence Livermore National Laboratory, Livermore, 1981.

48. Greenhut, G.K., and Bean, B.R., Aircraft measurements of boundary-layer turbulance over the central equatorial pacific ocean, Boundary Layer Meteorology, 20, 221-41, 1981.

49. Verma, S.B., and Rosenberg, N.J., Further measurements of carbon dioxide concentration and flux in a large agricultural region of the great plains of North America, J. Geophys. Res., (in press).
50. Bingham, G.E., Gillespie, C.H., McQuaid, J.H., Lawrence Livermore National Laboratory, and Dooley, D.F., Molectron Corp., A miniature, battery-powered, pyroelectric detector-based differential infrared absorption sensor for ambient concentrations of carbon dioxide, Ferroelectrics, 34, 15-19, 1981.

51. McClatchey, R.A., and Selby, J.E.A., Atmospheric attenuation of laser radiation from 0.76 to 31.25 µm, AFCRL-TR-74-0003, Air Force Cambridge Research Laboratories, Bedford, 3 January 1974.

52. McClatchey, R.A., and D'Agati, A.P., Atmospheric transmission of laser radiation: Computer code LASER, AFGL-TR-78-0029, Air Force Geophysics Laboratory, Hanscom AFB, 31 January 1978.

53. White, K.O., Watkins, W.R., Bruce, C.W., Meredith, R.E., and Smith, F.G., Water vapor continuum absorption in the

3.5 µm to 4.0 µm region, ASL-TR-0004, Atmospheric Sciences Laboratory, White Sands Missile Range, 1978.

54. Anderson, D.E., Verma, S.B., and Rosenberg, N.J., Eddy correlation measurements of CO_2, latent heat, and sensible heat fluxes over a crop surface, Boundary Layer Meteorology, 29, 263-72, 1984.

55. Hanst, P.L., Pollution: Trace gas analysis, in Fourier Transform Infrared Spectroscopy, Applications to Chemical Systems, 2, 79, edited by Ferrand, J.R., and Basile, L.J., Academic Press, New York, 1979.

56. Haught, K.M., High resolution atmospheric transmission spectra from 5 to 3 µm, NRL Report 8297, Naval Research Laboratory, Washington, D.C., 9 March 1979.

57. Herget, W.F., Measurement of gaseous pollutants using a mobile Fourier transform infrared (FTIR) system, SPIE 289, 1981; Fourier Transform Spectroscopy, 449, Society for Photo-Optical Instrumentation Engineers, 1981.

58. Baker, D.J., Steed, A.J., Ware, G.A., Offermann, D., Lange, G., Lauche, H., Ground-based atmospheric infrared and visible emission measurements, J. Atmos. & Terr. Phy., 47, 133-45, 1985.

59. Huppi, E.R., Rogers, J.W., and Stair, A.T., Jr., Aircraft observations of the infrared emission of the atmosphere in the 700 - 2800 cm^{-1} region, Applied Optics, 13, 1466, 1974.

60. Bracewell, R.N., The Fourier Transform and its Applications, McGraw Hill, New York, 1978.

61. Baker, D., Steed, A., and Stair, A.T., Jr., Developments of infrared interferometry for upper atmospheric emission studies, Applied Optics, 20, 1734, 1981.

62. Baker, D.J., Field-Widened Interferometers for Fourier Spectroscopy, in Spectrometric Techniques, 71, edited by Vanasse, G.A., Academic Press, New York, 1977.

63. Haycock, R.H., A Cryogenically-Cooled Field-Widened Interferometer for Rocketborne Near-Infrared Atmospheric Studies, paper presented at the 11th Annual Meeting of Upper Atmospheric Studies by Optical Methods, Lindau/HARTZ, West Germany, 30 August - 2 September 1983.

RECEIVED December 12, 1985

Chemically Modified Electrode Sensors

Richard A. Durst and Elmo A. Blubaugh

Organic Analytical Research Division, Center for Analytical Chemistry, National Bureau of Standards, Gaithersburg, MD 20899

Electroanalytical sensors based on amperometric measurements at chemically modified electrodes are in the early stages of development. The modes of modification can take many forms, but the most common approach at the present time is the immobilization of ions and molecules in polymer films which are applied to bare metal, semi-conductor, and carbon electrodes. Such surface-modified electrodes exhibit unique electrochemical behavior which has been exploited for a variety of applications.

This review gives a brief summary of the types of chemically modified electrodes, their fabrication, and some examples of their uses. One especially promising area of application is that of selective chemical analysis. In general, the approach used is to attach to the electrode surface electrochemically reactive molecules which have electrocatalytic activity toward specific substrates or analytes. In addition, the incorporation of biochemical systems should greatly extend the usefulness of these devices for analytical purposes.

In recent years, considerable effort has gone into the development of a new class of electrochemical devices called chemically modified electrodes. While conventional electrodes are typified by generally nonspecific electrochemical behavior, i.e., they serve primarily as sites for heterogeneous electron transfer, the redox (reduction-oxidation) characteristics of chemically modified electrodes may be tailored to enhance desired redox processes over others. Thus, the chemical modification of an electrode surface can lead to a wide variety of effects including the retardation or acceleration of electrochemical reaction rates, protection of electrodes, electro-optical phenomena, and enhancement of electroanalytical specificity and sensitivity. As a result of the importance of these effects, a relatively new field of research has developed in which the electrochemical behavior of the attached substances is being studied as well as the influence of these modified surfaces on the catalysis and inhibition of a variety of electrochemical processes. The functionalization of an electrode surface may ultimately benefit a diverse array of processes such as electro-organic synthesis, electrocatalysis, semiconductor stabilization, photosensitization, photoelectrochemical energy conversion, electrochromism, and selective chemical analysis.

To a large extent, the discovery and application of adsorption phenomena for the modification of electrode surfaces has been an empirical process with few highly systematic or fundamental studies being employed until recent years. For example, successful efforts to quantitate the adsorption phenomena at electrodes have recently been published [1-3]. These efforts utilized both double potential step chronocoulometry and thin-layer spectroelectrochemistry to characterize the deposition of the product of an electrochemical reaction. For redox systems in which there is product deposition, the mathematical treatment described permits the calculation of various thermodynamic and transport properties. Of more recent origin is the approach whereby modifiers are selected on the basis of known and desired properties and deliberately immobilized on an electrode surface to convert the properties of the surface from those of the electrode material to those of the immobilized substance.

Surface Immobilization Techniques

There are three principal approaches used for the immobilization of electroactive substances onto surfaces: chemisorption, covalent bonding, and film deposition.

In chemisorption, the electrochemically reactive material is strongly (and to a large extent irreversibly) adsorbed onto the electrode surface. Lane and Hubbard [4] were among the first to use this approach when they chemisorbed quinone-bearing olefins on platinum electrodes and demonstrated a pronounced effect of the adsorbed molecules on electrochemical reactions at the metal surface.

Covalent attachment schemes have been developed for both mono- and multi-molecular layers of electroactive sites on semiconductor, metal oxide, and carbon electrodes. Since its introduction by Murray and co-workers [5], covalent immobilization by organosilanes has become the most widely used technique for preparing modified electrodes. Other covalent linking agents have been used including cyanuric chloride [6], thionyl chloride [7], and acetyl chloride [8]. While it is sometimes possible to attach the electroactive substance directly to the electrode surface, covalent linking agents are much more commonly used because they allow the use of a greater variety of terminal functional groups than could be produced on the electrode surface alone. Covalent attachment is also usually more tolerant of exposure to different types of solvents.

Film deposition refers to the preparation of polymer (organic, organometallic, and metal coordination) films which contain the equivalent of many monomolecular layers of electroactive sites. As many as 10^5 monolayer-equivalents may be present [9]. The polymer film is held on the electrode surface by a combination of chemisorptive and solubility effects. Since the polymer film bonding is rather non-specific, this approach can be used to modify almost any type of

electrode material. Polymer films have been applied to electrode
surfaces by dip and spin coating, bonding via covalent linking
agents, electrochemical precipitation and polymerization, adsorption
from solution, and plasma discharge polymerization [10].

One of the major deterrents to the successful application of
electroanalytical sensors has been the lack of long-term stability of
the polymer films. At least three factors effect the stability of
these amperometric sensors. These factors are the mode of polymer
film attachment to the electrode surface (adsorption vs. covalent
bonding), solubility of the film in the contacting solution, and
finally, the mode of attachment of the catalyst in the polymer film
(electrostatic vs. covalent).

The stability of a covalently attached catalyst will be
significantly greater than a catalyst bound to the polymer film via
electrostatic forces. This is supported by experiments with ion
exchange films, where the electrostatistically bound electroactive
species can reversible exchange with the cation/anion in the contacting solutions [11-13]. This behavior does not occur with a
covalently-linked species.

To increase the stability and allow greater resistance to
polymer dissolution and degradation in swelling solvents, crosslinking procedures have been used. This has been accomplished in the
past by electrochemical and radiation initiated crosslinking [14-17].
Highly crosslinked polymer films have been formed by means of glow-
discharge polymerization [18]. A copolymer of vinylferrocene and
δ-(methacrylpropyltrimethoxy silane) was shown to form an insoluble,
yet electroactive, film on Pt via thermally induced siloxane crosslinking [19]. Recent work has demonstrated that the thermally and
photochemically initiated crosslinking of a commercially available
copolymer, poly(vinylpyridine/styrene) is feasible [20]. The crosslinking was thermally initiated by benzoly peroxide which undergroes
free-radical reaction with both the copolymer and a crosslinking
agent (triallyltrimellitate) which then couples to generate cross-
linked polymer macromolecules. The photochemical crosslinking
involves photo-excitation of a dye sensitizer (3-ketocoumarin) which,
when used in conjunction with a consensitizer (e.g., ethyl-p-d-
imethylaminobenzoate), acts to initiate the crosslinking reaction.
The stability of the film and film/electrode interface was evaluated
as a function of percent crosslinking agent added. With no crosslinking agent, these polymer films dissolved readily when placed in a
stirred methanol solution. With about two mole-percent of crosslinking agent, the film remained physically intact, with about thirty
percent weight loss due to leaching of uncrosslinked polymeric
materials.

The concept of using the functional groups of electrode surfaces
themselves to attach reagents by means of covalent bonding offers
synthetic diversity and has been developed for mono- and multi-layer
modifications. The electrode surface can be activated by reagents
such as organosilanes [5] which can be used to covalently bond electroactive species to the activated electrode surface. Recently,
thermally induced free-radical polymerization reactions at the surfaces of silica gel have been demonstrated [21]. This procedure has
been applied to Pt and carbon electrode surfaces. These thermally
initiated polymer macromolecules have the surface of the electrode as
one of their terminal groups. Preliminary studies indicate that the

covalent attachment gives a very stable electrode/polymer-film interface. The logical extension of this technique for the generation of a variety of copolymer and block copolymers can be imagined.

There are several reasons for the appeal of polymer modification: immobilization is technically easier than working with monolayers; the films are generally more stable; and because of the multiple layers redox sites, the electrochemical responses are larger. Questions remain, however, as to how the electrochemical reaction of multimolecular layers of electroactive sites in a polymer matrix occur, e.g., mass transport and electron transfer processes by which the multilayers exchange electrons with the electrode and with reactive molecules in the contacting solution [9].

Mediated Electrocatalysis

Electrocatalysis at a modified electrode is usually an electron transfer reaction, mediated by an immobilized redox couple, between the electrode and some solution substrate which proceeds at a lower overpotential than would otherwise occur at the bare electrode. This type of mediated electrocatalysis process can be represented by the scheme:

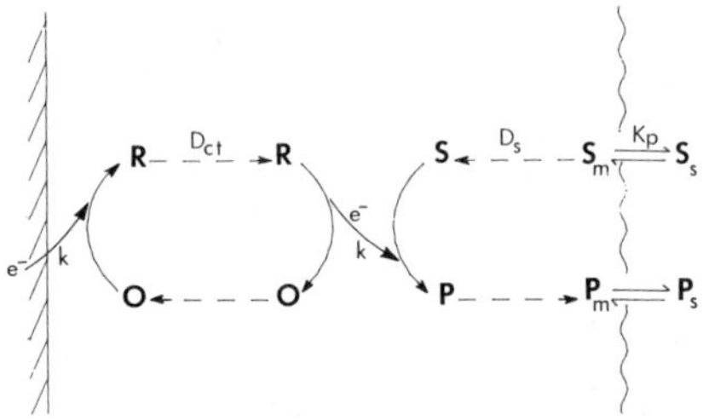

Figure 1.   Mediated   electrocatalysis   at   a   polymer-modified
            electrode: charge and mass transport processes.

In this scheme, the substrate, S, which is irreversibly reduced (or electroinactive) at the bare electrode is transported across the polymer film-solution interface (partition coefficient, K_p) and diffuses into the polymer film (diffusion coefficient, D_s). The electrocatalyst or mediator, R/O, undergoes rapid heterogeneous electron transfer at the electrode surface and charge is propagated through the polymer film at a rate given by the charge-transport diffusion coefficient, D_{ct}. This coefficient may be composed of one or more components including a self-exchange reaction between electrocatalytic sites ("electron hopping"), site diffusion (polymer-chain motion), counterion diffusion and electromigration. In the bulk of the film, the substrate and mediator undergo a homogeneous electron transfer regenerating the oxidized form of the mediator and forming the reaction product, P. If the rate of electron transfer between the electrode and mediator is faster than the rate of the mediator-substrate reaction, then the substrate will undergo electrolysis at a potential near the surface formal potential of the mediator couple. The catalytic redox potential also depends on the rate constant for the mediator-substrate reaction. This type of electrocatalytic scheme has been the focus of numerous modified electrode studies.

This approach can be further extended to photoelectrochemical reactions at modified semiconductor electrodes. In such cases the immobilized substance(s) may serve several functions: mediation of the redox process, photosensitization of the semiconductor, and photocorrosion protection.

Techniques for Modified Electrode Study

To characterize the properties of molecules and polymer films attached to an electrode surface, a wide variety of methods have been used to measure the electroactivity, chemical reactivity, and surface structure of the electrode-immobilized materials [9]. These methods have been primarily electrochemical and spectral as indicated in Table I. Suffice it to say that a multidisciplinary approach is needed to adequately characterize chemically modified electrodes combining electrochemical methods with surface analysis techniques and a variety of other chemical and physical approaches.

Polymer-Film Electrodes

Chemical modification of electrode surfaces by polymer films offers the advantages of inherent chemical and physical stability, incorporation of large numbers of electroactive sites, and relatively facile electron transport across the film. Since the polymer films usually contain the equivalent of one to more than 10^5 monolayers of electroactive sites, the resulting electrochemical responses are generally larger and thus more easily observed than those of immobilized monomolecular layers. Also, the concentration of sites in the film can be as high as 5 mol/L and may influence the reactivity of the sites because their solvent and ionic environments differ considerably from dilute homogeneous solutions [9].

In most polymer films, the electroactivity of the redox centers depends on the ionic conductivity of the film. This is usually achieved either by the penetration of supporting electrolyte ions through pores in the film or by the presence in the film of numerous fixed charge sites plus mobile counterions. In many cases, solvent permeation into the polymer facilitates ionic penetration and mobility. If the polymer film possesses electronic, rather than ionic, conductivity, the electron-transfer reactions will most likely occur at the polymer/solution interface and the advantage of a three-dimensional reaction zone will be reduced.

Most of the redox centers in a polymer film cannot rapidly come into direct contact with the electrode surface. The widely accepted mechanism proposed for electron transport is one in which the electroactive sites become oxidized or reduced by a succession of electron-transfer self-exchange reactions between neighboring redox sites [22]. However, control of the overall rate is a more complex problem. To maintain electroneutrality within the film, a flow of counterions and associated solvent is necessary during electron transport. There is also motion of the polymer chains and the attached redox centers which provides an additional diffusive process for transport. The rate-determining step in the electron site-site hopping is still in question and is likely to be different in different materials.

Table I. Techniques for Modified Electrode Study

Electrochemical Methods

- cyclic voltammetry
- differential pulse voltammetry
- chronamperometry and chronocoulometry
- chronopotentiometry
- coulostatics
- AC voltammetry
- rotated disk and ring-disk voltammetry

Spectral Methods

- x-ray photoelectron and Auger electron spectroscopy
- transmission spectroscopy at optically transparent electrodes
- reflectance spectroscopy
- photoacoustic and photothermal spectroscopy
- Raman spectroscopy
- inelastic electron tunneling spectroscopy
- secondary ion mass spectrometry
- second harmonic generation
- ellipsometry
- ultrasoft x-ray fluorescence spectroscopy

Other Methods

- film thickness and roughness
- scanning electron microscopy
- elemental analysis
- contact angles
- radiolabels

Analytical Applications

The relative fragility and preparative difficulty associated with monolayer-modified electrode surfaces hampered significant analytical progress for some time, and it was not until polymer-film electrodes were developed that the utility of modified electrodes in analysis could be demonstrated.

One approach utilizes the polymer-modified surface as a preconcentrating surface in which the analyte or some reaction product is collected and concentrated by reactive groups attached to the electrode.   The preconcentrated analyte is then measured electrochemically.   Ideally, the collection process will be selective for the analyte species of interest.   If not, the analyte must be electroanalytically discriminated from other collected species.   The capacity of the polymer film on the modified electrode should be sufficient to prevent saturation by the analyte, and the electrochemical response during measurement should provide good sensitivity to the collected analyte [9].

Electrostatic binding [11] may provide another very useful approach to preconcentration analysis.   Enhancement of the redox ion concentration in the ion-exchange polymer volume should permit very sensitive analysis when combined with an appropriate electroanalytical method [12,13].   However, the sensitivity of the ion-exchange equilibrium to the sample solution electrolyte composition and concentration and the necessity of having a multiply charged analyte ion may limit the usefulness of the electrostatic binding approach.

Bioanalytical Sensors

Many biological redox systems undergo very slow heterogeneous electron transfer at electrodes and consequently exhibit quasi-reversible or irreversible electrochemical behavior.   One approach to circumvent this problem is to add to the solution an electroactive species, called a mediator, which acts as an electron shuttle to provide redox coupling between the electrode and the redox center in the biological compound [23].   In principle, the enhancement of the electrochemical behavior of the biocomponent can be accomplished by attachment of the mediator directly to the electrode surface. This approach has been demonstrated, for example, by Wrighton and co-workers for the electrocatalysis of cytochrome c at platinum electrodes modified with various organosilane polymers [24,25].   Schemes using immobilized-mediator electrodes will probably be extended to a variety of previously intractable biological redox reactions, both for analytical purposes and for fundamental studies of the biological couples.

Electrode modification by the attachment of various types of biocomponents holds considerable promise as a novel approach for electrochemical (potentiometric, conductometric, and amperometric) biosensors.   Potentiometric sensors based on coupled biochemical processes have already demonstrated considerable analytical success [26,27].   More recently, amperometric biosensors have received increasing attention [27,28] partially as a result of advances made in the chemical modification of electrode surfaces.   Systems based on

enzymatic and immunological reactions are the most promising as a
consequence of their high specificity and the possibilities for chemical signal amplification.   In addition, microorganisms, plant
leaves, and plant and animal tissues have been successfully used in
potentiometric and oxygen electrode-based biosensors [26,28-30].

Chemical amplification of the biosensor signal can be achieved
by a variety of ingenious schemes.   Perhaps the most innovative
approach uses liposome-encapsulated electroactive marker compounds
which are released via complement-mediated lysis (rupture) of the
sensitized liposomes (lipid vesicles) [31,32] or red blood cell (RBC)
ghosts [31].   The electroactive marker, such as quaternary ammonium
ion in the potentiometric approach [31,33] or glucose using the
amperometric glucose oxidase/oxygen sensor [32], is loaded into the
liposomes or RBC ghosts which have been sensitized by the incorporation of antigens or haptens for the recognition of specific immuno-
agents.   In the presence of complement, the immunological reaction
(formation of an antigen-antibody complex) initiates a series of
enzymatic reactions which activates complement and results in the
lysis of the liposome thereby releasing the encapsulated markers to
be sensed by the electrochemical transducer.   In a recently reported
non-electrochemical scheme [34], liposomal lysis was produced by the
competitive action of hapten conjugates with cytolytic agents. This
homogeneous liposome-mediated immunoassay required neither complement
nor sensitized liposomes.   Furthermore, enzyme markers were encapsulated and, when released, the kinetic rate of their activity was
monitored photometrically.

Since none of the liposomal immunoassay approaches described in
the scientific literature thus far took advantage of surface immobilization techniques, one could envision a double-amplification biosensor in which surface modification plays an important role [35].
For example, consider a dehydrogenase enzyme marker system which
requires an electroactive cofactor such as NAD^+.   In the enzymatic
reaction scheme:

$$S + NAD^+ \xrightarrow{\text{SDH}} P + NADH + H^+$$
$$\underset{-e^-}{\longleftarrow}$$

substrate S is converted to product P via liposome-released substrate
dehydrogenase SDH and the cofactor NAD^+.   The NADH which is produced
can be electrochemically reoxidized to NAD^+ and the oxidation current
is a measure of the extent of enzymatic reaction.   If we further consider the membrane of the enzyme-containing liposome to be sensitized
by the incorporation of an antibody Y, then in the presence of the
analyte (antigen) A, a complex A_2Y forms which activates complement C
and results in liposome lysis.   The overall double-amplification process is illustrated in Figure 2.   In this example, the liposomes and
cofactor molecules are coimmobilized onto the electrode surface and
the bathing solution contains complement and a high concentration of
the substrate.   It should be noted that this is a generic approach in
which only the immunological reaction, i.e., the liposome sensitizer,
is changed according to the desired analyte recognition.

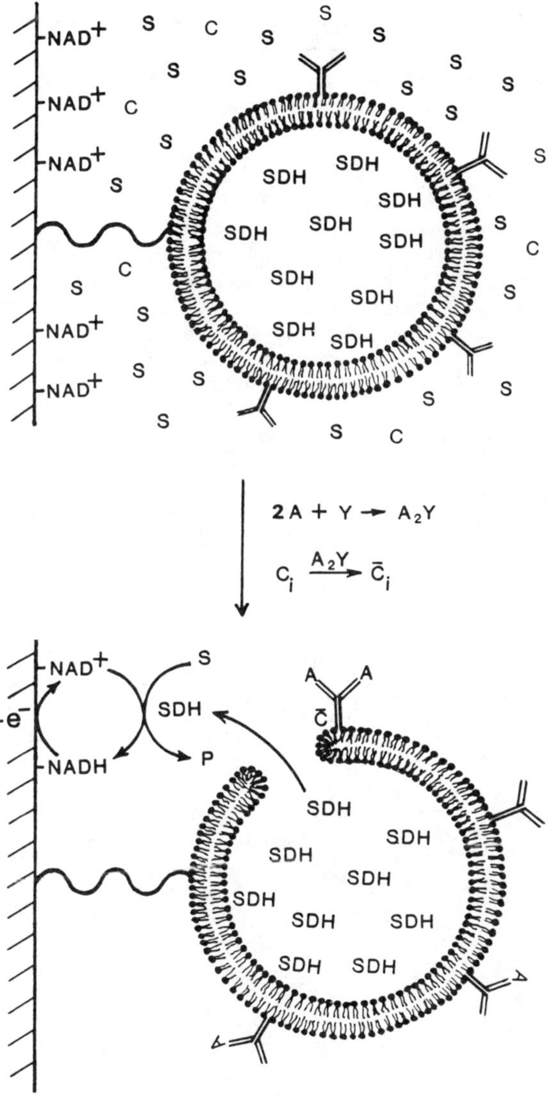

Figure 2. Immune lysis of a sensitized liposome. Immobilized antibody-sensitized liposome undergoes complement-induced lysis. Released enzyme catalyzes substrate-product reaction with concomitant reduction of immobilized cofactor. The cofactor is electrochemically reoxidized and the current is related to the analyte concentration. (See text for discussion.) Symbols: A, analyte (antigen); Y, antibody; S, substrate; SDH, substrate dehydrogenase; P, product; C, complement, $\bar{C}$, activated complement; NAD^+, nicotinamide adenine dinucleotide; NADH, reduced NAD^+.

The possibilities of these approaches, considering the large numbers of potentially useful biochemical reactions, are enormous. It is expected that significant advances will occur in the field of biosensor development in the near future, especially when newer biotechnological processes and chemical modification approaches are adapted to sensor development.

Conclusion

This brief review attempts to summarize the salient features of chemically modified electrodes, and, of necessity, does not address many of the theoretical and practical concepts in any real detail. It is clear, however, that this field will continue to grow rapidly in the future to provide electrodes for a variety of purposes including electrocatalysis, electrochromic displays, surface corrosion protection, electrosynthesis, photosensitization, and selective chemical concentration and analysis. But before many of these applications are realized, numerous unanswered questions concerning surface orientation, bonding, electron-transfer processes, mass-transport phenomena and non-ideal redox behavior must be addressed. This is a very challenging area of research, and the potential for important contributions, both fundamental and applied, is extremely high.

Literature Cited

1. Burke, R. T.; Blubaugh, E. A.; Yap, W. T.; Durst, R. A.   J. Electroanal. Chem. 1983, 159, 28-293.
2. Yap, W. T.; Blubaugh, E. A.; Durst, R. A.; Burke, R. T.   J. Electroanal. Chem. 1984, 160, 73-78.
3. Burke, R. T.; Blubaugh, E. A.; Yap, W. T.; Durst, R. A.   J. Electroanal. Chem. 1984, 177, 77-88.
4. Lane, R. F.; Hubbard, A. J.   J. Phys. Chem. 1973, 77, 1401; ibid. 1973, 77, 144.
5. Moses, P. R.; Wier, L.; Murray, R. W.   Anal. Chem. 1975, 47, 1882.
6. Lin, A. W. C.; Yeh, P.; Yacynych, A. M.; Kuwana, T.   J. Electroanal. Chem. 1977, 84, 411; A. M. Yacynych and T. Kuwana, Anal. Chem. 1978, 50, 640; M. J. Dautartas; J. F. Evans; and T. Kuwana; ibid. 1979, 51, 104.
7. Watkins, B. J.; Behling, J. R.; Kariv, E.; Miller, L. L.   J. Am. Chem. Soc. 1975, 97, 3549
8. Rocklin, R. D.; Murray, R. W.   J. Electroanal. Chem. 1979, 100, 271.
9. Murray, R. W.   Chemically Modified Electrodes, in Electroanalytical Chemistry, A. J. Bard, Ed., 1983, Vol. 13, Marcel Dekker, NY.
10. Murray, R. W.   Acc. Chem. Res. 1980, 13, 135.
11. Oyama, N.; Anson, F. C.   J. Electrochem. Soc. 1980, 127, 247.
12. Szentirmay, M. N.; Martin, C. R.   Anal. Chem. 1984, 56, 1898.
13. Nagy, G.; Gerhardt, G.; Oke, A.; Adams; R. N.; Moore, R. B.; Szentirmay, M. N.; Martin, C. R.   J. Electroanal. Chem. (In Press).
14. Shaw, B. R.; Haight, G. P., Jr.; Faulkner, L. R.   J. Electroanal. Chem., 1982, 140, 147.

15. Kaufman, F. B.; Schroeder, A. H.; Patel, V. V.; Nichols, K. H. J. Electroanal. Chem. 1982, 132.

16. Abruna, H. D.; Denisevich, P.; Umara, M.; Meyer, T. J.; Murray, R. W. J. Am. Chem. Soc. 1981, 103, 1.

17. DeCastro, E. S.; Smith, D. A.; Mark, J. E.; Heineman, W. R. J. Electroanal. Chem. 1982, 138, 197.

18. Doblhofer, K.; Nölte, D.; Ulstrup, J. Ber. Bunsenges. Phys. Chem. 1978, 82, 403.

19. Nakahama, S.; Murray, R. W. J. Electroanal. Chem. 1983, 158, 303.

20. Blubaugh, E. A.; Bushong, W. C.; Shupack, S. I.; Durst, R. A. J. Electroanal. Chem., Submitted for Publication.

21. Fery, V. N.; Laible, R.; Hamann, K. Die Angewandte Makromolekulare Chemie 1973, 34, 81.

22. Kaufman, F. B.; Schroeder, H.; Engler, E. M.; Kramer, S. R.; Chambers, J. Q. J. Am. Chem. Soc. 1980, 102, 483.

23. Fultz, M. L.; Durst, R. A. Anal. Chim. Acta 1982, 140, 1.

24. Lewis, N. S.; Wrighton, M. S. Science 1981, 211, 944.

25. Chao, S.; Robbins, J. L.; Wrighton, M. S. J. Am. Chem. Soc. 1983, 105, 181.

26. Rechnitz, G. A. Anal. Chem. 1982, 54, 1194A-1200A.

27. Czaban, J. D. Anal. Chem. 1985, 57, 345A-356A.

28. Karube, I.; Suzuki, S. Ion-Selective Electrode Rev. 1984, 6, 15-58.

29. Kobos, R. K. TRAC 2, 1983, 154-157.

30. Thompson, M.; Krull, J. J. TRAC 3, 1984, 173-178.

31. Shiba, K.; Umezawa, Y.; Watanabe, T.; Ogawa, S.; Fujiwara, S. Anal. Chem. 1980, 52, 1610-1613.

32. Umezawa, Y.; Sofue, S.; Takamoto, Y. Anal. Lett. 1982, 15, 135-146.

33. D'Orazio, P.; Rechnitz, G. A. Anal. Chem. 1977, 49, 2083-2086.

34. Litchfield, W. J.; Freytag, J. W.; Adamich, M. Clin. Chem. 1984, 30, 1441-1445.

35. Durst, R. A. "Chemically Modified Electrode Sensors for Biocomponents," Presented at the Fourth Scientific Session on Ion-Selective Electrodes, Matrafüred, Hungary, October 10, 1984.

RECEIVED October 31, 1985

15

Coated-Wire Ion-Selective Electrodes

L. Cunningham and H. Freiser

Strategic Metals Recovery Research Facility, Department of Chemistry, University of Arizona, Tucson, AZ 85721

Coated wire ion selective electrodes were first developed in 1971, and comprise a film of PVC or other suitable polymeric matrix substrate containing a dissolved electro-active species, coated on a conducting substrate (generally a metal, although any material whose conductivity is substantially higher than that of the film can be used.) Electrodes of this sort are simple, inexpensive, durable and capable of reliable response in the concentration range of 10^{-1} M to 10^{-6} M for a wide variety of both organic and inorganic cations and anions. The principles on which these electrodes are based, as well as their application to a variety of analytical problems, will be discussed.

When faced with the problem of a trace level determination of an inorganic ion, the analyst often considers use of an ion-selective electrode (ISE) for the species of interest. This approach is advantageous because of the speed and ease of ISE procedures in which little or no sample is required. Further, they possess wide dynamic ranges, and are relatively low in cost. These characteristics have inevitably led to sensors for several ionic species, and the list of available electrodes has grown substantially over the past two decades. In most cases, the traditional barrel configuration has been utilized. However, the large size of this type of ISE along with the requirement that it be used in a nearly upright position renders it somewhat cumbersome to use and unnecessarily expensive.

In our laboratory, these disadvantages have been overcome with the development of the coated wire electrode (CWE). This sensor, having response characteristics equal to and occasionally better than conventional types, is only 1-2 mm in diameter (further size reduction can be easily achieved), can be used at any angle, and costs only a few pennies to make. Indeed, they can be considered "disposable", though with proper handling lifetimes of over six

months have been realized. During the course of CWE investigations
here, the list of analyte species has been lengthened to include not
only most common inorganic ions of interest, but also organic species which are anionic or cationic under appropriate solution conditions (see Table 1). This paper will review this research from the
inception of the CWE to the present.

The first electrode of this type was based on the Ca^{++}-didecyl-
phosphate/dioctylphenyl phosphonate system ($\underline{1}$). An effective Ca^{++}
selective CWE resulted when a 6:1 mixture of 5% PVC in cyclohexanone
and 0.1 M. Ca didecylphosphate in dioctylphenylphosphonate was dried
on the end of a platinum wire. Favorable comparison of this electrode response characteristics against the commercial counterpart
(Table 2) encouraged further studies with other membrane components.
The Ca^{++} electrode response relied upon the complexation of aqueous
Ca^{++} by didecylphosphate dispersed in the organic, or membrane,
phase. In a similar manner, incorporation of methyltricaprylam-
monium (Aliquat 336S) salts in polymer membranes produced CWEs for
their respective anions ($\underline{2}$). A 60%(v/v) solution of Aliquat 336S in
decanol was first converted to the desired anionic form via shaking
with 1 M aqueous solution of the appropriate Na^{+} salt. A 10:1
mixture of PVC in cyclohexanone and this decanol solution was then
used to coat copper wires by repeated dipping and drying until a
small bead completely encapsulated their ends. Listed in Tables 3
and 4 are anionic species for which Aliquat based electrodes were
prepared. Applications ranged from critical micelle determination
using laurylsulfonate sensors ($\underline{3}$), analyses of atmospheric NO_x
pollutants with nitrate electrodes ($\underline{4}$), and assay of phenobarbital
tablets using a phenobarbital anion CWE ($\underline{5}$).

In many cases, poly(methyl methacrylate) or epoxy resin could
be substituted for PVC with retention response. This, along with
absence of the traditional internal reference electrode, raised
fundamental questions surrounding the charge conduction mechanism
occurring in the membrane and at the polymer-substrate interface.
Calculation of activation energies from the temperature dependence
of conduction suggested that an electronic mechanism was operative,
such as that observed in organic semiconductors ($\underline{6}$). Later studies
of the pressure dependence of conduction gave strong evidence for
ionic conduction because much larger activation volumes than could
be expected from an electronic mechanism were obtained ($\underline{7}$, $\underline{8}$). As
such, the existence of some redox couple at the substrate-polymer
interface probably functions as an "internal reference". This
hypothesis is further reinforced when one considers that conditional
standard potentials shift by significant and reproducible amounts
from one type of metal substrate to another.

Our attention next turned to the development of cation selective electrodes in order to develop methods for protonated alkylammonium ions. Initial studies in this area were aimed at improving
selectivity among similarly charged cations by utilizing a mobile
exchange site, facilitating membrane response to changing counter
ions ($\underline{9}$). These membranes were comprised of dinonylnaphthalene
sulfonic acid (DNNS), a lipophilic anionic extractant, dissolved in

Table 1. Coated Wire Electrodes

<u>Lipophilic Cation-Based</u>

Halide: Cl^-, Br^-, I^-, CNS^-

Oxyanion: NO_3^-, ClO_4^-

Organic Anion: $RCOO^-$, RSO_3^-

Amino Acid:

$$H_2N-C(-R)-C{\overset{\displaystyle O}{\underset{\displaystyle OH}{}}}$$

<u>Neutral Carrier-Based</u>

K^+

<u>Dinonylnapthalenesulfonate-Based</u>

Quaternary Ammonium Ions

Drugs of Abuse, e.g., PCP, speed, methadone

β-Adrenergic Drugs, e.g., acebutalol

Ca-Blockers, e.g., Verapamil

Phenothiazines, e.g., chloropromazine

Table 2. Selectivity Coefficients, $k_{i,j}^{pot}$,[a] of Various Divalent Cations [2]

Interferent	Orion Electrode	Coated Wire Electrode
Ni^{2+}	0.026	0.0039
Cu^{2+}	0.24	0.15
Mg^{2+}	0.033	0.014
Ba^{2+}	0.016	0.0036
Sr^{2+}	0.029	0.021
Pb^{2+}	0.23	1.86
Zn^{2+}	1.44	32.3

[a]Calculated from $\Delta E = 30 \log (1 + k_{i,j}^{pot} a_i / a_{Ca})$

Table 3. Response Characteristics of Coated Wire Electrodes

Electrode	Slope, mV/log a	Concn. Range of Linear Response, M	Useful Concn. Range, M
Perchlorate	58	10^{-1}–10^{-4}	10^{-1}–10^{-4}
Chloride	55	10^{-1}–10^{-4}	10^{-1}–10^{-4}
Bromide	59	10^{-1}–10^{-3}	10^{-1}–10^{-4}
Iodide	60	10^{-1}–10^{-4}	10^{-1}–10^{-4}
Thiocyanate	59	10^{-1}–10^{-3}	10^{-1}–10^{-4}
Oxalate	28[a]	10^{-1}–10^{-4}	10^{-1}–10^{-4}
Acetate	50[a]	10^{-1}–10^{-3}	10^{-1}–10^{-4}
Benzoate	53[a]	10^{-1}–10^{-3}	10^{-1}–10^{-4}
Sulfate	28	10^{-1}–10^{-3}	10^{-1}–10^{-4}
Salicylate	53[a]	10^{-1}–10^{-3}	10^{-1}–10^{-3}
Phenylalanine	54[a]	10^{-1}–$10^{-2.6}$	10^{-1}–10^{-3}
Leucine	52[a]	10^{-1}–$10^{-2.6}$	10^{-1}–10^{-3}

[a] log C

Table 4. Selectivity Coefficients, $k_{i,j}^{pot}$, for the Coated Wire Electrodes Compared to Those for the Liquid-Membrane Electrodes [3]

| Electrode | Chloride | Interfering Anion | | Miscellaneous |
		Nitrate	Sulfate	
Perchlorate	0.004 (0.18)[a]	0.028 (0.12)	<0.001 (<0.001)	ClO_3^-, 0.039 (0.20)
Chloride	...	2.0 (3.0)	0.12 (0.061)	Br^-, 1.2 (2.7)
Bromide	0.19 (0.40)	2.0 (1.1)	0.020 (<0.001)	I^-, 14.5 (3.6)
Iodide	0.0048 (0.064)	0.11 (0.23)	<0.001 (<0.001)	Br^-, 0.056 (0.22)
Thiocyanate	<0.001 (<0.001)	<0.046 (0.067)	0.001 (0.018)	I^-, 0.34 (0.42)
Benzoate	1.3 (0.17)	1.3 (0.48)	0.15 (<0.001)	Salicylate, 1.4 (0.29)
Salicylate	<0.001 (<0.001)	0.42 (0.030)	<0.001 (<0.001)	m-OH benzoate, 0.22 (0.13)
Oxalate	51 (80)	...	1.3	OAc^-, 11(11)
Sulfate	16 (100)	30 (820)	...	
Phenylalanine	0.8 (1)	2 (1.6)	0.04 (0.1)	Gly 0.025 (0.040) Leu, 0.13 (0.40)
Leucine	0.56 (0.50)	0.50 (0.30)	0.020 (0.025)	Gly, 0.032 (0.063) Val, 0.25 (0.25) Phe, 2.0 (1.6)

[a] Value of $k_{i,j}^{pot}$ for liquid membrane electrode is in parenthesis.

a PVC membrane plasticized with dioctylphthalate. Extremely high
selectivity was observed for alkylammonium ions over common inorganic ions tested (Table 5). Among organic species, selectivity
increased regularly with the number of carbon atoms of the interferents tested. This indicated great promise for sensors of phar-
maceutical interest, since many such compounds are high molecular
weight protonated amines in the physiologic pH range. Currently,
DNNS based CWEs are made by dissolving the amine of interest in a 5%
PVC in tetrahydrofuran (THF) which is also 0.5% in DNNS and 4.5% in
plasticizer, usually dioctylphthalate. This solution is then used
to coat the end of a copper wire which is elsewhere insulated with
non-plasticized PVC. Following conditioning in a 10^{-4} to 10^{-3} M
solution of the analyte, the electrodes are ready for use.

<u>EMF and Selectivity Measurements</u>

Early on, it was anticipated that many repetitive calibrations and
EMF measurements would be carried out in the evaluation of a large
quantity of electrodes. The first microcomputer-based automated
titration system utilizing high level software (CONVERS) hastened
these studies (<u>10</u>), as did a more recently constructed minicomputer
system (Figure 1). Typical results are shown in Figure 2, where a
set of five protriptyline CWEs were calibrated simultaneously (<u>11</u>).
Graphic side-by-side comparison of different electrode calibrations
was also useful in establishing structure-selectivity relationships.
     In solvent extraction studies of various anions using Aliquat
336S, values of K_{ex} were obtained that showed a one-to-one correspondence between K_{ex} and electrode selectivity coefficients,
$k_{i,j}^{pot}$. This latter value is calculated from EMF responses of the
electrodes sampled in the presence and in the absence of an interfering ion using the following relationship:

$$E = E^{o} + \frac{59.2}{n} (a_i + k_{i,j}^{pot} a_j^{n/z_j})$$

where n and z_j are charges of the primary and interfering ions,
respectively, and a_i and a_j are their activities. Experimentally,
computer-generated solutions of specific ratios of a_i/a_j are made,
followed by correction of the "new" activity of ion "i" due to
addition of ion "j".
     Systematic selectivity studies of DNNS based electrodes were
carried out in precisely this fashion (<u>13</u>). Using tributylammonium
as a primary ion, log $k_{i,j}^{pot}$ values were determined for various substituted alkylammonium ions. As was the case for Aliquat-based
electrodes for organic species, selectivity improved considerably
with increasing molecular weight. These observations are summarized
in Figures 3 and 4. In Figure 3, log K_D values, as calculated by

Table 5.   Selectivity Coefficients for the
DNNS-Based Electrodes [9]

| | | | DoTA$^+$-ISE | |
| | | | CWE | Conven-tional |
Inter-ferants	Na$^+$-ISE	TBA$^+$-ISE	CWE	Conven-tional
DoTA^{+b}	$>10^{5a}$	5.3	(1)	(1)
TBA^{+c}	$>10^{5a}$	(1)	0.18	0.21
DTA^{+d}		0.54	0.084	0.085
TPA^{+e}		0.02	0.031	0.038
TEA^{+f}	4.7×10^3	4.4×10^{-4}	1.1×10^{-4}	1.5×10^{-4}
TMA^{+g}	30	1.2×10^{-4}	$<10^{-4}$	$<10^{-4}$
Ag$^+$	22			
K$^+$	6.5	$<10^{-4}$	$<10^{-4}$	$<10^{-4}$
NH$_4^+$	3.1	$<10^{-4}$	$<10^{-4}$	$<10^{-4}$
Na$^+$	(1)	$<10^{-4}$		
Li$^+$	0.76		$<10^{-4}$	$<10^{-4}$
H$^+$	0.46	$<10^{-4}$	$<10^{-4}$	$<10^{-4}$
Pb^{2+}		$<10^{-4}$	$<10^{-4}$	$<10^{-4}$
Ca^{2+}		$<10^{-4}$	$<10^{-4}$	$<10^{-4}$
Mg^{2+}	0.69	$<10^{-4}$	$<10^{-4}$	$<10^{-4}$

^{a} $k_{i,j}^{pot}$ varies greatly with concentrations of primary and interfering ions. ^{b}Dodecyltrimethylammonium. ^{c}Tetrabutylammonium. ^{d}Decyltrimethylammonium. ^{e}Tetrapropylammonium. ^{f}Tetraethylammonium. ^{g}Tetramethylammonium.

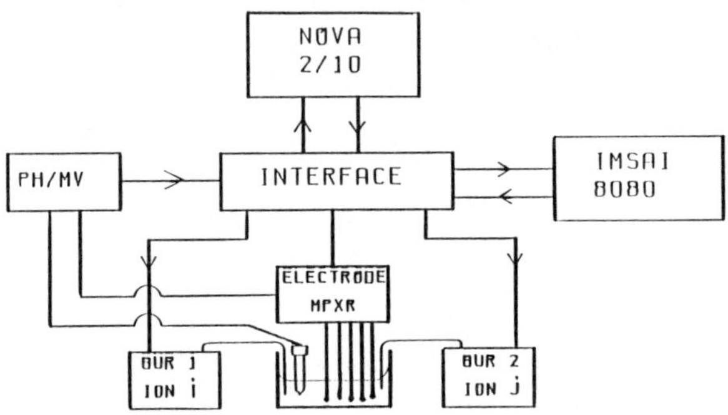

Figure 1. Computer-controlled potentiometric analysis system.

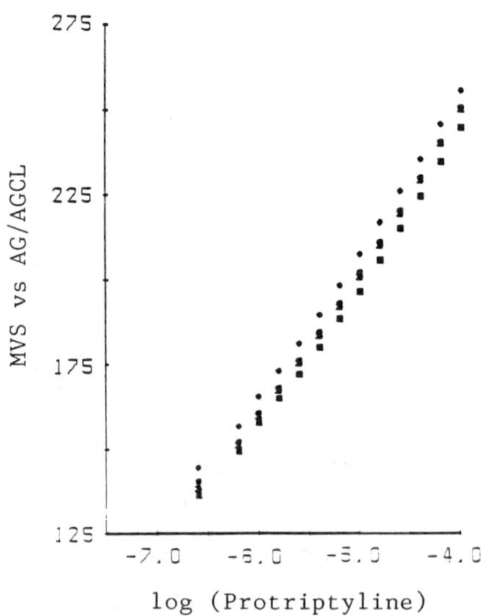

Figure 2. Calibration curve of five different protriptyline electrodes.

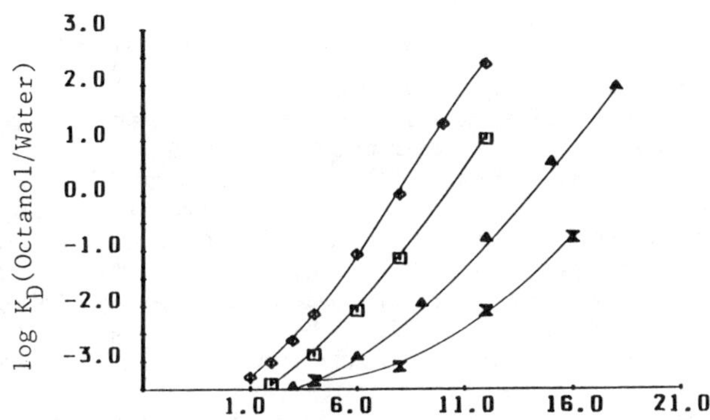

Figure 3.   Log K_D vs. carbon number for various alkylammonium ions:
◇ = primary ammonium ions, ▣ = secondary ammonium ions,
△ = tertiary ammonium ions,   ⬙ = quaternary ammonium
ions.

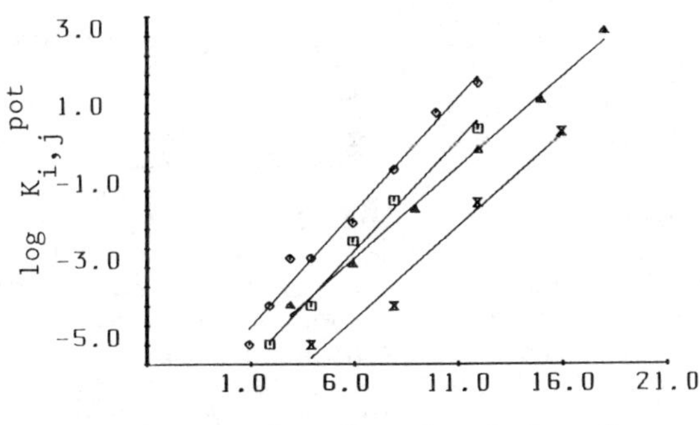

Figure 4.   Log $k_{i,j}^{pot}$ for tributylammonium CWEs vs. carbon number for
various alkylammonium ions: ◇ = primary ammonium ions,
▣ = secondary ammonium ions, △ = tertiary ammonium ions,
⬙ = quaternary ammonium ions.

the correlation technique of Hansch and Leo ($\underline{14}$), are plotted against ammonium on carbon number. When compared to Figure 4, where the corresponding log $k_{i,j}^{pot}$ values for a set of TBA^+ CWEs against these ions are plotted, one sees the utility in considering solvent extraction parameters in predicting electrode selectivity behavior. Indeed, plots of log $k_{i,j}^{pot}$ vs. log K_{ex} are linear (cor = 0.98) though quaternary ammonium ions displayed a greater slope (2.1 vs 1.1) (Figure 5) than did the less substituted ions tested. This may be due to the faster increase in steric hindrance as successive methyl groups are added to the more symmetric quaternary ions.

The effect of increased interference with increasing molecular weight of interferents could be minimized when heteroatoms and polar functional groups are present. Their effect, though, would be to increase the hydrophilicity of the molecule, thereby decreasing the selectivity coefficient. On the other hand, an electrode prepared for a highly-substituted compound may show significant response to a lower-molecular weight less-substituted ammonium ion. These predictions were realized in a study of N-ethylcyclohexylammonium ions, where the log $k_{i,j}^{pot}$ values calculated from measured EMF responses of tributylammonium CWEs decreased with increasing hydrophilic character of the added functionality. In Table 6, these values are compared to log K_D values calculated by the Hansch method. These arguments lead to another consideration, that is among a similar set of interfering ions, the electrode for the primary ion highest in lipophilicity would display greatest selectivity. Electrodes for aliphatic ammonium ions of carbon number twelve were prepared and selectivity coefficients measured for each set vs. the other two. The selectivity order was found to be dodecyl-ammonium > dihexylammonium > tributylammonium (see Table 5).

In addition to the greater magnitude of the primary amine distribution constant, less sterically hindered nitrogens would result in a larger value of K_{IP}, the ion-pairing constant of the protonated ammonium ion with the DNNS anion. Therefore, among various CWEs, the family curves obtained for tributylammonium (Figure 4) would shift either positively for more hydrophilic primary ions, or negatively for more hydrophobic ones.

These results were consistent with subsequent studies of CWEs for the drugs shown in Figure 6 ($\underline{11}$). Selectivity orders were in excellent agreement with K_D values calculated for these compounds.

Addition of hydrophilic groups tended to decrease selectivity of even high molecular weight compounds, though sensitivity seemed to be independent of this effect. In general, though, relative selectivity among a group of CWEs selective to different drugs is readily predicted from calibration curves. As shown in Figure 7, the most selective electrodes (see Table 7) generally yield a higher value for the calibration curve y-intercept. This supports the previous assertions, that is more lipophilic ions will produce a high membrane potential and, therefore, a larger value of $E_{CWE}-E_{REF}$.

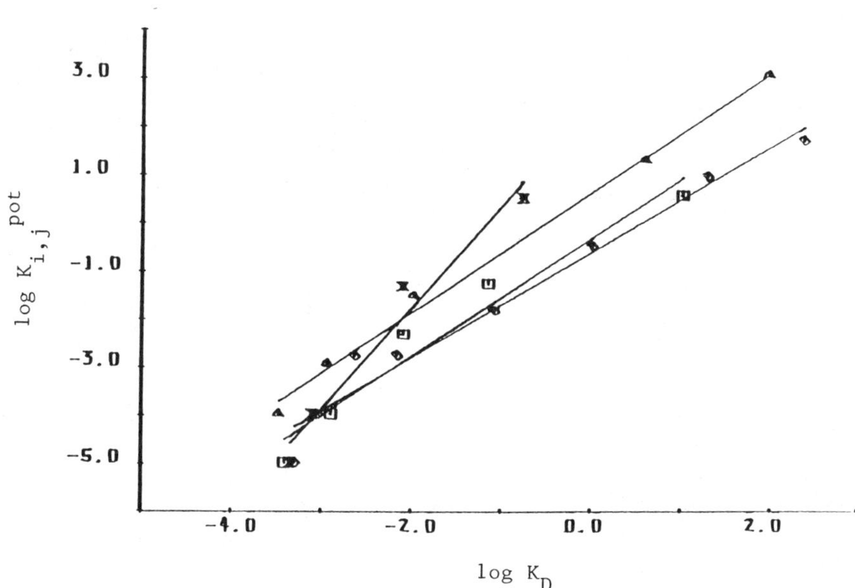

Figure 5.   Log $k_{i,j}^{pot}$ for tributylammonium CWEs vs. Log K_D for various alkylammonium ions: ◇ = primary ammonium ions, ⊡ = secondary ammonium ions, △ = tertiary ammonium ions, ⬙ = quaternary ammonium ions.

Figure 6.   β-Adrenergic and calcium blocking drugs.

Table 6. Log $k_{i,j}^{pot}$ and Log P Values for Substituted
Cyclohexylamines at 25° C

Compound	Log $k_{i,j}^{pot\,a}$	Log K_D (Hansch)
N-Ethylcyclohexylamine	-1.97	-1.36
N-2-Hydroxyethylcyclohexylamine	-2.35	-3.98
N-2-Cyanoethylcyclohexylamine	-2.12	-2.16
N-3-Aminopropylcyclohexylamine	-2.22	-3.52

^{a}TBA$^+$ coated-wire electrode. The concentration of interfering ion
was 10^{-1} M with a background level of 10^{-4} M TBA$^+$.

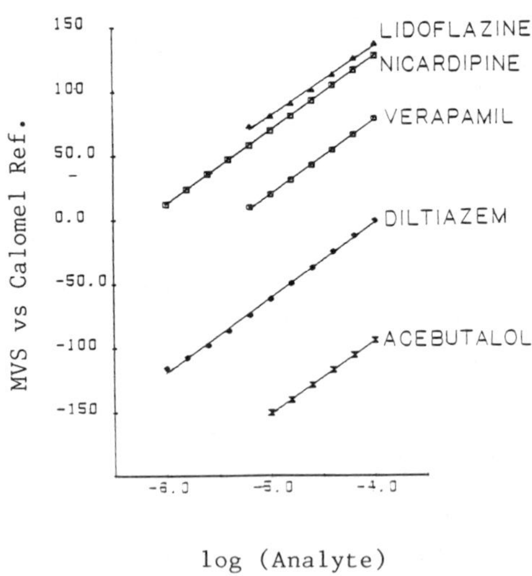

Figure 7. Calibration curves for drugs shown in Figure 6.

Table 7.   Log $k_{i,j}^{pot}$ and Log P Values for Substituted
Cyclohexylamines at 25 $^{\circ}$C

Drug	Log $k_{i,TBA}^{pot}$	+/−	St. Dev.	Log K_D
Acebutalol	2.44	+/−	0.2	−0.2
Dilthiazem	0.71	+/−	0.02	2.3
Lidoflazine	0.30	+/−	0.02	5.3
Verapamil	−0.32	+/−	0.02	5.9
Nicardipine	−0.89	+/−	0.02	6.3

Recent studies by other workers have shown that similar membrane systems give potentiometric response to changes in solution pH (15). Preliminary investigations in this laboratory have shown that DNNS-based drug electrodes behave differently depending upon the incorporated drug. The magnitude of the pH responses were such that over a sufficiently wide range of solution pH, shifts of 0.02 units would only cause concentration errors of 1.0%. Future work in this area, besides developing CWEs for additional drugs and compounds of interest, will be aimed at improving the sensitivity of these devices. Current studies of interfacial processes occurring at liquid-liquid interfaces should provide insight into the fundamental mechanisms responsible for CWE behavior with regards to detection limits, selectivity, and varying pH response.

This research was supported by a grant from the Office of Naval Research.

Literature Cited

1. Cattrall, R. W.; Freiser, H. _Anal. Chem._ 1971, 43, 1905.

2. James, H; Carmack, G. D.; Freiser, H. _Anal. Chem._ 1972, 44, 1842.

3. Fujinaga, T; Okazaki, S.; Freiser, H. _Anal. Chem._ 1974, 46, 1842.

4. Kneebone, B. M.; Freiser, H. _Anal. Chem._ 1973, 45, 449.

5. Carmack, G. D.; Freiser, H. _Anal. Chem._ 1977, 49, 1577.

6. Carmack, G. D.; Freiser, H. _Anal. Chem._ 1973, 45, 1976.

7. Carmack, G. D.; Freiser, H. _Anal. Chem._ 1975, 47, 2249.

8. Carmack, G. D.; Freiser, H. _Anal. Chem._ 1977, 49, 767.

9. Martin, C. R.; Freiser, H. _Anal. Chem._ 1980, 52, 562.

10. Martin, C. R.; Freiser, H. _Anal. Chem._ 1979, 51, 803.

11. Cunningham, L.; Freiser, H. _Anal. Chim. Acta_ 1982, 139, 97.

12. James, H. J.; Carmack, G. D.; Freiser, H. _Anal. Chem_ 1972, 44, 853.

13. Cunningham, L; Freiser, H. _Anal. Chim. Acta_ 1981, 132, 43.

14. Hansch, C.; Leo, J. L. "Substituent Constants for Correlation Analysis in Chemistry and Biology"; Wiley: New York, 1979.

15. Schulthess, P.; Shijo, Y.; Pham, H. V.; Pretsch, E.; Amman, D.; Simon, W. _Anal. Chim. Acta_ 1981, 131, 111-116.

RECEIVED January 21, 1986

Chemical Sensing Using Near-IR Reflectance Analysis

David L. Wetzel

Department of Grain Science and Industry, Kansas State University, Manhattan, KS 66506

Near-infrared reflectance analysis is particularly well suited to chemical sensing because it operates on "as is" samples and yet has chemical specificity. The absorptions observed originate from vibrations of a relatively few chemical groups whose overtones and combination bands appear in the near infrared region. These groups are commonly found in natural and synthetic materials and their quantitation is possible at major component levels. Chemical sensing by near-infrared techniques does not usually require complete scanning since information for quantitation is found in select wavelength responses incorporated into a mathematical analytical expression. Correlation transformation between laboratory pre-analyzed samples (training set) and the optical data collected allows statistical wavelength selection and assignment of regression coefficients. The result is a select calibration for an analyte in a particular matrix or commodity. Quantitative chemical sensing occurs when the built-in computer of the near-infrared analyzer simultaneously calculates multiple component concentrations by solving prediction equations which use multiple wavelength intensity ratioed data for individual samples. Similarly, qualitative sensing information by discriminant analysis is obtained by use of a discrete wavelength multiterm function. Preselected wavelengths and dedicated preprogrammed microcomputers are ideal for numerous sensing and ultimately control functions.

Chemical sensing applied to "as is" material requires an analytical system which is tolerant of the sample as it exists. Thus, the analyte must be determined in the matrix in which it is found. This precludes concentration steps, dilution steps, sample purification or workup of any sort and particularly makes sophisticated analytical separations prior to monitoring impractical.

Direct absorption spectroscopic sensing, in some cases, fulfills the above requirements. In many cases, however, lack of specificity for determination of indvidual analytes in a complex matrix would appear to be an almost insurmountable obstacle to this approach.   If the sample "condition" to be sensed has light scattering and/or spectroscopic absorption characteristics it may be possible to make multichannel (multiwavelength) measurement work.

Through statistical means and correlation transformation of the spectroscopic data, quantitative chemical sensing is possible. This is true, however, only in cases which permit good matrix (background) correction and optical measurement condition correction which allow linear analytical response in a reasonable range.

In near-infrared reflectance, the matrix correction is made possible by the additivity of each weighted contribution for that wavelength in the spectrum being observed. The low absorptivity in the near-infrared contributes to this necessary "additivity" and eliminates, also, the necessity of sample dilution prior to sensing.

Unlike classical analytical spectroscopy performed on liquids or dilute solutions of analytes, diffuse reflectance measurement in the near-infrared must deal with a composite effect of spectroscopic absorption and scattering from the analyte and the matrix in which it is found. Differences in refractive indices of the sample material, specular reflection and observance of relatively small differences are all dealt with in this technique.

Flow through spectroscopic cells for liquids have been used for decades for ultraviolet, visible, infrared and fluorescence on-line spectroscopic monitoring. The same is true for monitoring gases. Fluids, as the name implies, present only minor sample transport problems but dealing with granular materials presents a challenge. Color meters have been used for granular materials for decades for clinical, industrial and other applications.

Unfortunately there are limited cases where samples can be analyzed directly without prior reaction with special reagents. The responses of intense electronic spectra are not necessarily linear even if spectroscopic resolution provides selectivity, hence quantitation is difficult. Similarly, autofluorescence, coupled with selective excitation and observation wavelengths has been useful. Fluorophores, selectively coupled to a portion of the sample, have allowed the art of fluorescence microscopy to be refined and add the dimension of chemical selectivity (1). Addition of reagents prior to measurement detracts from the practicality of either fluorescence or absorbtion for sensing purposes. Autofluorescence of finely divided solids in a viscous suspension has been used. Quantitation, in this case involving wheat milling fractions, has been enhanced by statistical data treatment methods (2). These include partial least squares, factor analysis, or principle component analysis. Such data fitting techniques provide compensation for measurement or sample matrix conditions which would otherwise compromise the linearity of the sensing response.

To achieve quantitative sensing by near-infrared reflectance techniques the wavelengths most responsive to the analyte must be selected. The analytical equation produced by regression should provide a weighted term to account for differences in relative absorptivities. Method development including wavelength selection, training set selection and equation testing will be treated in a latter section of this paper. Sample presentation and an example

Table I. Uses of Near-Infrared Sensing

Sensing Function	Action
on/off	switching (alarm, TV channel change)
go/no go	accept/reject (too wet to store)
semi-quantitation	ingredient specifications (blend formulation)
quantitation	product specifications (buy/sell)

of a working solid sample on-line near-infrared sensing system will also be detailed as a case history.

Comparison of Near-Infrared Reflectance Analysis with Classical Spectroscopic Techniques

Characteristics specific to near-infrared (diffuse) reflectance analysis vs classical spectrometry:

- near-infrared compared to mid-infrared
- diffuse reflectance compared to transmittance
- neat samples compared to solution spectrometry
- correlation techniques for wavelength selection
- data base for empirical analytical coefficients

The near-infrared region of the spectrum that uses a lead sulfide detector spans approximately 1100 to 2500 nanometers. Excellent classical spectroscopic exploration and review of this region preceeded its current practical use (3-5). The vibrational overtone and combination absorption bands in this region are broad, they are overlapping with each other especially for solids and the absorptions are extremely weak. This is in comparison to the mid infrared which has sharp well defined bands which are quite often resolved from each other, at least for pure compounds, and which have a high intensity. At first consideration, these disadvantages of insensitivity brought about by low absorptivity and the lack of resolution that one finds with broad bands would seem to be a devastating blow to the near-infrared technique. The fact is, however, that the weak absorptivity in this region of the spectrum is an advantage because the necessity to dilute a solid sample is avoided (6). The absorbance of a major component in a neat, granulated sample will, in fact, appear in the range of 0.4 - 1.0 absorbance units and non-linearity due to extremely strong absorbtion is usually not a problem. Another advantage of the weak absorbtion is that when doing diffuse reflectance, it becomes necessary to correct for the specular reflectance component. The low absorption, coupled with a relatively constant index of refraction among members of a sample set makes the mathematics of correction for the specular component of reflectance quite simple.

A linear multiterm spectroscopic response as a function of the concentration of the analyte is an achievable goal (7). However, additivity of the spectroscopic response at the same wavelength of one component to the other component of the sample matrix is absolutely essential. This is due to the fact that with a neat sample, infinite dilution does not exist as in solution spectrophotometry for which classical laws were developed. Since the bands are broad, isolating an analyte band from the bands of the matrix is virtually impossible. Therefore, additivity is essential to make quantitation possible. Unless one is applying this technique to pure chemicals, there is nearly always a serious background problem. Although the problem is serious, it can be dealt with readily. Spectral subtraction is not used but rather a statistical technique is employed for this purpose. The coefficients on each term in the equation will, in fact, include the background correction for other materials present in the sample and in a number of samples constituting a set, the magnitude of each of these terms and the ratio of one of these coefficients to another is determined by empirical means (8). Multiple linear regression is one of the primary tools. With the multiple wavelength expression, it is in fact possible to have a non-linear effect which would appear in a single term by itself, be offset by a non-linear effect in a different term in the opposite direction (9). Although perfect compensation cannot always be predicted, adherence to this expression of the analyte concentration is always tested for and standard error of determination is calculated on the basis of testing the equation, complete with a sample preparation and measurement technique, on a set of known samples. This test is, in fact, a part of the analytical method development for any purpose.

In granular solids or in analysis of liquids or slurries in which a considerable amount of particulate material exists, the scattering effect attenuates the optical signal in addition to the absorption. Scattering back from the body of the sample toward the surface produces the intensity to be measured as diffuse reflectance. Scattering also controls the depth of penetration of the sample as well as does its absorptivity (10). The complexity of these two factors acting at once is difficult to predict a _priori_. This is another reason why the empirical method and the empirical equation coefficients produced by a training set are essential.

Unlike transmission spectroscopic techniques involving true solutions or pure liquids, the diffuse reflectance becomes angle dependent for incident radiation and for purposes of collecting diffusely reflected radiation preferentially over the specularly reflected radiation. The specular component is minimized by experimental design, if possible, but the unavoidable specular contribution requires correction. Not only does the presence of the specular component simply increase the total amount of radiation hitting the detector, but it obscures the finding of information from the intensity of the signal. As the specular component approaches metallic reflection, it can, in fact, produce an intensity contribution opposite in sign to that of diffuse

reflectance. This needless to say completely ruins chemical
quantitation or at least, adversely affects the desired linear
relationship. The chemical sensing by near-infrared reflectance
analysis thus involves correlation spectroscopy as opposed to
classical, near-infrared instead of mid-infrared and diffuse
reflectance scattering considerations instead of simple transmission.

Correlation Techniques For Wavelength Selection

Diffuse reflectance differs from classical transmission in which no
particulate matter exists to scatter the beam of radiation. It is
necessary to contrast correlation spectroscopy (correlation
analytical techniques based on spectroscopic measurements) to a
classical, one wavelength, monochromatic application of Beer's law.
The use of multiple wavelengths produces a multiterm analytical
equation in reflectance R of the general type:

$$\% \text{ Analyte} = z + a \log 1/R_1 + b \log 1/R_2 + c \log 1/R_3 + ...$$

In such an equation, at least one term, is used as an indicator
wavelength. The indication evidenced by absorption could be
positive or negative. An equation may use the wavelength at
which the analyte absorbs. A positive coefficient will exist on
that term. In a closed system, such as wheat, which is constituted
of approximately 70% starch, with the other 30% divided between
protein and moisture and a few minor components, a lower quantity
of an absorber such as starch could be an inverse indicator of the
presence of protein. It is common when measuring protein in a
high starch system, to utilize a wavelength at which the carbo-
hydrate absorbs and to employ a negative coefficient on that
term. Thus, those two terms have essentially a push pull effect
and add to the sensitivity. It is nearly always essential to have
at least one reference term, to show the overall intensity level of
the baseline. Use of reference wavelengths provides some
mathematical assistance to avoid baseline shift due to scattering
effects from different particle size in the case of a granular
sample or a different particle population in the case of a slurry.
Two wavelengths may be chosen close together which would indicate a
slope. In such a case, the difference between absorbance at those
two wavelengths would appear in the analytical expression.
    In other cases, a baseline corrected peak height for a parti-
cular absorber may be employed as a term in the equation. In such
a case, the wavelength difference on either side of a peak maximum
will affect the contribution of that complex term. That increment
or gap, in fact, under such circumstances becomes a part of the
calibration. It is as important a contribution to the calibration
as the coefficients on the wavelength terms. In this correlation
spectroscopy, classical band assignments are not always possible.
Little specific near-infrared literature exists in advance of most
applications and it is not always possible to predict which wavelengths will produce the best linearity and the best sensitivity for
a given analytical problem. In the empirical approach a variety of
statistical treatments have been attempted. By far the most

commonly used is multiple linear regression. This can be done with limited spectroscopic data obtained from filter instruments or with hundreds of data points obtained from scanning instruments. Multiple linear regression may be performed on a "step up basis" where one begins with the wavelength most highly correlated and subsequently adds other highly correlated wavelengths as well as a reference wavelength. In the step up procedure when addition of further wavelength data does not improve the correlation or reduce the standard error of calibration, then no further wavelengths are used. Reverse stepwise multiple linear regression has also been used, particularly when dealing with a filter instrument which is capable of measurement at a limited, relatively small, number of wavelengths. An example would be to perform a multiple linear regression on a sample set which is greater in size than the number of filters and to repeat the process by discarding the wavelength which had the least impact on the previous regression. Typically, the Student t-test on each individual term would be the criterion for eliminating that term. After a series of steps, for example, the 19 initial wavelengths used could be reduced down to the point at which the correlation degrades significantly, or the standard error of calibration increases significantly, then the reverse stepwise process would be stopped.

It is also possible with computers to perform many regressions sequentially and automatically. With less than 100 data points on a rapid computer, or less than 20 on a slow computer, it is possible to routinely regress all possible combinations of three wavelengths from a field of 100 or a field of 20, respectively, and select the best combination. Computer programs are also available for all possible combinations of five or more. However, at some point, since there is a geometric progression in the number of combinations as one adds additional terms, the computer time required is not necessarily warranted for the small improvement in the results. In addition to the multiple linear regression, other techniques such as partial least squares, factor analysis (principal components) and Fourier transform have been applied to near-infrared spectroscopic data. Qualitative sensing based on discriminant analysis data treatment also shows promise (11,12).

<u>Instrumentation For Near-Infrared Sensors</u>

Tungsten-quartz-halogen sources and lead sulfide detectors are common in most near-infrared chemical sensors. The exception to this is the use of independent light emitting diodes as individual monochromatic sources (Trebor Ind., Gaithersberg, MD) in the very near-infrared where silicon detectors can be used. With the exception of wavelength selection by a choice of LED sources interference filters are used in the most routine instruments to select the appropriate wavelengths from the continuum. The main difference in filter instruments are: 1. The mechanism for wavelength change, 2. The mechanism and timing associated with referencing the signal intensity reflected from the sample to the intensity off of a standard reflector, 3. The way in which diffusely reflected radiation is collected.

Wavelength (filter selection) is accomplished by moving a

discrete filter into the beam (perpendicular to the beam) for accumulating reflectance measurement at that wavelength before selecting the next wavelength with another discrete filter. The alternative approach involves a series of rapidly tilting filters which accumulate data at numerous wavelengths (depending on the angle to the beam) with each rotation cycle. With this mechanism, accumulation of data requires summation of intensities at each wavelength from multiple rotations (Figure 1).

In practice, reflectance is the ratio of intensity off of the sample to intensity off of a standard reflecting surface. A clean ceramic surface or a diffuse gold mirror are typical standard reflectors. Intensity referencing may be done at each discrete filter position immediately following the sample intensity measurement using a moveable mirror. The resulting ratio of intensities is stored in memory for each preselected wavelength before measurement at the next wavelength. In an alternate instrumental design the system stores reference intensities at multiple wavelengths with the standard reflector and collects sample intensities at multiple wavelengths on a separate optical pass with the sample in place. Grating monochromators with an f = 2.0 or better and interferometers are used for laboratory experimentation and method development in near infrared reflectance analysis but are unlikely to be used as routine chemical sensors.

Collection and measurement of diffusely reflected radiation scattered back from the sample is accomplished in several ways. A collimated beam of radiation from the filter strikes the sample surface normal to the plane of the surface. Reflected radiation intensity from the sample will vary depending on the angle of observation. It may also differ with direction particularly if the surface has texture. All optical detection systems attempt to minimize specular reflection and maximize diffuse reflection which reaches the detector. One system employs an integrating sphere to collect diffuse reflection at a maximum solid angle but excludes the cone surrounding the incident beam where specular reflection has the greatest intensity. Radiation entering the integrating sphere is collected on one of two detectors located near the base of the sphere. Another design uses a single detector at 45 degrees from the surface and rotates the sample to average out directional differences. Multiple detectors placed on four sides at an elevation of 45 degrees are also used. With all three systems the geometry of the collection angle and the resulting intensity is highly dependent upon the vertical positioning of the sample relative to the fixed detector and window.

Variation in sample height on a non contact optical unit can be a major analytical limitation. There are at least three American and three European near-infrared moisture monitoring systems which use water absorption at 1940 nm with one or more reference wavelengths. These are designed to be mounted above a conveyor belt with clearance between the optical window and the sample surface. Since water is a strong absorber the coefficients used in the analytical expression are small and the uncertainty associated by multiplying the sampling noise may be acceptable. For analytes where absorbance differences are less, the precision may not be sufficient.

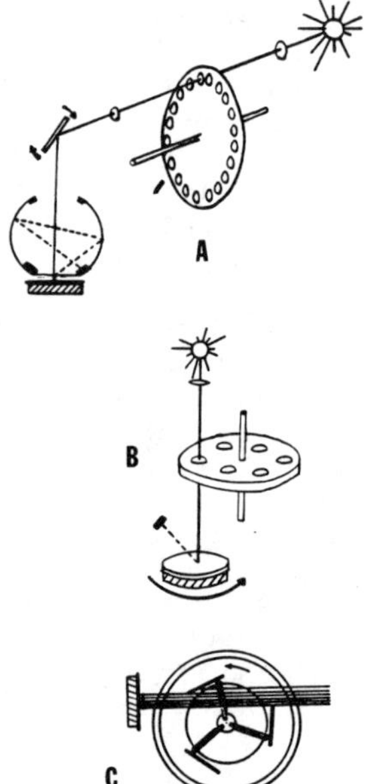

Figure 1   Optical systems for near-infrared reflectance,
A) turret mounted discrete filters, integrating sphere
and moveable mirror referencing.
B) turret mounted discrete filters, single detector and
rotating sample cup (Dickey-John, Anburn, IL).
C) tilting filter mount (continuously moving) with
detectors on four sides of sample cup.

Sample Presentation

<u>Solids</u>.   Conventional near-infrared reflectance analyzers use a variety of methods to position the sample into the incident collimated beam and collect reproducibly the diffusely reflected radiation to measure the absorption which takes place in the body of the sample traversed. Solid samples are ground with care to achieve reproducible and reasonably uniform granulation for calibration and analysis measurements. The overall scattering characteristics of the sample which shift the reflectance baseline and control the depth of sample penetration and opportunity for absorption become a part of the method and the empirical analytical equation (6,9). Solid sampling is summarized on the following table.

Table II.   Examples of Sample Presentation Systems For Solids

---

Closed quartz covered cup for powders
Open cup for pastes such as processed meat
Cavity adjacent vertical window for powder
Compacted cavity above inverted window for powder
Film or sheet of polymer

<u>Liquids</u>.   Liquid sample handling devices have been designed and supplied by instrument manufacturers in a variety of shapes and sizes. One approach for near-infrared measurement of liquids is to place a detector opposite the source to use the transmission mode. This is done in one design, regardless of the turbidity of the sample. Another design retains the use of the integrating sphere and its positioning for diffuse reflectance and uses a combination of absorption of radiation going in and back through the liquid before it is collected in the integrating sphere. This amounts to a succession of absorption by the sample, diffuse reflectance from a reflector on the bottom followed by absorption of the sample. Needless to say, the thickness of the cell is extremely important, it is temperature dependent and a good deal of control is required. Such a system, and a sophisticated electronic control circuitry, is part of a liquid sample drawer which fits into a commercial instrument and is controlled by the firmware built into that instrument. This configuration of an instrument is marketed for industrial fluids and the dairy industry. For liquid milk samples, it is essential to employ a homogenizer prior to the introduction of the sample into the liquid drawer accessory to the instrument. The drawer is also thermostated and a pump is placed into the system to furnish samples to the optical system. The same system with a homogenizer has been used for other industrial samples which have characteristics similar to that of milk. Oils and previously melted fats can be measured neat in a heated, thermostatically controlled liquid drawer. Other liquid drawers have been fabricated without the programmable control for use with an external water bath

or other thermostated system. Some of these special liquid cells have been used at ambient temperature for alcohol in wine and for various components in beer and in wort during the process of fermentation. The liquid cell volume is rather small but the tubing leading into that can be tailored for the particular purposes. Large bore cells have been produced which can accomodate viscous fluids and those containing as much as 60-70% solids. Although diffuse reflectance in the near-infrared was never intended as a microtechnique, the necessity to make measurements on small amounts of samples has been dealt with by several people. Small volume cells involving 200 microliters of liquid have been produced. One approach to dealing with a small sample has been the design and fabrication of a focusing cup (13). In this case, the cup is transparent and the bottom of it is mirrored so that light which enters from the top is then focused back up toward the center part containing the sample. A reasonable percentage of the radiation incident on the sample exits the sample towards the collection optics of an InfraAlyzer 400.

Process Analysis

Grab samples can be pulled at strategic points in a process to determine the quality of the intermediate product, or from the raw material. Obviously, for the purposes of control, the intermediate product has the greatest potential for giving the processor usable and timely information. Instead of random sampling, it may be desirable to have periodic sampling, analyze these samples as rapidly as possible in an off-line procedure, transfer the information obtained directly to the personnel or to the control device used to take the appropriate processing action. A few examples of such processing quality parameter monitoring would include moisture prior to some processing step, protein retention after a separation process has occurred, purity of the product or intermediate from the material for which separation has been performed.   In the case of dry corn milling, one could use monitoring of the corn grits or flour to look for a residue of oil. Specifications call for a limit to the oil level in the grits. It is also possible to monitor the amount of product which was not recovered from the by-product. Such would be the case of monitoring for starch left on the fiber in either dry wheat milling or wet corn milling. In both cases, the starch thrown out with the by-product would be considered a loss and a mark of reduced efficiency in the process. For condensation polmers the process of polymerization can be followed in some cases by the disappearance of terminal groups found in monomers which are incorporated then into the polymer. The extent of cross linkage and end capping has also been measured for processing control purposes. Unsaturation during the hydrogenation processing of oil has been followed and quantitated by near-infrared. Fermentation broths used to produce pharmaceuticals have been monitored for activity during the fermentation process by near-infrared. Starch gelatinization or degree of cook has also been monitored in an off-line manner. There are other examples of process monitoring analyses, but they are too numerous to mention here. This discussion will concentrate on examples of actual on-line systems used since the timeliness of on-line analysis gives the operator a chance to perform a processing control function better than off-line techniques.

<u>On-Line</u> <u>Monitoring</u> <u>of</u> <u>Liquids</u> <u>and</u> <u>Slurries</u>

On-line liquid systems operating at ambient temperature have been in use for a few years for monitoring alcohol in the wine industry, particularly in France, by near-infrared combined transmission reflectance techniques. Beer, as a finished product, has been monitored by pumping debubbled beer from a sidearm transfer pipe. This has been pumped through a flow cell and measurements taken on regular intervals from the sidearm sampling loop. In the dairy industry, milk has been pumped through a homogenizer into the liquid cell and measured in a stop flow mode. Also, in the brewing industry, wort has been monitored by pumping samples into a liquid sample drawer during the process of fermentation. In wet corn milling, the protein in the liquor from the centrifuges can be determined by pumping it through a liquid sampling device. From a high temperature high pressure hydrogenation reaction vessel, in the edible fats and oils industry, it is potentially possible through a sampling loop, to introduce hot oil into the liquid sampling device even in the presence of the catalyst particles. More on-line liquid and slurry measurement systems will no doubt be devised as needed. Most of these use a stop flow procedure with a sample loop. Whereas the sampling is intermittent and reflects what is flowing through the pipe or circulating through the reaction vessel at the time, such systems involve, nevertheless, true on-line procedures.

<u>On-Line</u> <u>Monitoring</u> <u>of</u> <u>Granular</u> <u>Materials</u>

It is difficult to generalize concerning the flow characteristics of granular materials. Each material has its own transport characteristics based on cohesive and adhesive forces, density, size of the particles, shape and perhaps other unknown characteristics. For a material of a given composition, humidity represents a day to day or hour to hour variable which could supercede the other characteristics. For on-line sensing of granular materials, the main challenge is a suitable transport system. It is necessary not only to get the sample into the optical beam, but to do it reproducibly, and to reproduce the amount of compaction each time. Complete removal of the sample is necessary so there will be no memory effect and new material can continually progress in the path of the light beam. In addition to the work at Kansas State University in our laboratory and pilot mill, which will be described later, other adaptations of commercial benchtop instruments to on-line monitoring are noted below. There is at least one system in the Netherlands adapted for dealing with dry milk powder. In addition, there are several systems in Europe where an instrument is placed at the end of a flour milling process to examine the final flour for protein, moisture and purity (14,15). Based on the information available from these industrial installations, in each case, the optical sensing head is tethered by electrical cable sufficiently short so that the mainframe instrument is housed in the production facility.

In addition, in each of these cases, the measurement is done in a stop flow mode where a discrete filter instrument is involved and the sample is removed by force using either an air blast or suction or both. These involve commercially available or modified table top instruments of Percon, Hamburg, West Germany and Technicon, Tarrytown, NY, USA. One contact type of system in Hungary (Focus Engineering) uses a diffuse reflectance illumination/observation window in the bottom of a conveying spout to monitor protein in precrushed sunflower to ensure optimum protein level of the resultant sunflower meal at the end of the oil extraction process. The repetitive, averaging, rapid scanning instruments of Pacific Scientific (Silver Spring, MD) have been adapted to on-line use in the transmission and the diffuse reflectance mode (16, 17). In the former, a sample fills a chute between the filtered source and the detectors. In diffuse reflectance a fiber optic probe has been used as an alternative to a vertical illumination and diffuse reflectance detection window where samples are placed in contact by a four vane fill-read-dump vertical rotary system.

KSU System of Near-Infrared Reflectance Monitoring in a Pilot Flour Mill

The following items need to be considered for a successful on-line monitoring system and its economic use in a processing facility:

- A transport system
- Key quality factor selection
- An optical sensing head
- Interfacing to the computer module
- A custom data handling system
- A design of potential action to affect the process based on the monitoring data

Hardware. The sample transport system must be designed for the flow characteristics of the material being handled. For material with certain granularity, with certain cohesive forces, one system will work when another material may require some sort of agitation to promote flow. The optical sensing head must be mounted on the transport system. The transport system described here is to monitor samples in typical metal spouting in the process of vertical transport which is fed by gravity after the material has been elevated by a pneumatic lift. This is completely different than mounting an optical head a number of inches away from a moving conveyor belt. With a moving conveyor belt, one is apt to have a different elevation of the product and different distances from the sensing head will change the optical geometry significantly. A slight vertical shift in the position of the sample will change the solid angle of collection and totally change the sensitivity and the calibration constants. That particular feature is possibly the biggest limitation with conveyor belt mounted monitors. The remote optical sensing head is tethered to the parent computer in this case the Technicon InfraAlyzer Model 400 mainframe. There are peripherals to the central computer connected through an RS232 interface as will be described later. In addition to the central computer and its peripherals, a data feedback system to the processing plant is essential. In the case of the flour mill, it was necessary for the data feedback system to be low voltage and explosion proof.

The task required of the central hardware includes controlling the remote data acquisition. The central hardware must receive a raw signal from a remote sensor and process the signal for each measurement. It is desirable also for mass storage of data, such as would be obtained in one day's run or the run of one operating shift in the processing plant. Software is necessary for averaging of sequential process data points. In an on-line system, it is impossible to have absolutely every single reading be a perfect one. There is potential for cavitation in the cell. It is possible that we could get one completely erroneous reading and rather than be overly influenced by that, it is necessary to do some averaging. A typical averaging would be a running mean. We have chosen a running mean of five. Five readings are obtained with our system at the present time in less than 2.5 minutes. We plan to reduce this to approximately 1.25 minutes. Other tasks of the central hardware include output of the averaged data to a printer and transmission of processed data to the control area for action either by a process control computer or by human operators in charge of the processing system. Figure 2 illustrates the Infralyzer 400 mainframe connected via an RS232 interface to a Hewlett Packard 85 data acquisition computer. From the HP 85 through an HPIL interface loop, the data is sent back to the processing plant where it is received by a handheld calculator HP 41CX (18). Peripherals: dual microflexible disc storage and ThinkJet printer (Hewlett Packard models 9121D and 2225A) are connected to the HP 85 by way of an HPIB interface card.

The tasks of the remote hardware include transmitting power to the lamp of the sensing head and control of the mechanical function of the sensing head. In this case the mechanical function involves action of a stepping motor to change filters or to move a mirror and allow referencing. Thermal control must also be provided for the chamber containing the filters. Other functions of the remote hardware are the in-plant readout system, which must communicate with the central hardware system. The central hardware system will transmit real time data to plant control personnel or to a process control computer. In the plant, a calculator or other peripheral display receives what is being transmitted from the central hardware unit in the lab. The optical sensing head is attached to the transport system. The custom built transport system (Figure 3) involves a three inch aluminum spouting that is typically found in flour mills. To this has been welded a vertical 2"x4" piece of extruded aluminum. The spouting is typically at an angle of repose of approximately 70° or more from ground level and the vertical extrusion is dropped from that spouting. It is connected at the bottom through a round tube perpendicular to the rectangular extrusion which then feeds back to the spouting by way of a screw conveyor (19). Bolted to the extrusion is the optical sensing head, its quartz window is then flush with the hole milled in the extruded aluminum to accomodate that observation port. Opposite the window of the sensing head is an inspection port so it is possible to make sure that there has been no accumulation of materials on the optical

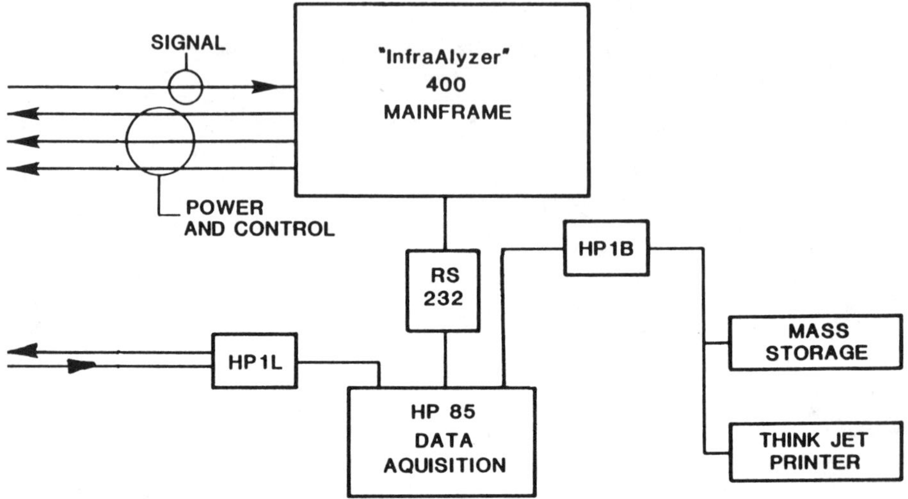

Figure 2   Block diagram of central computer, associated peripherals and I/O.

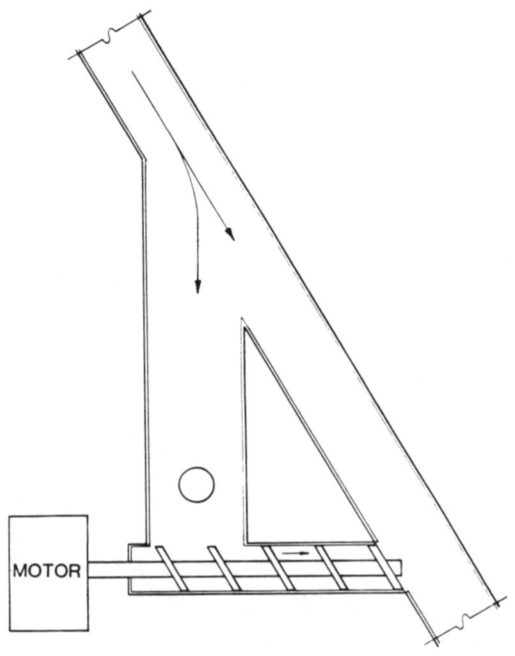

Figure 3   Sample transport system for flour milling intermediate product quality sensing.

window or if cleaning would be necessary, it could be reached
without disassembling the whole system.

The selection of the height of stock above the observation
window depends upon the material being observed. It is desirable to
have enough stock above the window to produce a relatively constant
compaction against the window but it is undesirable to have an
unnecessarily high head of stock because this produces a lag between
what is being observed and what is flowing above. The distance from
the observation port to the conveyor screw was chosen to be just
high enough that the flow in the conveyor screw would not produce
cavitation in the window area. A hydraulic motor is used to drive
the screw.

<u>Remote Readout</u>. The readout system in the plant, in this case, on
the mill floor or in the mill office, is as follows. An HPIL loop
which emanates from the HPIL interface to the HP85 control computer
extends back to the mill a distance of 108 feet. In the mill, a
handheld HP41CX calculator is plugged into the loop. An HPIL video
interface is also plugged into the loop. From the video interface
coaxial cable leads to a video monitor located in the mill office.
These monitors can be multiplexed and can be in as many locations as
deemed necessary. Furthermore, it is possible to have multiple
recepticles for HP41CX data display. This handheld device can be
carried on the operator's belt and can be plugged into a port
on any floor of the operating mill to obtain a current readout.
Figure 4 shows a block diagram of the entire system, including the
instrument room, the operation floor of the pilot flour mill, and
the location of the video monitor in the mill office. An external
HP85 computer can be used to control the Infralyzer 400 mainframe
cycle   if a patchbox is placed between the Infralyzer 400 mainframe
and the RS232 interface. The patchbox includes connection to the
standard RS232 output and a second connection loop to an internal
switching point. It is then possible to program the HP85 to time
control the sequence of data acquisiton. Instead of being driven
by the firmware of the InfraAlyzer 400, the HP85 can be programmed
for any appropriate time function, depending on the flow of product
or on the necessity to acquire data. With a 3 filter readout, for
example the minimum time would be thirteen seconds. When data is
acquired at more than 3 wavelengths, with an instrument of this
type, a longer period is required. Another variation on hardware is
elimination of the HP85 and adaptation of the handheld HP 41CX
calculator for control, readout and driving peripherals. This is
possible with an RS232/HPIL interface attached to the external
control interface patchbox. This handheld calculator can actually
be at the remote area and still initiate the measurement cycle of
the InfraAlyzer 400 in its central laboratory location.

<u>Software</u>. To devise an in plant, on-line system from table top
laboratory equipment involves some software as well. The tasks of
the software include the control of the sensing device, this is
driven by the Technicon InfraAlyzer 400 firmware. Initiation of the
cycle can be operated in one of three methods. Method A – would be
to use the autocycle firmware of the Technicon 400. Method B would
be external HP85 programmed time cycle.   Method C would be remote

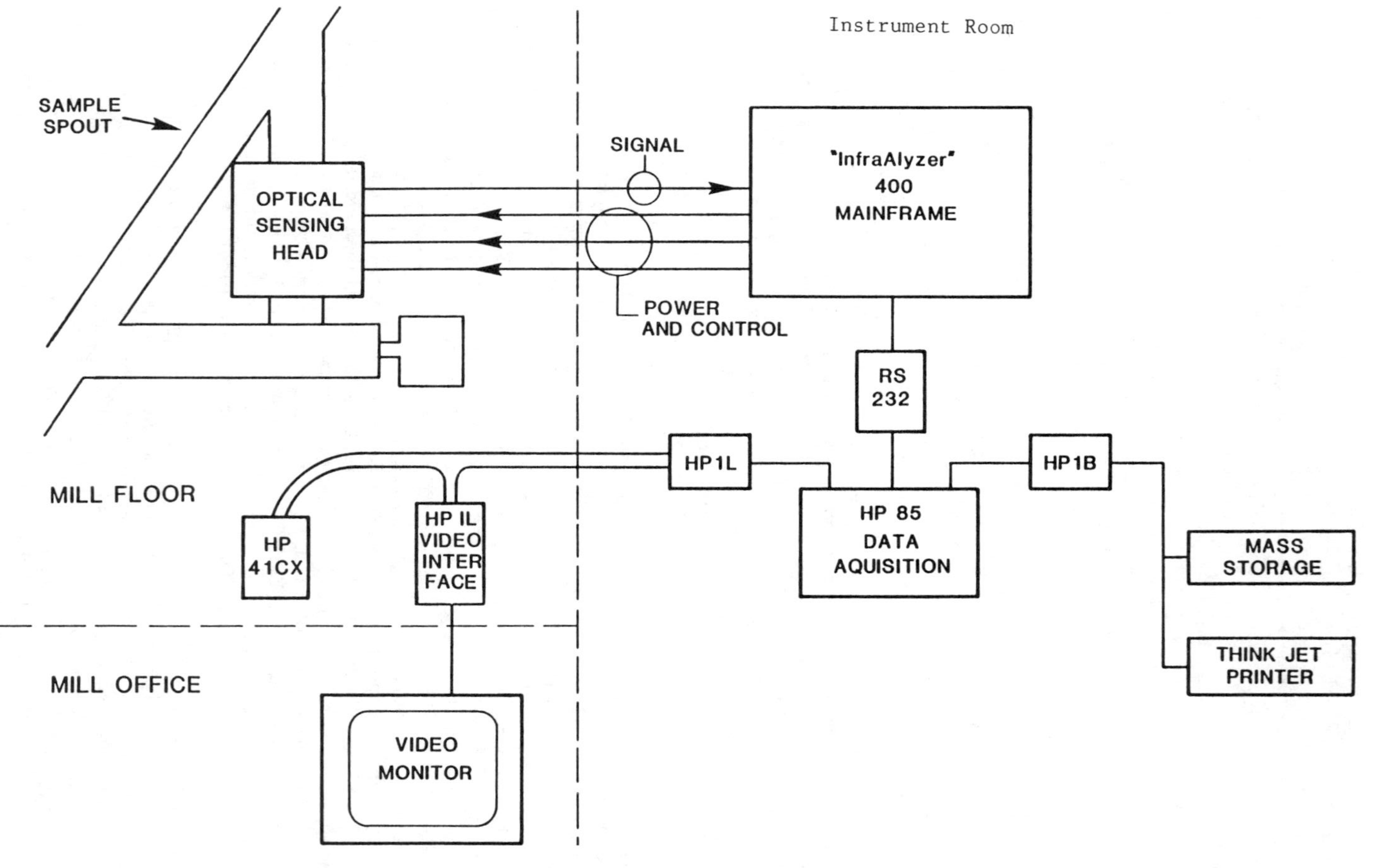

Figure 4   Block diagram of remote on-line system.

HP41CX programmed time cycle.   Another task of the software is to
process the raw data.   The InfraAlyzer 400 firmware can be used as
is for this task.   For on-line monitoring, it has been previously
mentioned, it is necessary to average the data.   The averaging
(running mean of five readings) is done on the HP85 which also
takes care of storage, local output and remote output.   The remote
output display required HP41CX programming in order to access the
signal coming from the HPIL loop and to display it as desired.
     Calibration of an on-line system is not a trivial endeavor.
Ideally,   calibration constants and if necessary, wavelength
selection should be performed on the actual sensing head with its
geometry, with the material in a flowing mode or stop flow mode,
depending on the design of the on-line system. Unfortunately, it is
not always possible to tie up the manufacturing facility to do this.
It may also not be possible to produce and run through the line
material which varies sufficiently enough to give one the breadth of
calibration needed.   If one is using an on-line moisture meter,
the selection of wavelengths is not a problem.   The analyte is
defined, its observation wavelengths are well-known and the wavelength selection can be done entirely off-line. The coefficients in
the analytical equation will be subject somewhat to the optical
geometry and to the particular filters and their densities in the
sensing head. In this case, the purity of a particular intermediate
product stream was to be monitored for several months of operation.
Samples were taken at the monitoring point and a large set was
accumulated.   By various classical techniques, it was determined
what the extremes were that had been observed through many
operations. The extremes from several runs were pooled, the pooled
ones of one extreme were called A, the pooled ones from another
extreme were called B and synthetic mixtures of A and B were made by
weight. Such a procedure is based on the assumption that with linear
combinations of the extremes it will be possible to represent the
natural intermediates which fall in between.   Assumptions like
this can be verified by using a calibration produced from a
synthetic set and applying it back to the intermediates and if
absolute numbers are not available for the intermediates, they can
at least be ranked to see if the ranking agrees with ranking by
another method.   Figure 5 shows that ideally it is desirable to
monitor the causal quality parameters.   However, it may also serve
the same purpose to measure a parameter which runs parallel or is
proportional, even if it is an inverse relationship.   There are a
number of options available.   In making synthetic binary mixtures,
one must be very careful that the primary difference between A and B
is not moisture content, which has strong absorbing peaks, or
granulation.   As for the moisture content, it is possible to avoid
incorporating moisture wavelengths into your analytical equation.
As for granulation, this can be observed from optical data of
the parent stocks of A and B.

Implementation.   The rationale for control using monitoring of
quality between A, defined as good quality and B, defined as poor
quality, is as follows.   If the stock going through the monitor is
like A (good quality) then the process would be routed by a certain
pathway or a certain control of a particular machine.   If the stock

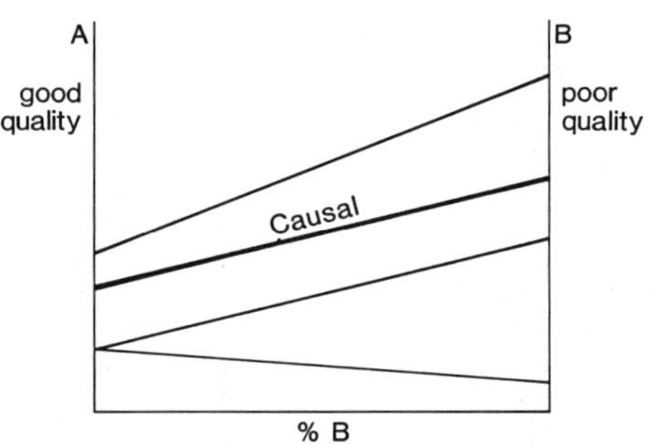

Figure 5 Parameter variation of binary mixture.

is more like B, then it would go either by an alternate pathway or a slighlty modified function of the next machine.   It is necessary to choose the limits of percent B for the control functions and rapidly analyze percent B with the remote on-line monitor.   Timeliness of analysis is the key to process control.   The system described is a potential means to that end.

From the economic and practicality standpoint, it should be obvious that it is essential to find a key monitoring point.   If there is little variation of stock at that point, there is little reason to monitor.   If a considerable amount of variation is observed, whether this comes about by natural variables in the mill operation or by natural variables brought about from different batches of raw materials, or whether this is brought about by a pathological condition from something going wrong with a piece of machinery, then there is a potential for the system to be out of control and the time lag for finding this may be long, and the losses may be great.   The penalties of not detecting this change may vary greatly from producing a small amount of off-grade product to having inefficient production for the period of out of control operation.   In a 24 hour process, this could mean 10 to 16 hours of inefficient or off-grade production, if no correction is made.   The second thing to consider in evaluating practicality is what can be done if changes are observed?   The obvious question is what would the supervisor do to the process if he were aware of the change? What the supervisor would do can be programmed into a process control computer which could automatically make a correction. Control action requires some kind of a movement of a lever to divert stream of product,   opening of a valve, opening of an airjet, controlling of a motor, controlling of the final adjustment of pieces of equipment.   The control action course is dictated by the particular process and the economics associated with implementing it.   The economics associated with payback must be calculated on an individual basis.   Rapid, timely, on-line monitoring provides the processor with an option.   It is naive to assume that the entire efficiency of a complex process could be controlled by just one monitoring point and one automated corrective action.   It is more likely that multiple monitoring points would be necessary and multiple intercorrelated correction processes would have to be driven.   All of this translates into the potential of considerable expenditure and the necessity for parallel sensoring systems net-worked into a central computer.   There is also the possibility of multiplexing the several sensing heads to one mainframe instrument in order to share the function of that mainframe.   Typical economic data and variables observed in a flour mill are presented in a later discussion.

<u>Monitoring System Development</u>.   The sequence of developing the system just described involved seven steps.   The first involved gathering fundamental knowledge of the infrared spectroscopic characteristics of wheat components (Figure 6). Secondly, experience was obtained in quantitation of constituents of individual streams in spite of their heterogeniety and varied chemical composition. Optical sorting has enabled us to reduce the standard error of analysis by a factor of 2 by sorting into subsets according to

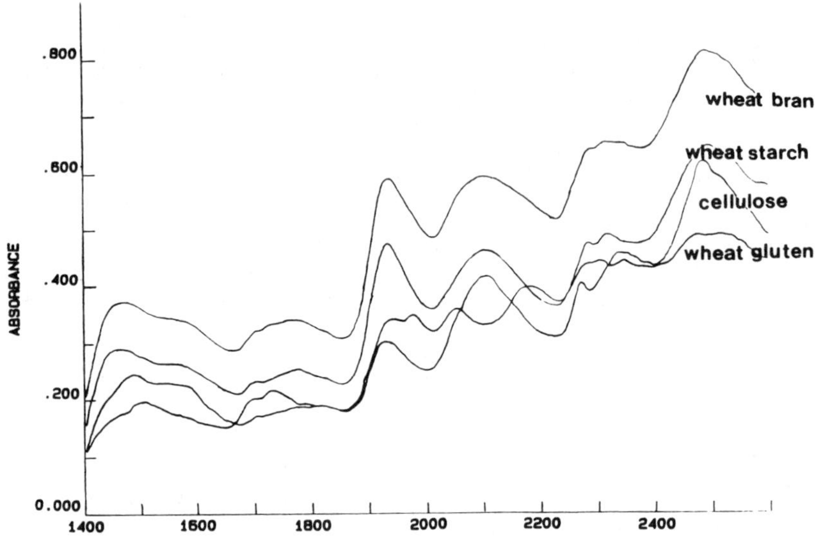

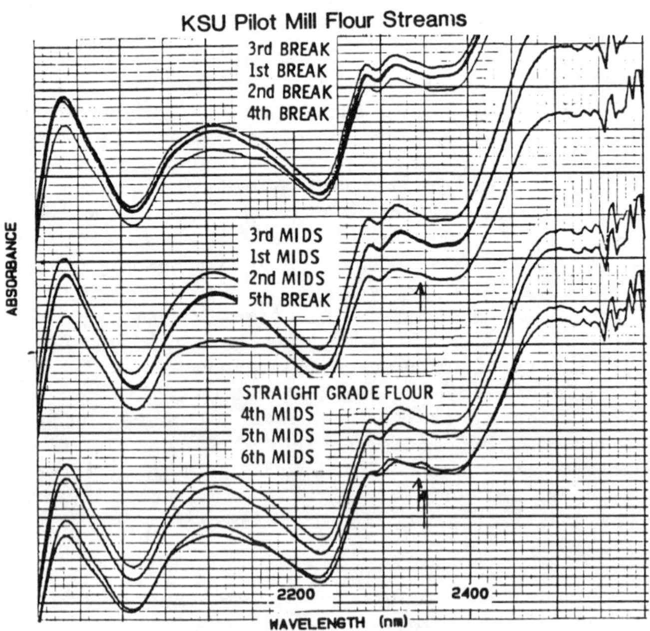

Figure 6   Near-infrared spectra of wheat fractions (Wetzel, D. L.; Mark, H. "Spectroscopic Determination of Cellulose as Criterion of Flour Purity with Respect to Bran", American Association of Cereal Chemists 62nd Annual Meeting, San Francisco, CA 1977).

median particle size. Where we have an overall calibration which gives us a standard error of 0.283% protein for example, out of roughly 12% protein stock, the standard errors are going to be reduced to 0.15% by a sorting process. That is to say, the range of calibration is broken up either according to the range of the analyte or according to characteristics of the matrix and individual calibrations are done then a sorting process can direct the computer to apply the most appropriate calibration which will give you the tightest fit and the best analytical results. With samples as diverse in composition and heterogeniety as different flour streams in a mill, this was required for protein analyses, but it was not required for moisture analysis (20). For the purity monitoring which we described in this case history, one test point and the range of variability at that test point had to be obtained by collecting samples under various conditions of operation over a period of time in order to establish the variables in granulation size, composition and purity. Off-line calibration was done involving specific intermediate products. A table top instrument was modified for remote on-line use by separation of optical and computing modules and extension of cables. The transport system was designed, fabricated and tested in the mill. The overall system, including the 400 mainframe, all the peripherals and all the remote systems was installed, debugged and tested during mill operation. Finally, special milling processes were run with the on-line monitor system in place (21). This was used for regular scheduled runs of the mill. The economics of monitoring were tested by experimental pilot mill runs where deliberate changes were made in the mill and deliberate corrective actions were taken. To assess the overall effect, numerous samples were collected of 23 product (flour) streams and analyzed by conventional methods. The profile of the mill was determined to generate the economic data. Examples of such profile data are shown on Figure 7 where a maximum of low ash product is desirable and Figure 8 where the weighted summation of products having two values (obtained from Figure 7) differs for processing with and without control action.

Figure 9 illustrates response as a function of time. The parameter monitored in this case was protein. In this case, a different batch of material was sent through to test the system's ability to respond to that new material. One economic note, in the case of one of our flour milling experiments shows that where it is possible to produce 58% of the more valuable product before adjustment after the correction of processing, this increased to 63%. The value of this product (patent flour) is about 125% the value of the secondary product (clear flour) so a shift of 5% of material from the category of lower value to one of higher value represents an important step in processing efficiency. For a flour mill operating at 1500/cwt per day, a 5% shift represents 120 to 165 dollars per day or 44-60 thousand dollars per year. The spectra showed on Figure 10 are four successive readings at 9:15, 10:15, 10:50, 11:15 of milling operations the same day. These were samples collected and scanned on a scanning laboratory instrument. The shift of the baseline indicates a drastic change in granularity. Such drastic changes in granularity over relatively short periods of time show that any calibration would have to be sufficiently robust

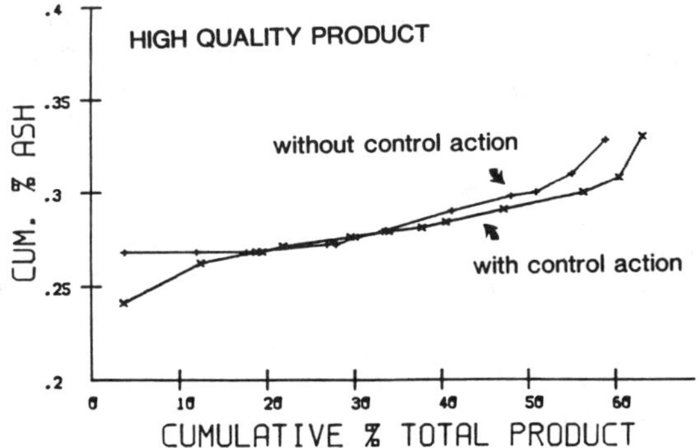

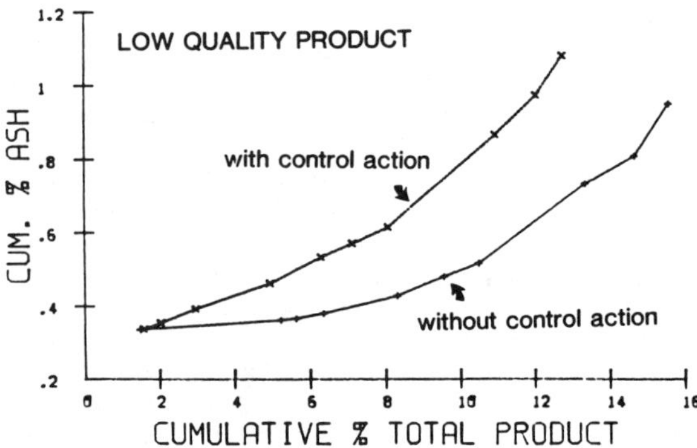

Figure 7　Miller's (cumulative ash) curves for production purity and extraction with control action (more high value, low value) without control action (less high, more low value).

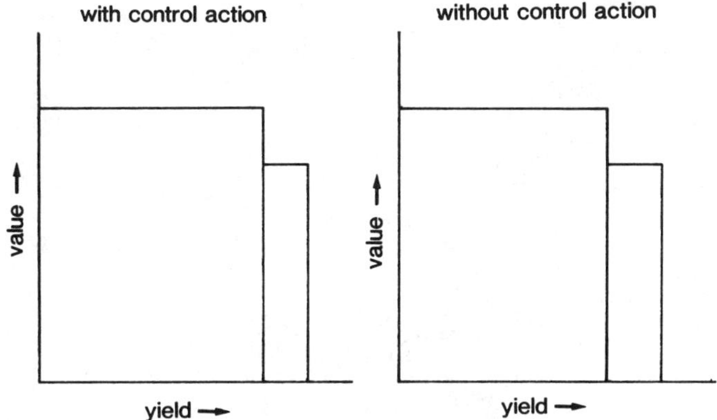

Figure 8   Distribution of milling products by value and yield (area denotes summation of product value).

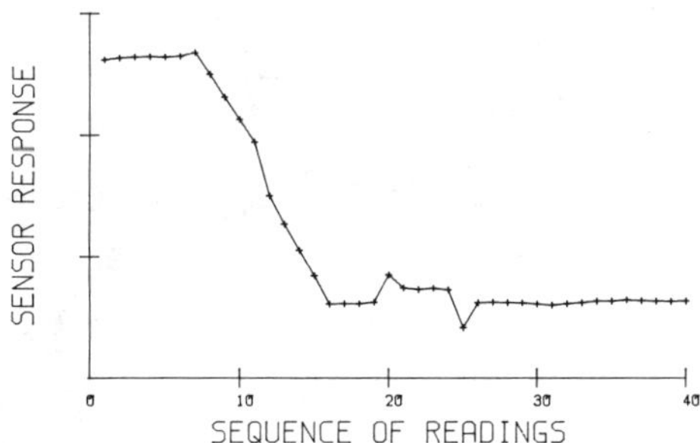

Figure 9   Response of running mean of sensor to change in protein.

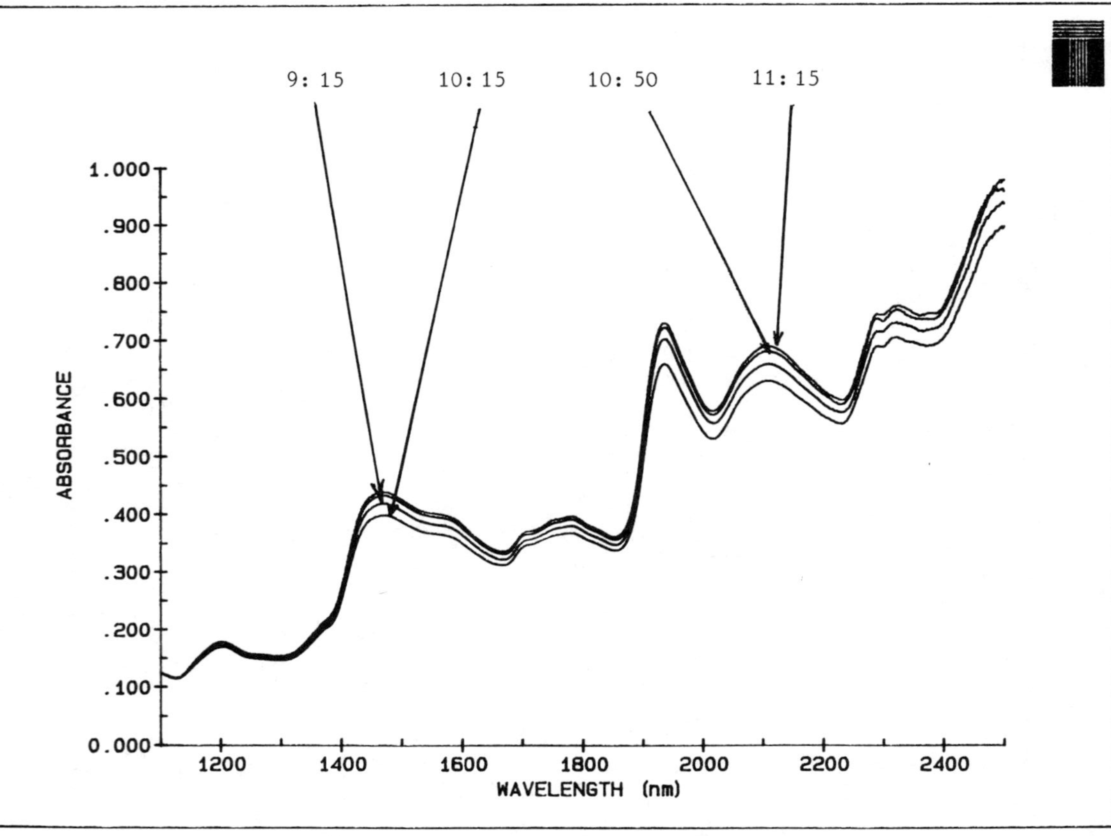

Figure 10   Near-infrared reflectance spectra of intermediate products at four times in the same run.

to withstand this or would have to have built into it a warning to tell you that at a given time, the reading you are getting may be outside the calibration range. Such variations in sample were not encountered in our first five and one half months of operation but the results shown on the previous diagram occurred when the miller decided to try something different, such a simple thing as turning off the air which provided turbulence in a particular purifier resulted in gravity taking over where you normally had turbulence and after five and one half months of successful operation with a particular calibration, suddenly a change in physical character- istics of the sample put the product outside of our calibration range. When the unexpected happens, that has to be added to the experience bag and incorporated so there will be less unexpected things occurring.

Contribution No. 86-178-B from the Kansas Agricultural Experiment Station.

Literature Cited

1.   Fulcher, R. G. Food Microstructure. 1982, 1, 176-75.
2.   Jensen, SV. A.; Munk, L.; Martens, H. Cereal Chem. 1982, 59(6), 477-84.
3.   Kaye, W. Spectrochim. Acta. 1954, 6, 257.
4.   Kaye, W. Spectrochim. Acta. 1955, 1, 181.
5.   Whetsel, K. B. Appl. Spectrosc. Rev. 1968, 2(1), 1.
6.   Wetzel, D. L. Anal. Chem. 1983, 55, 1165A.
7.   Ben-Gera, I.; Norris, K. H. Israel J. Agr. Res. 1968, 18, 125.
8.   Honigs, D. E; Hieftje, G. M.; Hirschfeld, T. Appl. Spectroscopy. 1984, 38, 844.
9.   Stark, E. W. "Interaction Between Data Treatments and Calibration Techniques in NIRA". 1985 Pittsburgh Conference on Analytical Chemistry and Applied Spectroscopy. New Orleans, LA. Feb. 1985; Paper 1096.
10.. Wetzel, D. L. "Physical Sample Characterization by Granulation Sorting from Diffuse Reflectance Measurement in the Near-Infrared". Proc. Third Ann. Users Conference for NIR Researchers. Pacific Scientific, Silver Springs, MD. Feb. 1984.
11.   Rose, J. R. "Discriminant Analysis of NIR Research", 10th Annual Federation of Analytical Chemistry and Spectroscopy Societies meeting, Philadelphia, PA, Sept. 1983; Paper 26.
12.   Mark, H. L. Anal. Chem. 1985. 57, 1449-56.
13.   Hirschfeld, T. "Near Infrared Trace and Microanalysis", 1985, Pittsburgh Conference on Analytical Chemistry and Applied Spectroscopy. New Orleans, LA, Feb. 1985; Paper 1093.
14.   Bolling, H.; Zwingelberg, H. Getreide Mehl u. Brot, 1984, 38, 3.
15.   Osborne, B. "Progress of NIRA in Milling and Baking". Proc. of the Seventh International Symposium on Near- Infrared Reflectance Analysis, Technicon Industrial Systems, Tarrytown, NY. July 1984.

16.   Stokes, L. "An Economical and Practical Evaluation of an
       On-Line Instrument for controlling Protein in Soybean Meal",
       American Association of Cereal Chemists 64th Annual Meeting,
       Washington, DC, Oct. 1979; Paper 145.
17.   Cooper, P. J. "Developing and Testing On-Line NIR Methods for
       Process Control", 1985 Pittsburgh Conference on Analytical
       Chemistry and Applied Spectroscopy. New Orleans, LA, Feb.
       1985; Paper 852.
18.   Wetzel, D. L., Levin, J. A. and Posner, E. S. "On-Line
       Process Monitoring: Hardware Software and Economics". Proc.
       of the Eighth International Symposium on Near Infrared
       Reflectance Analysis, Technicon Industrial Systems,
       Tarrytown, NY, August 1985.
19.   Wetzel, D. L. "On-Line Monitoring for Control of Process".
       Proc. Seventh International Near-Infrared Reflectance
       Symposium, Technicon Industrial Systems, Tarrytown, NY, July
       1984.
20.   Wetzel, D. L., Wehling, R. L. and Stalnaker, J. B.
       "Individual Millstream Analysis by Near Infrared Reflectance",
       Annual Meeting of the American Association of Cereal Chemists,
       Chicago, IL, Sept. 1960.
21.   Posner, E. S. and Wetzel, D. L. "NIR for control of a flour
       mill", Annual meeting of the Association of Operative Millers.
       Cleveland, OH, April 1985.

RECEIVED January 24, 1986

ENVIRONMENTAL SENSORS

Electrochemical Sensors, Sensor Arrays, and Computer Algorithms

For Detection and Identification of Airborne Chemicals

Joseph R. Stetter

Energy and Environmental Systems Division, Argonne National Laboratory, Argonne, IL 60439

Recent developments in the field of sensing airborne chemicals using electrochemical sensors and sensor arrays are reviewed. Such systems detect, identify, and quantify potential chemical hazards to protect the health and safety of workers and citizens. The application discussed in this review article is single chemicals at part-per-million levels in air. The sensor system consists of an array of sensors used in four modes of operation, and the data are interpreted by a computer algorithm. Pattern recognition techniques are being used to understand the information content of the arrays and to focus future experimental work.

This paper provides a brief review and technical perspective on recent developments in the field of sensing airborne chemicals using electrochemical sensors and sensor arrays. Selective detection of gases and vapors is central to solving the many industrial and societal problems surrounding hazardous chemicals. For example, inexpensive sensors that can detect, identify, and quantify otherwise invisible hazards are needed to characterize source emissions, trace the transport of chemicals through the environment, measure levels of human exposure, design cost effective clean-up strategies, and protect the health and safety of both workers and citizens.

Solving these gas and vapor detection problems will require a variety of new sensors, sensor systems, and instruments. Field detection of airborne chemicals can be somewhat arbitrarily divided into three distinct situations. The first case is when a spill or leak results in a single compound occurring in air far in excess of its background concentration. The second case is when one or several trace constituent(s) occur in a complex background ("needle-in-the-haystack" problem). The third case is when a complete analysis is needed for all minor as well as major constituents of a complex mixture. The first case is the one specifically addressed by the approaches discussed in this review article. The second and third cases, although not discussed here, will be addressed in the near future by a portable, inexpensive sensor technology.

The three categories of gas detection problems mentioned above are extremely important in developing the initial guidelines for sensors and instruments. Each type of problem will require certain capabilities in the sensors and the algorithm required to interpret the data. However, it is not reasonable to expect a single sensor or instrument to be developed for each chemical or situation. One way to increase the information content of sensors is to use them in sensor arrays. This approach is contributing significantly to solutions for the detection problems presented by the three categories.

All sensor arrays are not equally promising. So, while the technology is being developed, how is one to judge which avenues to pursue and which to avoid? The discussions in the following three sections provide a useful framework for approaching this problem of sensor research and instrument development. Commentaries on recent designs for electrochemical sensors, sensor arrays, and algorithms for identification and quantification of airborne chemicals are presented from the above perspective.

Electrochemical Sensors

Amperometry can be used to detect and identify airborne chemicals, and devices based on this technology have existed for several years for measuring such gases as CO ($\underline{1}$), NO ($\underline{2}$), NO_2 ($\underline{2}$), H_2S ($\underline{3}$), alcohol ($\underline{4}$), and hydrazines ($\underline{5}$). Figure 1 is a schematic diagram of an electrochemical detector system. The sensors are operated at constant potential; thus, the general analytical technique can be categorized as chromoamperometry or constant-potential amperometry.

The sensor consists of six major parts (see Figure 1): filter, membrane, working or sensing electrode, electrolyte, counter electrode, and reference electrode. Each part influences the overall performance characteristics of the sensor. Choosing construction materials and sensor geometry is critical and has a profound influence on the accuracy, precision, response time, sensitivity, background, noise, stability, lifetime, and selectivity of the resulting sensor. The relationships among materials of construction, sensor geometry, and performance characteristics of these sensors are still poorly defined. Although a thorough discussion of these relationships is beyond the scope of this article, the response mechanism of an amperometric electrochemical cell should be discussed.

The response of an amperometric gas sensor can be described by the following eight steps:

1. Introduction of the chemical to the sensor through the filter,

2. Diffusion of the reactant across the working electrode membrane,

3. Dissolution of the electroactive species in the electrolyte,

4. Diffusion of the chemical to the electrode-electrolyte interface,

5. Adsorption onto the electrode surface,

6. Electrochemical reaction,

7. Desorption of the products, and

8. Diffusion of the products away from the reaction zone,

Any of these steps can be rate limiting, thus determining the ultimate sensitivity and response characteristics of the electrochemical sensor. The parameters most frequently observed to influence the characteristics of these sensors include sample flow rate, working electrode composition, electrolyte, membrane type, and electrochemical potential of the sensing electrode. By controlling these parameters during design, the sensor engineer can achieve the desired sensor response characteristics.

The eight reaction steps in the sensor model include a variety of chemical and physical processes, all of which are influenced by the system components shown in Fig. 1. The sensor is usually designed so that the kinetics of the physical processes (i.e., mass transport by diffusion) are limiting, but it is possible to construct sensors that exhibit performance characteristics limited by the kinetics of the chemical/electrochemical processes.

Step 2 is usually limited by the permeability of the membrane. In certain sensor designs, the membrane is eliminated to avoid this step. Step 4 refers to the diffusion of the solvated gas in the electrolyte to the electrode-electrolyte interface. Diffusion in liquids is often considerably slower than diffusion across a membrane. If the sensing electrode is flooded with electrolyte, the response is slow because the gas must diffuse through the electrolyte before reaching the reaction surface.

Typically, increasing the volumetric flow rate through the sensor increases the mass transport of the analyte to the working electrode, thereby increasing the observed signal (i.e., sensitivity). High flow rates also decrease sensor response time. Sensor response with changing flow rate has been discussed (1, 6). The selectivity of the sensor can also be improved by controlling the electrochemical potential of the working electrode. For example, proper selection of a Au electrode potential will allow the determination of NO_2 in the presence of NO (2, 7).

The construction materials of each sensor part will influence its operating characteristics, as illustrated in the following examples. Choosing a Au rather than a Pt electrocatalyst for the sensing electrode allows for selective determination of H_2S in the presence of CO (8). Using a charcoal filter in combination with a

CO sensor allows detection of CO in the presence of hydrocarbons and other adsorbable contaminants. The membrane is usually chosen for its ability to protect the sensing electrode. However, if it has low permeability to air, the sensor will have a slower response time. The electrolyte and counter electrode have also been reported as influencing selectivity and device performance in the determination of hydrazines (5) and NO_2 (9), respectively. Finally, materials of construction are typically Teflon and high-density plastics like polypropylene because such materials must be compatible with reactive gases and corrosive electrolytes.

Amperometric sensors are important in portable instrument design because they are relatively small, inexpensive, and lightweight, and use very little power to generate significant signals. They can detect part-per-million (ppm) levels of electrochemically active gases and vapors, can be engineered to have significant selectivity, and can be operated over a wide range of temperatures. These excellent operating features make electrochemical cells particularly suited for portable instruments constructed with electrochemical cells. With the proper choice of filters, flow rates, and potentials, they can provide stable and meaningful measurements in a variety of field situations.

Sensor Arrays

Electrochemical sensors respond to a limited number of chemicals, and each one responds with limited selectivity. One way to overcome these limitations, or at least to improve the sensor's capabilities, is to construct sensor arrays (10, 11). Information is created in a sensor by its reaction to a chemical stimulus. This reaction of the sensor to the type of chemical and its concentration creates an analytical signal that is decoded into an electronic signal by the sensor. In other words, the electronic signal contains the recorded information. More specifically, if chemicals are to be identified on the basis of their relative electrochemical reactivities, then the sensor array must be designed to accomplish this task. It must contain sensors that record a different electrochemical response for each chemical species. Conversely, the chosen sensors for the array must only decode those chemicals exhibiting differing reactivities on the given sensors. Thus, to achieve the most effective method, the sensor array and the analytical problem must be considered together.

The objective is to perform both qualitative and quantitative analyses simultaneously. Each sensor yields a specific type and amount of information. However, if sensors are combined, the ratios of sensitivities among different sensors provide new analytical information. Thus, the information content of the arrays is greater than that developed by an individual sensor (12).

A sensor array consisting of four electrochemical sensors and two hot wire filaments has recently been evaluated (13-15). The array was constructed to both identify and quantify a toxic vapor in a few minutes using a portable, battery-operated, lightweight instrument. The electrochemical sensors are combined with hot catalyst filaments to allow the electrochemical sensors to respond to a broad range of chemicals. Gases are drawn over the filament and into the electrochemical sensor, where pyrolysis fragments created by the hot filaments are detected by the electrochemical sensors. Electroactive gases and vapors are sensed directly, and nonelectroactive gases and vapors (e.g., benzene) are detected after pyrolysis over the heated filament ($\underline{10}$).

In this manner, a nearly universal and very nonselective detector is created that is a compromise between widespread response and high selectivity. For example, the photoionization detector (PID) can detect part-per-billion levels of benzene but cannot detect methane. Conversely, the flame ionization detector (FID) can detect part-per-billion levels of methane but does not detect chlorinated compounds like CCl_4 very effectively. By combining the filament and electrochemical sensor, all of these chemicals can be detected but only at part-per-million levels and above. Because most chemical vapors have toxic exposure limits above 1 ppm (a few such as hydrazines have limits below 1 ppm), this sensitivity is adequate for the initial applications. Several cases of electrochemical sensors being used at the sub-part-per-million level have been reported ($\underline{3}$, $\underline{16}$). The filament and electrochemical sensor form the basic gas sensor required for detecting a wide variety of chemicals in air, but with little or no selectivity. The next step is to use an array of such sensors in a variety of ways (modes) to obtain the information required to perform the qualitative analysis of an unknown airborne chemical.

The four electrochemical sensors were carefully chosen and have two working electrodes of Au and two of Pt. One Au and one Pt electrode are operated at anodic potentials to facilitate oxidations, and the other two are operated at cathodic potentials to facilitate reductions. When an electroactive gas passes through this array, the half-wave potential of the chemical species is not measured. However, by comparing the signals from the sensors in the array, one can tell whether the half-wave potential is above or below 1.0 V vs. the standard hydrogen electrode (nhe). Thus, the array signals from these sensors, while not measuring the thermodynamic half-wave potential, do provide a set of chemical parameters related to the vapor's electrochemical properties. Hence, the term "chemical parameter spectrometry" was chosen to describe this technique.

Under microprocessor control, a constant concentration of the unknown vapor (e.g., 100 ppm benzene in air) is passed through the array, and the steady state signal from the four sensors is read with (1) the Pt filament at 900°C, (2) the Rh filament at 900°C, (3) the Rh filament at 1000°C, and (4) no filaments on. This procedure generates 16 signals for each gas or vapor as shown in Figure 2 for two representative chemicals, benzene and cyclohexane. The largest signal is set equal to 1.0, and the other channels are scaled accordingly, making the histogram concentration independent in the linear range of the sensors. A wide range of compounds has been investigated at concentrations of 3 ppm to several hundred parts per million, each producing a unique response patterns ($\underline{14}$). Once the

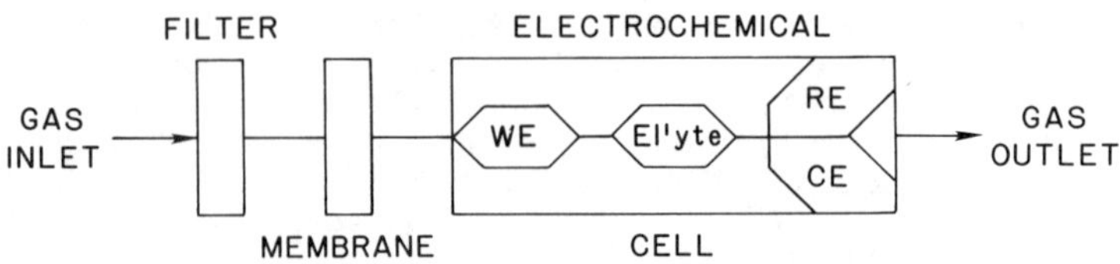

Figure 1 Conceptual model of an amperometric gas/vapor sensor (WE = working electrode, El'yte = electrolyte, RE = reference electrode, and CE = counter electrode).

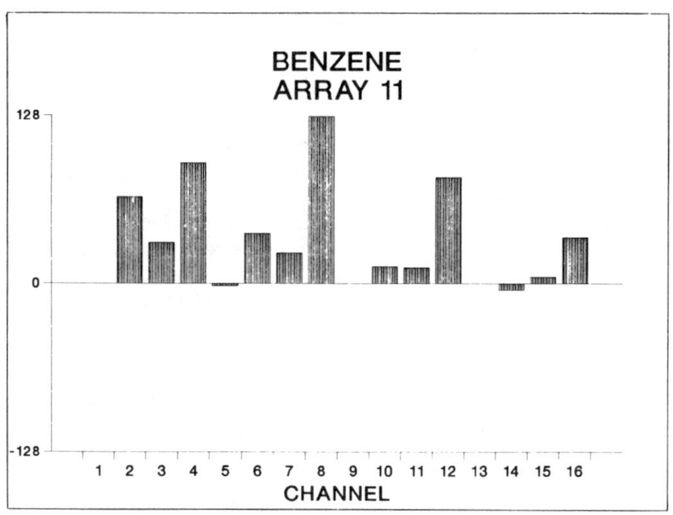

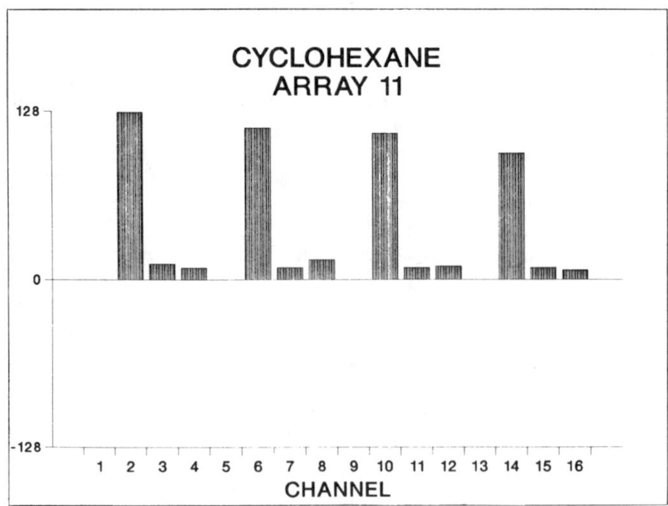

Figure 2 Normalized response of sensor array to benzene and cyclohexane.

histogram is known for a given compound, it can be used like an infrared spectrum or fingerprint to identify the chemical.

During operation of this sensor array, significant information is gathered by the four electrochemical sensors about the electrochemical activity of the chemical and its pyrolysis products.   Comparing the sensor readings taken with the Rh filament at 900°C vs. 1000°C reveals whether formation of electroactive products is a highly activated or a nonactivated process. Similarly, comparing sensor readings for the Pt vs. Rh filaments provides a comparison of the catalytic activity of Pt vs. Rh for the pyrolytic formation of electro-oxidizable or electro-reducible compounds.   Thus, sensor readings taken together provide a set of parameters related to the specific electrocatalytic and catalytic properties of the individual chemicals.

An automated instrument incorporating this sensor array and operating procedure was constructed (15) to evaluate and demonstrate its utility for gas detection.   Using microprocessor control, the instrument sensors could be zeroed, calibrated, or operated in any one of three modes of operation -- "univ(ersal)," "select," or "ident(ify)."   In universal mode, the instrument displays a simple bar graph of any contaminant present in the air to which this sensor array responds (14), and this graphic display can be used to locate leaks.   In the identify mode, approximately 500 cm^3 of vapor/air mixture trapped in a sample bag is analyzed (i.e., a histogram is generated).   After analysis, the user is given a report of the gas identity and concentration, the percentage of the short-term exposure limit (STEL), number and name of other compounds in the device library that are "close" to the unknown, and the "euclidean" distance of the unknown from the closest library entry.

These latter two pieces of information relate to the reliability of the analysis performed by the unit.   The intent of this exercise has been to demonstrate that the array device can provide sufficient information to identify a chemical hazard and to decide whether the analysis should be believed.

<u>Algorithms</u>

The system algorithm is the heart of the array device's ability to provide the needed information rapidly and in a useful format.   The quantitative information can be easily obtained from the strongest linear data channel using traditional calibration techniques.   More subtlety is needed to obtain the qualitative information described previously.   How that is done is the subject of the following section.

The simplest form of pattern comparison (euclidean distance in 16-dimensional space) was used to design and build an operational portable gas monitoring unit (15).   With the limited computer power of a portable instrument, any one of about a dozen gases could be identified in less than one minute of computational time.   This algorithm was evaluated using a data set for repeated runs of 16 different chemicals in 2 different sensor arrays (14).   The results indicated that the arrays were correct 52 of 62 and 40 of 48 times, respectively, using the averages of the data sets as reference libraries. Using a weight vector that emphasized channel 16 improved the percent correct identifications to 97.5%. The euclidean distance of the individual pattern vector from its library entry is related to the correctness of the identification. This demonstration of a useful array confirms the practical information content of the array and the confidence with which this information can be routinely extracted from the data set.

To interpret the signals generated by sensor arrays, computerized data-handling techniques are required. Although "fingerprints" can often be recognized by the trained eye as unique, such qualitative information must be translated rapidly and efficiently into a quantitative measure of how unique the fingerprint is. A data set containing 22 individual chemical responses was evaluated using pattern recognition techniques (17). Such pattern recognition algorithms are very useful in the quantitative evaluation of the information content of a particular array or data set. In this study, correlation was found among most pairs of data channels: only five uncorrelated channels were found. However, even though the data channels exhibited high correlation, the patterns were sufficiently unique to identify each of the 22 compounds in the data set. These results provided the first quantitative measure of the "uniqueness" of the information generated by this array and method (17).

The information content in such systems is potentially large and should be sufficient not only to identify single chemicals in air, but also to identify mixtures of chemicals. To resolve mixtures of N compounds unambiguously, N independent parameters yielding N simultaneous independent equations are needed. An earlier work (11) suggests a practical approach to this problem. The minimum number of parameters P required to identify compounds N on the basis of the presence or absence of significant signals in various channels is:

$$2^P - 1 \geq \sum_{I=1}^{A} \frac{N!}{(N-I)!I!}$$

where the compounds may be present in combinations of up to A components. For example, for N = 100 chemicals in mixtures where A = 3 or 4, then P (number of independent channels) must be $\geq$18 or $\geq$22, respectively.

To identify any one or any combination of thousands of chemicals with a sensor array is a prodigious task. Experimentally, such identification will demand a large number of sensors and operating modes, resulting in extended measurement times. Designing a system for 99.99% of the expected or most frequent occurrences would result in high reliability and a simpler data-gathering system (sensor array). The available algorithms and the sensor array are a closely connected team. The array must be designed to collect the information that contains the unambiguous answer to the sensing problem, and the algorithm must extract this information rapidly and accurately. If properly designed, the algorithm will require a minimum amount of computational time on a small computer and will minimize the design requirements of the sensor array. While the algorithm that interprets the sensor response must be efficient, it must also extract the desired information unambiguously. This requirement is a substantial challenge when working with elaborate arrays, large data sets, and complicated responses (e.g., mixtures of chemicals).

As previously mentioned, computer programs can be used to evaluate the "uniqueness" of the data set produced by a given array for a given set of chemicals. This capability is extremely important because algorithms can be designed and used to measure array redundancy (11), thereby providing the information required to custom design arrays to solve specific problems. In other words, computer-aided-design (CAD) of arrays is possible. Further, because the difference between known and unknown patterns can be rapidly calculated, portable instruments can indeed be used for rapid identification in the field. Finally, it must be stressed that pattern recognition does not create information that the sensors did not generate. However, computer algorithms can often make the information contained in such data sets more obvious and certainly easier to display and use. Thus, the focus for hardware is to design arrays that generate more information, and the role of algorithms is to guide instrument development and evaluate the success of the experiment.

Conclusions and Future Work

Experiments to date have shown that a portable instrument incorporating a thoughtfully chosen array of sensors can detect, identify, and quantify a wide variety of chemicals in air. Also, pattern recognition techniques are being used to understand the information content of the arrays and to focus future experimental work. Development of smaller, more sensitive, and more reliable electrochemical sensors will expand the applications of the system described here.

Progress in the application of sensor arrays to gas analysis will be made through increasingly independent data channels using novel combinations of sensors and operating modes. Computational approaches will be modified to suit specific types of sensor arrays and to make economical use of computational space for portable instrument applications. The primary challenges of the near future will be to solve the "needle-in-the-haystack" problem and to proceed to complex mixture analysis using a plurality of sensor responses.

Acknowledgments

Work performed for the U.S. Coast Guard through an agreement with the U.S. Department of Energy, under Contract DTCG23-84-F-20045.

Literature Cited

1. Bay, H. W.; Blurton, K. F.; Sedlak, J. M.; Valentine, A. M.
    Anal. Chem. 1974, 46, 1837.
2. Blurton, K. F.; Sedlak, J. M. Talanta. 1976, 23, 811.
3. Stetter, J. R; Sedlak, J. M.; Blurton, K. F. J. Chromatog.
    Sci. 1977, 15, 125.
4. Blurton, K. F.; Bay, H. W.; Lieb, H. C.; Oswin, H. G. Nature.
    1972, 240, 52.
5. Stetter, J. R.; Blurton, K. F.; Valentine, A. M.; Tellefsen,
    K. A. J. Electrochem. Soc. 1978, 125, 1804.
6. Stetter, J. R.; Tellefsen, K. A.; Saunders, R. A.; DeCorpo, J.
    J. Talanta. 1979, 26, 799.
7. Zaromb, S.; Stetter, J. R.; Attard, G. J. J. Electroanal. and
    Interfacial Electrochem. 1983, 148, 289.
8. Blurton, K. F.; Sedlak, J. M. Talanta. 1976, 23, 445.
9. Stetter, J. R.; Cromer, R. B. U.S. Patent 4 302 315, 1981.
10. Stetter, J. R.; Zaromb, S.; Findlay, M. W., Jr. Sensors and
    Actuators. 1985, 6, 269.
11. Zaromb, S.; Stetter, J. R. Sensors and Actuators. 1984, 6,
    225.
12. Hirshfeld, T.; Callis, J. B.; Kowalski, B. R. Science. 1984,
    116, 312.
13. Stetter, J. R.; Zaromb, S.; Penrose, W. R.; Findlay, M. W.;
    Otagawa, T.; Sincali, A. J. Proc. Hazardous Materials Spills
    Conf., J. Ludwigson, ed., 1984.
14. Stetter, J. R.; Penrose, W. R.; Zaromb, S.; Christian, D.;
    Otagawa, T. Proc. 2nd Annual Tech. Seminar on Chem. Spills,
    1985.
15. Stetter, J. R.; Penrose, W. R.; Zaromb, S.; Stull, J. O. Anal.
    Instrum. 1985, 21, 163.
16. Rogers, P. M.; Stetter, J. R.; Lau, Y-K; Cromer, R. B. Proc.
    JANNAF Environmental and Safety Subcommittee, 1980, pp. 1-22.
17. Stetter, J. R.; Jurs, P. C.; Rose, S. L.; Stull, J. O. Proc.
    1985 JANNAF Safety and Environmental Protection Subcommittee,
    1985.

RECEIVED December 4, 1985

Amidoxime-Functionalized Coatings for Surface Acoustic Wave Detection of Simulant Vapors

Neldon L. Jarvis[1], John Lint[2], Arthur W. Snow[3], and Hank Wohltjen[4]

Chemistry Division, U.S. Naval Research Laboratory, Washington, DC 20375–5000

The objective of the work was to prepare coatings for a surface acoustic wave (SAW) sensor that would have a chemical specificity and sensitivity toward phosphonate ester agent simulants. Butadiene-acrylonitrile-acrylamidoxime terpolymer coatings of varying composition were prepared and evaluated for chemical sensitivity toward methanesulfonyl fluoride (MSF) and dimethyl methylphosphonate (DMMP) simulants. Chemical reactivity was monitored by infrared spectroscopy, and physical adsorption by SAW frequency measurements. The amidoxime functional group did not react rapidly and quantitatively with the MSF vapor. Both simulants were physically adsorbed by the coating the degree of which is correlated with solubility parameters, vapor pressures and glass transition temperatures.

The preparation of thin film coatings with highly sensitive and specific absorption for particular vapors is a critical requirement for the development of chemical electronic microsensors as point detectors. In this work, coatings which have a specific interaction with simulants for nerve agents are being investigated using a SAW device. The SAW device detects extremely small gravimetric changes in a coating by registering a frequency shift in the resonance of the piezoelectric substrate. Previous work from this laboratory has employed a model reactive polymer-vapor system based on the Diels-Alder reaction between poly(ethylene maleate) and cyclopentadiene to learn how the SAW sensor responds to reactive and non-reactive vapors (1). In addition to discrimination between physisorption and

[1]Current address: Chemical Research and Development Center, U.S. Army, Aberdeen Proving Grounds, MD 21010
[2]Current address: 5536 Fillmore Ave., Alexandria, VA 22311
[3]Author to whom correspondence should be addressed.
[4]Current address: Microsensor Systems, Inc., P.O. Box 90, Fairfax, VA 22030

chemisorption, it was also learned that good sensitivity is dependent on the polymeric film matrix being above its glass transition temperature for efficient permeation of the vapor. In this work the objective is to incorporate a functional group, that has specific reactivity toward nerve agent simulants, in the thin film coating of a SAW device and to evaluate the detector's response to the simulant's vapor. The amidoxime was selected as the reactive functional group and methanesulfonyl fluoride (MSF) (MSF has been patented as a simulant for the nerve agent GB (2)) and dimethyl methylphosphonate (DMMP) were selected as reactive and non-reactive agent simulants. (Caution: MSF is toxic (anticholinesterase activity) and should be handled with gloves and gas-tight syringes in an efficient hood.) The synthesis, characterization and SAW detector response are described in this report.

<u>Experimental</u>

All reagents and solvents were of reagent grade quality, purchased commercially and used without further purification unless otherwise noted. Infrared spectroscopic data were obtained with a Perkin-Elmer 267. Glass transition temperatures were determined by differential scanning calorimetry with a Dupont 990 Thermal Analyzer and a 910 Differential Scanning Calorimeter. Polymer solubility parameters were determined by measuring weight percent swelling in solvents of varying solubility parameter (water, 23.4 $(cal/ml)^{1/2}$; methanol, 14.5; ethanol, 12.7; 2-propanol, 11.5; t-butyl alcohol, 10.6; acetone, 10.0; tetrahydrofuran, 9.5; ethyl acetate, 9.5; carbon tetrachloride, 8.6; cyclohexane, 8.2; diethylether, 7.4; benzene, 9.2; methylene chloride, 9.7; n-pentane, 7.0) after 12 hr. immersion. SAW devices were coated by solvent evaporation from a dilute polymer solution placed on the device surface. Inert atmosphere dilution of simulant vapors and SAW frequency measurements have been described previously.(3)

The butadiene (80%)-acrylonitrile(17%)-acrylamidoxime(3%) terpolymer was prepared by reacting 2.5g butadiene(80%)-acrylonitrile(20%) copolymer (Aldrich) in 150 ml xylene with 1.61g hydroxylamine hydrochloride in 12 ml n-butanol (freed of HCl immediately before addition by method of Hurd (4) by dropwise addition under nitrogen. Reaction time and temperature were 23 hr. and 90-95°C. The polymer product was worked up by slowly adding the reaction mixture to 500 ml ether with rapid stirring. The precipitated polymer was allowed to settle, the supernate decanted and the polymer resuspended and washed with two 100 ml portions of ether followed by vacuum drying. Yield 2.04g (81%). Conversion of nitrile functional groups to amidoxime groups was 15% by infrared absorption.

The butadiene(55%)-acrylonitrile(38%)-acylamidoxime(7%) terpolymer was prepared by reacting 2.00g butadiene(55%)-acrylnitrile(45%) copolymer in 50 ml tetrahydrofuran with 1.46g hydroxylamine hydrogen chloride (freed of HCl (5)) in 12 ml n-butanol at 70°C for 18 hrs. under nitrogen. The product was worked up analogously to the 3% amidoxime terpolymer. Yield and conversion to amidoxime were 2.02g (91%) and 15%, respectively.

The butadiene(55%)-acrylamidoxime(45%) terpolymer was prepared by reacting 2.00g butadiene(55%)-acrylonitrile(45%) copolymer in 50 ml tetrahydrofuran with 2.92g hydroxyl amine hydrogen chloride (freed

of HCl ($\underline{5}$)) in 25 ml n-butanol at 75-80°C for 24 hrs. under nitrogen.
The product was worked up analogously to the 3% amidoxime terpolymer.
Yield and conversion to amidoxime were 2.45g (96%) and 100%, respec-
tively.

Results and Discussion

The selection of the amidoxime functional group was based on several
considerations. As a reasonably well understood organophosphonate CW
agent specific reaction, the nucleophilic phosphorylation reaction
was chosen. It was desired to employ a neutral, highly nucleophilic
functional group that could readily be covalently attached to a
uniform thin-film-forming polymeric matrix and undergo a phosphorylation reaction with an agent vapor. Of the common nucleophilic
groups studied ($\underline{5\text{-}8}$) (hydroxamic acids, oximes and phenols) most
require a basic solution medium since the conjugate anion is the
reactive species. The amidoxime group appeared to be a possible
exception. In its neutral state the amidoxime is reported to have a
stronger nucleophilicity in acylation ($\underline{9}$) and phosphorylation ($\underline{10}$)
reactions than does the oxime. At room temperature amidoximes are
readily acylated by acid halides or anhydrides ($\underline{11}$). Possible
schemes for the chemistry of this reaction with MSF are depicted in
Figure 1. It is also worth noting that the amidoxime is a powerful
hydrogen bonding group, and hydrogen bonded complexes can also be
envisioned. That the functional group be reactive in its neutral
state is necessary if the polymeric matrix to which it is attached is
to maintain a permeable rubbery character. From a consideration of
synthetic accessibility, the nitrile group of a butadiene-acrylonitrile rubber may be converted to an amidoxime by reaction with
hydroxylamine (Equation 1).

$$\text{(Equation 1)} \tag{1}$$

The relative initial ratio of acrylonitrile to butadiene and
degree of conversion of nitrile to amidoxime are directly related to
the resultant film's solubility parameter and glass transition
temperature. Ideally, the concentration of amidoxime functional
groups would be maximized while the coating's solubility parameter is
matched to the vapor to be detected and the glass transition temperature is kept below room temperature. In practice, the conversion
limitations are set by the reaction conditions of limited polymer
solubility, reaction temperature and time. Three terpolymers of
varying butadiene, acrylonitrile and amidoxime compositions were
prepared as indicated in Table 1.

The infrared spectra of the butadiene-acrylonitrile copolymer
and butadiene-acrylonitrile-acrylamidoxime terpolymers are presented
in Figure 2. The amidoxime specific bands appear at 3480 and 3380
cm^{-1} (NH$_2$ stretching) and at 1660 cm^{-1} (C=N stretching) ($\underline{12}$). The
glass transition temperatures and solubility parameters of the
corresponding polymers are also presented in Table 1. As the
acrylamidoxime content increases from 3 to 7 to 45 mole percent, the

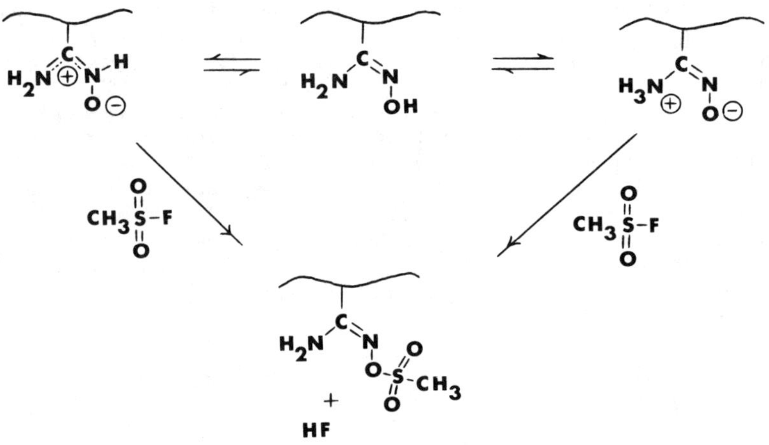

Fig. 1   Possible Reaction Schemes for Amidoxime with Methanesulfonyl Fluoride.

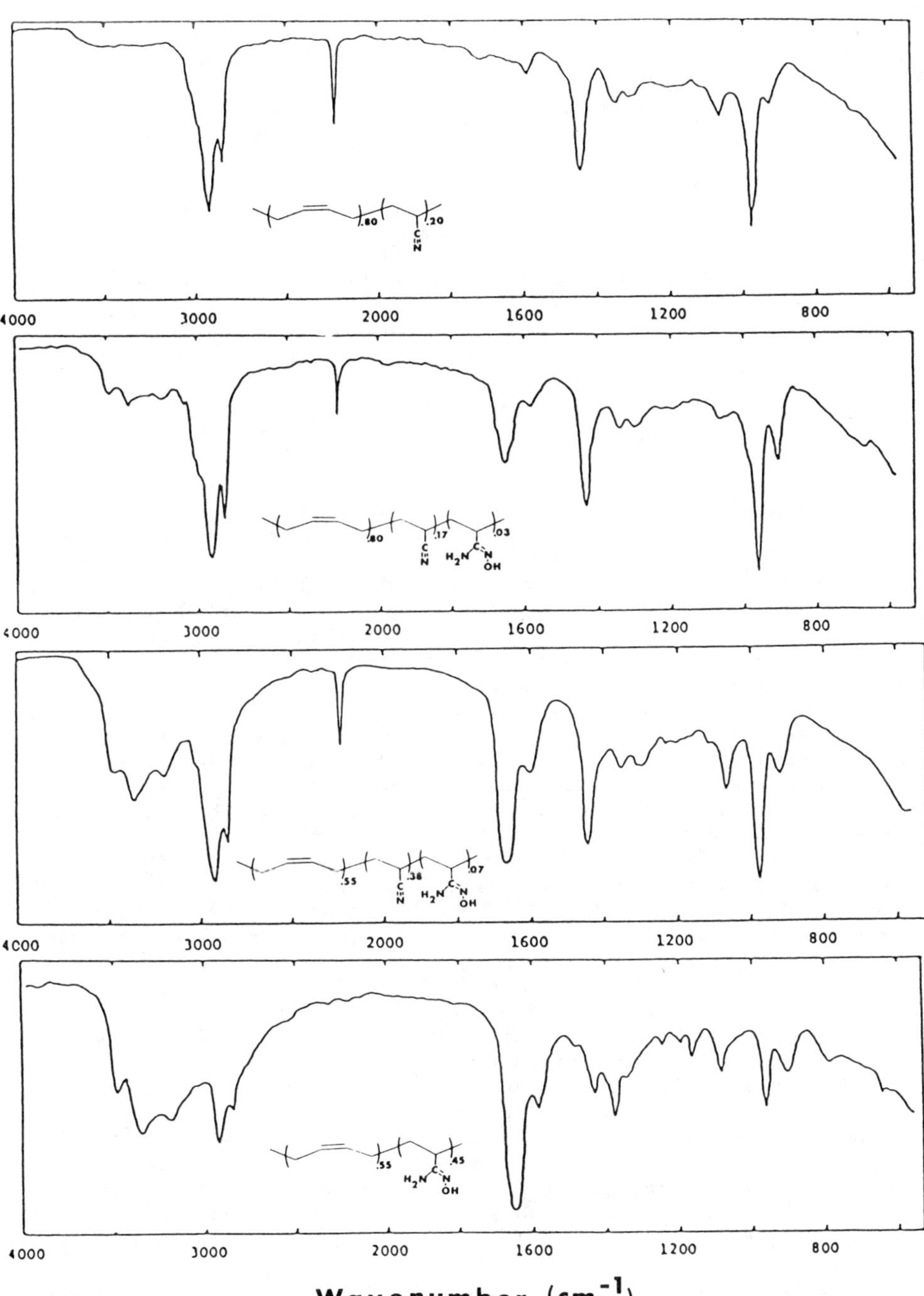

Fig. 2   IR Spectra of Butadiene-Acrylonitrile-Amidoxime Copolymers. Inserted structures indicate copolymer compositions but not microstructures.

corresponding Tg increases from -35 to 19 to 50°C, and solubility parameter increases from 9.3 to 9.7 to 16 $(cal/cm^3)^{1/2}$.

TABLE 1

Composition, Solubility Parameter and Glass Transition
Temperature of $(Butadiene)_x$-$(Acrylonitrile)_y$-$(Acrylamidoxime)_z$
Terpolymers

Terpolymer	x(%)	y(%)	z(%)	$\delta(cal/cm^3)^{1/2}$	Tg(°C)
1	80	17	3	9.3	-35
2	55	38	7	9.7	19
3	55	0	45	16	50

The interaction of the acrylamidoxime polymer coatings with the MSF and DMMP simulant vapors was investigated by infrared spectroscopy and by the SAW device response.

The infrared experiment consisted of placing a coating on a sodium chloride disc by solvent evaporation from a terpolymer solution and exposing it to a saturated simulant atmosphere in a closed container for one hour. The NaCl disc supported film was then immediately transferred to an infrared spectrometer, and the transmission spectrum obtained. The spectra of the exposed 7% amidoxime terpolymer together with the spectra of the liquid simulants are presented in Figure 3. In the case with MSF, the spectrum of the exposed polymer is mostly an additive superpositioning of the spectra of the unexposed polymer and the liquid simulant. The shaded bands at 1210 and 1400 cm^{-1}, which correspond to the O=S=O symmetric and asymmetric stretching (12), are particularly pertinent. If the fluoride ion were nucleophilically displaced, a 50 cm^{-1} shift of these bands should result (12). This indeed may be the source of the weak new band at 1178 cm^{-1}, but the reaction between the acrylamidoxime functional group and the MSF vapor is neither rapid nor quantitative. In the case with DMMP, a similar predominant physisorption is also observed as indicated in Figure 3.

Since the SAW device is extremely sensitive to small gravimetric changes caused by adsorption of vapors in supported films, there is still potential for these coatings as thin film detection elements. Thin films of the amidoxime terpolymers were placed on SAW devices by solvent evaporation. The device was then challenged with 500 to 3000 ppm concentration levels of simulant vapor in a purified air carrier continuous flow exponential dilution gas system. Device responses were recorded as negative resonant frequency shifts as a function of time. Typical data for 1000 ppm challenges of MSF and DMMP are presented in Figure 4. The initial 200 seconds, where no response occurs, is used to establish a baseline prior to syringe injection of the challenge vapor. As the simulant vapor is swept from the dilution flask to the device, a negative frequency shift is observed as the vapor adsorbs on the SAW coating. With passage of time and dilution of simulant in the flowing air, desorption of the simulant

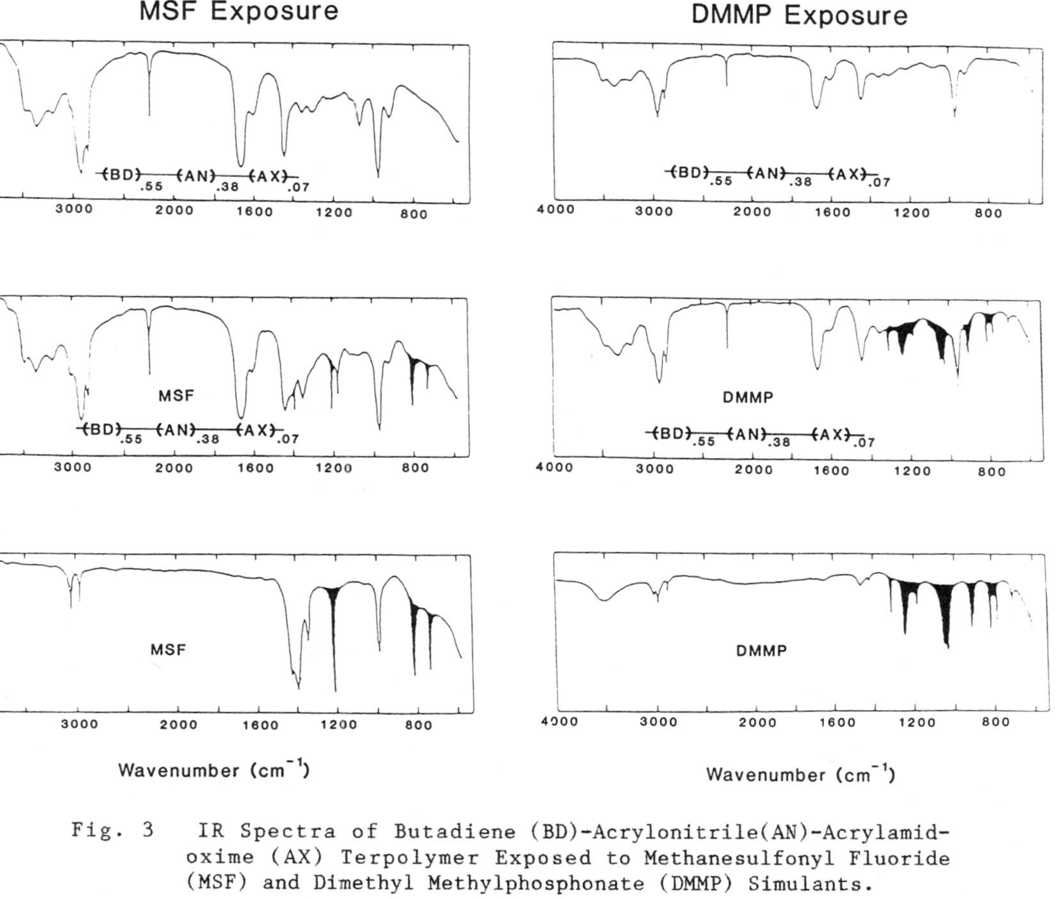

Fig. 3   IR Spectra of Butadiene (BD)-Acrylonitrile(AN)-Acrylamid-oxime (AX) Terpolymer Exposed to Methanesulfonyl Fluoride (MSF) and Dimethyl Methylphosphonate (DMMP) Simulants.

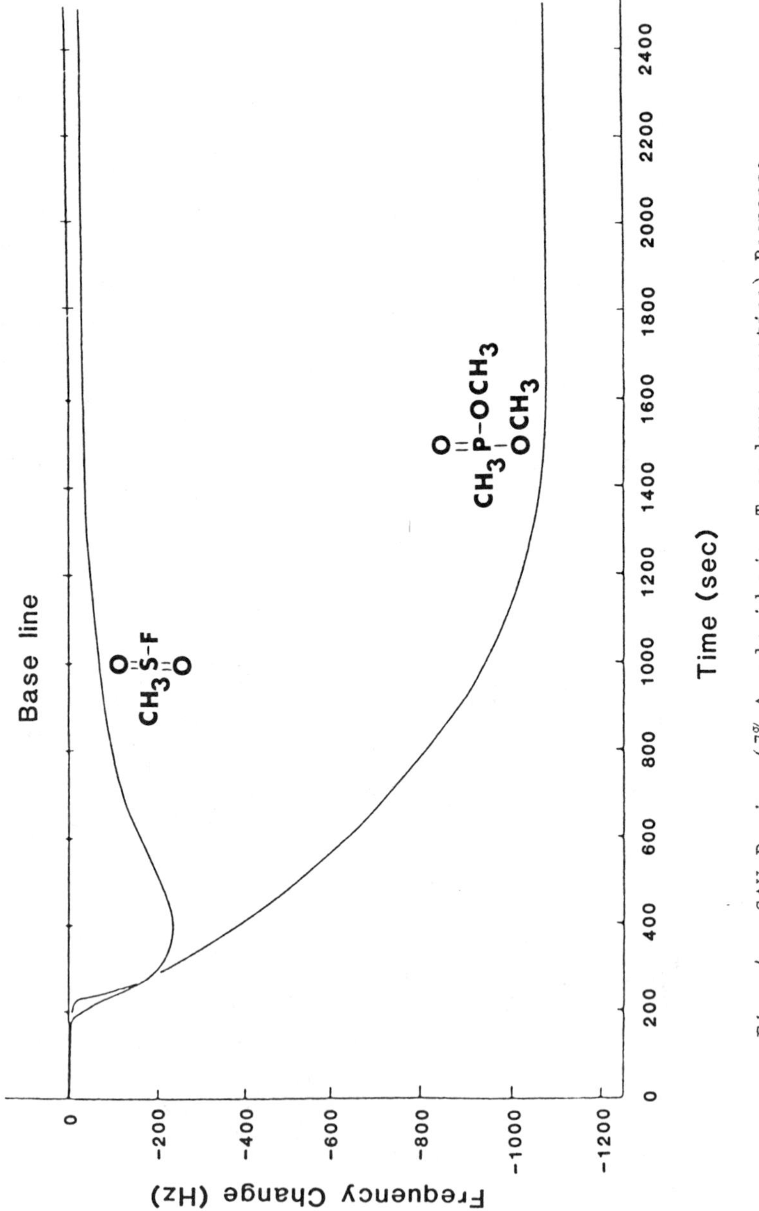

Fig. 4   SAW Device (7% Acrylamidoxime Terpolymer coating) Response to 1000 ppm MSF and DMMP.

may occur depending on the strength of the coating-simulant interaction and vapor pressure of the simulant similar to the situation
observed in gas chromatography. The MSF, which is more volatile than
the DMMP, desorbs much more rapidly and produces a smaller frequency
shift. The sensitivity of the coated device may be assessed by
plotting the maximum frequency shift against the simulant vapor
concentration that would maximally be obtained in the exponential
dilution flask. This data is presented in Figure 5 for the three
amidoxime terpolymers. Two important observations are: (a) the 7%
amidoxime terpolymer is the most sensitive to either MSF or DMMP
while the 45% amidoxime terpolymer is the least sensitive and (b) the
order of sensitivity to MSF and DMMP is reversed on progression from
the 7 to 45% amidoxime terpolymer. The large drop in overall
sensitivity correlates with the glass transition temperature increasing above room temperature. As the increasing concentration of
amidoxime functional groups causes the room temperature character of
the coating to change from rubbery to glassy, the vapor permeation
rate decreases rapidly with an accompanying decrease in gravimetric
sensitivity. The order of sensitivity between MSF and DMMP correlates with the solubility parameter match of the coating with the
simulant and with the relative vapor pressure of the simulant (13).
The solubility parameters for DMMP (calculated from heat of vapor-
ization and molar volume data (14)) and MSF (15) are 10.5 and 11.3
$(cal/cm^3)^{1/2}$, respectively. The closer the match between simulant
and coating solubility parameters, the more compatible will be the
thermodynamic interaction for adsorption; and the lower the simulants
vapor pressure, the slower the desorption rate and greater the
accumulation of adsorbed simulant. The 3 and 7% amidoxime terpolymers are rubbery coatings and more closely match the DMMP
solubility parameter while the 45% amidoxime is a rigid glassy
coating and is a better match to the MSF solubility parameter.

<u>Conclusions</u>

In summary, terpolymers of butadiene, acrylonitrile and acrylamidoxime of varying comonomer contents were prepared by reacting
butadiene-acrylontrile copolymers with hydroxylamine. Infrared
spectroscopy indicated that incorporation of the amidoxime functional
group in the terpolymer coating was successful but that a selective
absorption of the chemical simulant MSF based on a specific reactivity was not achieved. Instead, both the physical and the chemical
simulants were absorbed on the basis of their physical properties
(*i.e.* solubility parameter and vapor pressure). When thin films of
these terpolymers are coated on a SAW device, the adsorption of the
chemical simulant MSF and physical simulant DMMP is readily detectable and varies with the coating's glass transition temperature and
solubility parameter although no rapid or quantitative chemical
reaction was observed between the amidoxime functional group and the
MSF vapor. Correlations are observed between the SAW detector
sensitivity and the glass transition temperature of the coating and
between the solubility parameter match of the coating and the
simulant vapors. The objective of employing functional group
chemistry for discrimination between chemical and physical simulants
was not achieved. The issue of coating specificity defined as a
unique absorption of a particular vapor and exclusion of all others

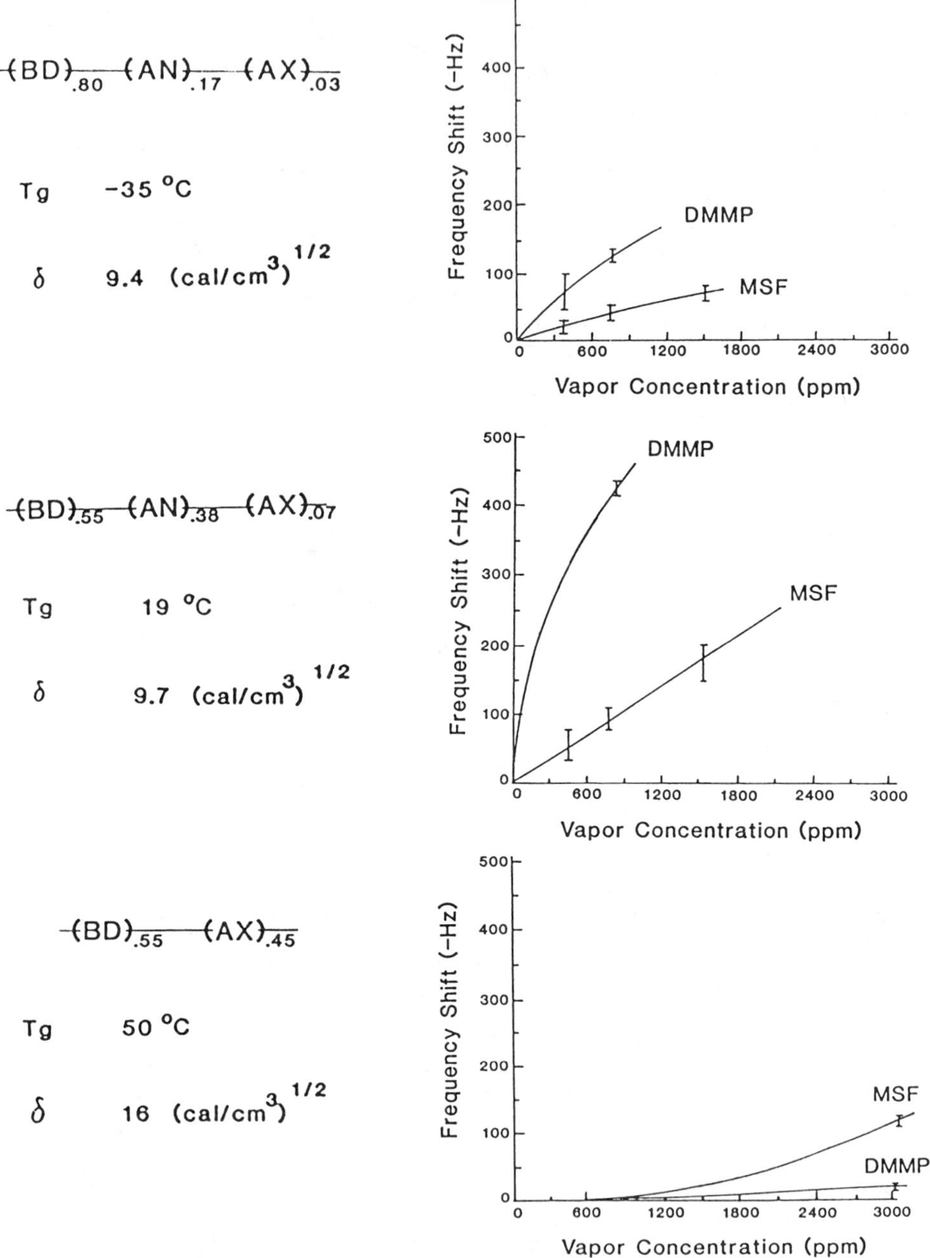

Fig. 5   MSF and DMMP Vapor Concentration Dependence of SAW Device
Coated with 3,7 and 45% Acrylamidoxime Terpolymers and
Correlation with Terpolymer Glass Transition Temperature and
Solubility Parameter.

has yet to be successfully addressed. Clearly, more elegant chemistry than simple functionalization of polymeric coatings is necessary if a single SAW measurement is to be employed for detection of particular vapors in the presence of interference vapors.

Acknowledgment

This work was supported by the U.S. Army Chemical Research and Development Center.

Literature Cited

1. Snow, A.W.; Wohltjen, H., Anal. Chem., 1984, 56, 1411.
2. Poziomek, E.J.; Crabtree, E.V., U.S. Patent 3,972,783, 1976.
3. Wohltjen, H., Proceedings Report No. ARCSL-SP-83030, Chemical Defense Research Conference, U.S. Army Chemical Research and Development Center, 1982.
4. Hurd, C.D.; Browstein, H.J., J. Am. Chem. Soc., 1925, 47, 67.
5. Hackley, B.E.; Plapinger, R.; Stolberg, M.; Wagner-Jauregg, T., J. Am. Chem. Soc., 1955, 77, 3651.
6. Davies, D.R.; Green, A.L., Disc. Faraday Soc., 1955, 70, 269.
7. Epstein, J.; Michel, H.O.; Rosenblatt, D.H.; Plapinger, R.E.; Stephani, R.H.; Cook, E., J. Am. Chem. Soc., 1964, 86, 4959.
8. Snow, A.W.; Wohltjen, H.; Jarvis, N.L.; Dominguez, D., NRL Memorandum Report 5050, 1983.
9. Aubort, J.D.; Hudson, R.F., J. Chem. Soc. Chem. Comm., 1969, 1342.
10. Bunton, C.A.; Cerichelli, G., J. Org. Chem., 1979, 44, 1880.
11. Eloy, F.; Lenaers, R., Chem. Rev.,1962, 62, 155.
12. Avram, M.; Mateescu, GH.D., "Infrared Spectroscopy", Wiley-Interscience, New York, NY, 1972, pp. 296-298.
13. Wohltjen, H.; Snow, A.W.; Lint, J.; Jarvis, N.L., Proceedings Report No. CRDC-SP-84014, Chemical Defense Research Conference, U.S. Army Chemical Research and Development Center, 1984.
14. Kosolapoff, G.M., J. Chem. Soc., 1955, 2964.
15. Snow, A.W., manuscript in preparation.

RECEIVED November 14, 1985

19

Selective Response of Polymeric-Film-Coated Optical Waveguide Devices to Water and Toxic Volatile Compounds

J. F. Giuliani, Neldon L. Jarvis[1], and Arthur W. Snow

Chemistry Division, U.S. Naval Research Laboratory, Washington, DC 20375-5000

Several polymeric coatings have been exposed to a number of compounds including one nerve agent simulant, water and two organic solvents. These polymeric materials were coated as thin films onto the exterior surface of a glass capillary tube forming the sensing element of an optical waveguide vapor detection system. A correlation can be made between the measured magnitude of the waveguide signals and the vapor pressures of the condensed vapors sensed. Three of the five polymer films tested showed extremely large responses to the simulant dimethyl methyl phosphonate (DMMP), and one of these films could detect DMMP vapor concentrations below the 20 ppm level.

The use of total internal multiple reflections of an optical beam at a glass organic film interface has been amply demonstrated as a highly sensitive and efficient probe for monitoring surface physical and chemical reactions. We have successfully designed and developed a reliable compact optical waveguide vapor detection system which is now routinely used in the laboratory testing and evaluation of all types of organic film coatings (1). Up to the present time, highly conjugated organic films (i.e dyes) which display color changes when exposed to certain chemical warfare (CW) simulants and other toxic vapors have been investigated. Invariably, although they were quite reactive to these vapors, some of them nonetheless exhibited instability and irreversibility over long periods of exposure to the vapor detected. These shortcomings result primarily from their susceptibility to degradation from certain wavelengths of light and to extreme temperature and humidity conditions. Although, the metallophthalocyanine and porphyrin dyes (2) show considerable optical and thermal stability, they nevertheless produce an irreversible color reaction (i.e. chemical change) when repeatedly exposed, for example, to a CW vapor.

We are at present beginning to investigate in more detail, as an alternative, polymeric materials as possible CW agent coatings for the optical waveguide system. Among their advantages over dyes, are

[1]Current address: Chemical Research and Development Center, U.S. Army, Aberdeen Proving Grounds, MD 21010

the relative ease of synthesis and coating procedures, film homo-
geneity and most importantly, their lack of optically active chromo-
phoric groups, which eliminates photochemical degradation when
exposed to visible radiation.

Experimental

A complete description of our optical waveguide detection system can
be found in reference (1). The only important modification to the
previously described detector is the replacement of the dye coatings
with polymeric coatings, which are prepared by dissolving the polymer
in a suitable solvent and then dipping a pre-cleaned glass capillary
into the solution in a single pass. After the dipping operation, the
solvent quickly evaporates, leaving a thin uniform film on the
capillary surface. The thickness of a typical film is on the order of
a micron.
     The polymers coated onto the outer capillary surface are listed
in Table I.
Three of the five materials are commercial products and two were
synthesized at NRL. Their structural formulas and refractive indices
(n) are also included in the Table. All these polymers produce a
transparent and uniform film on the waveguide surface.
     Figure 1 shows a schematic diagram of our experimental setup for
generating and detecting the saturated vapors from a number of
liquids including water. For these experiments the liquids are
placed in a flask and air is bubbled into the flask by means of a
regulated water aspirator system. A constant air flow of 0.3
liter/min is maintained for all liquids studied. The vapor-air is
next thoroughly mixed in a second flask before flowing into the
optical waveguide detector and 2 CC sampling chamber. The saturated
vapors presumably condense onto the polymeric film producing a
refractive index change at the condensed vapor-polymeric surface
interface. This interfacial optical modification in turn generates a
voltage difference which is amplified and displayed on a normal strip
chart recorder. Before each vapor test, a dry air baseline is run to
give a recorder baseline reference voltage. The testing procedure
consisted in exposing each polymer coating separately to a series of
vapor-generating solvents which included water, acetone, benzene and
the nerve agent simulant dimethyl-methyl-phosphonate (DMMP). Several
dry air/vapor cycling run's were made for each film to determine
whether the particular polymer studied, exhibited reversibility or
deteriorated with repeated exposure to a particular vapor.

Experimental Results

Figure 2 shows a d.c. recorder retracing of a typical set of dry
air/saturated vapor cycling responses of the optical waveguide coated
with the polymer, poly-epichlorohydrin (PEH) and exposed cyclically
to benzene vapors. Aside from the clearly detected electrical
signals above the dry air baseline reference, their amplitudes appear
to show apparent reversibility over the four exposure cycles indicated in the figure. Also, the rise time and decay of these signals
are rather symmetric and indicate a film response time of less than
one minute.

Table I. Polymer Coatings Tested

COATING	STRUCTURE	n_D	SOURCE
Polyisoprene (PI)	$+CH_2-C(CH_3)=C(H)-CH_2+$	1.52	Commercial
Polyepichlorohydrin (PEH)	$+OCH_2\ CH(CH_2CL)+$	1.5	Commercial
Polyfluoropolyol (PFP)	$+CH_2\ CHCH_2O-C(CF_3)(OH)-\phi-C(CF_3)(CF_3)-OCH_2\ CHCH_2O-C(CF_3)(OH)-CH_2CH=CH\ C(CF_3)\ O+$	1.4	NRL
Polyvinyl Pyrrolidone (PVP)	$+CH_2-CH+$ (N–C=O pyrrolidone ring)	1.53	Commercial
Polyethylene Maleate (PEM)	$+OCH_2\ CH_2\ O\ C(=O)-C=C(H)(C(=O)H)+$	1.5	NRL

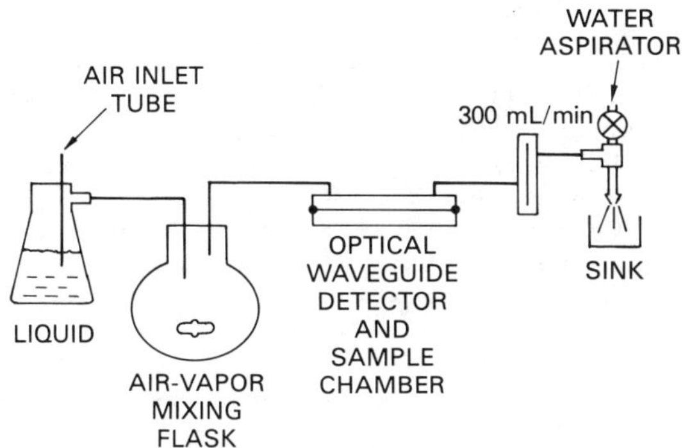

Figure 1. Schematic of the experimental set-up to produce and detect the condensed saturated vapors with the optical waveguide coated with several polymers.

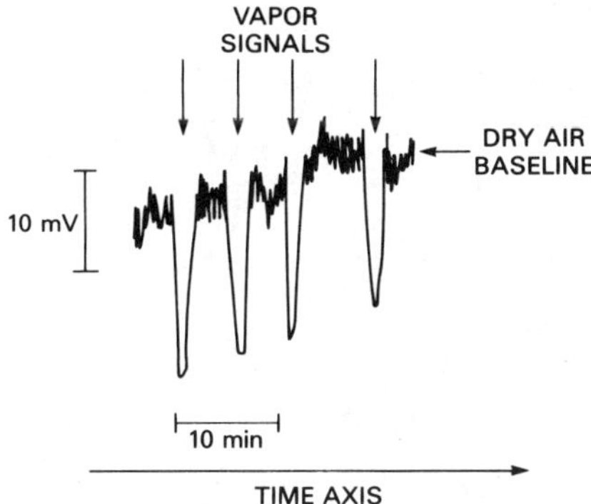

Figure 2. Typical electrical signals, retraced from the original optical waveguide data showing the cyclical response of the polymeric film PEH to dry air/condensed benzene vapors. Note particularly both the rapid response (less than 15 seconds) and amplitude reversibility.

A quantitative survey of all the polymer films tested against various saturated vapors is displayed as a bar chart in figure 3. Here, the measured response of the waveguide is normalized in the units of millivolts per millimoles per liter for each vapor molecule. It is readily seen that each polymer film responds differently to each vapor. However, with the exceptions of polyfluoropolyol (PFP) and polyisoprene (PI), all the polymers respond strongly to the nerve agent simulant DMMP. The shifts in the directions of some of the histograms are probably associated with the phenomena of frustrated total internal reflection (FTR) at the polymer film-glass interfacial boundary. Here such refractive index ratio changes and/or optical film thickness variations at the condensed vapor polymer boundary, can produce optical phase shifts resulting in either higher or lower reflections with respect to the dry air baseline. Film thickness variations can of course arise from the slight swelling of the polymer coating upon the diffusion of the condensed vapor molecules into the film after condensation has taken place (i.e. solubilization).

An interesting correlation between the apparent selectivity of the polymer film response to a series of condensed vapors can be obtained if the measured optical waveguide signal levels are plotted as a function of the vapor pressures in air for the series of vapors examined in this work. Figure 4 displays such a plot. It is immediately evident that the optical detection progressively increases with decreased volativity of the vapor detected. Thus for all coatings examined, the relatively high vapor pressures of acetone and benzene produce the smallest changes in signals (relative to dry air at 3 l/min) whereas the significantly lower vapor pressures of water and DMMP produce the largest signal levels. These phenomena had been observed by Snow, et al.(3-4) in studies of polymeric coatings with their surface acoustic wave microbalance device. The general correlation between volatility of the vapor and optical response displayed in figure 4, cannot at this time be explained with any degree of certainty, since many interacting variables contribute in the modification of the relative ratios of the refractive indices at the film-glass interface. For example, the amount of vapor effectively adsorbed on the film, and the specific solubility of each polymer to a particular condensed vapor are all unknown quantities in which either one or both could significantly produce the observed refractivity changes detected by our device. An interesting anomaly can be seen for the response of the polymers, PI and PFP which do not respond as strongly to DMMP as the other films, but nevertheless produce signal levels higher for DMMP than for benzene and acetone.

A more informative presentation of the curves depicted in figure 4 may be obtained if this same data is replotted on a log-log scale. This is shown in figure 5. The waveguide response versus vapor pressure suggests that the polymer films may be grouped into three response categories. Group 1 is formed by PVP and PEM. Note that these films respond almost in a straight-line behavior over the vapor pressure range represented by the solvents employed. Group 2 on the other hand, represented by PFP and PI films appear to respond noticeably in a highly nonlinear fashion with high selectivity and sensitivity at high vapor pressures but becoming much less selective at low vapor pressures. The polymer PEH appears to represent an intermediate case between the two extreme groupings. From the small

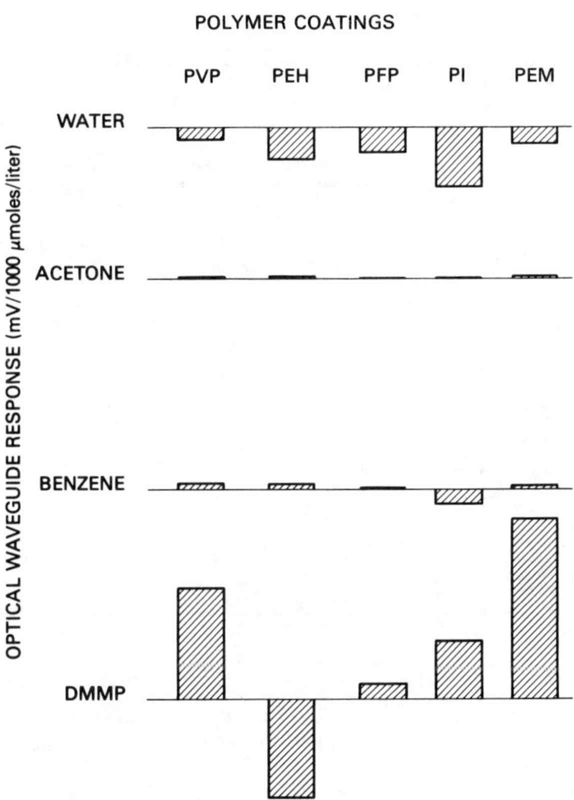

Figure 3.   Bar chart display, summarizing the optical waveguide detector coated with several polymeric film coatings which are exposed to various structured vapors.   The height of the rectangles reflect electrical measurements normalized for each vapor on a per mole basis for comparison.

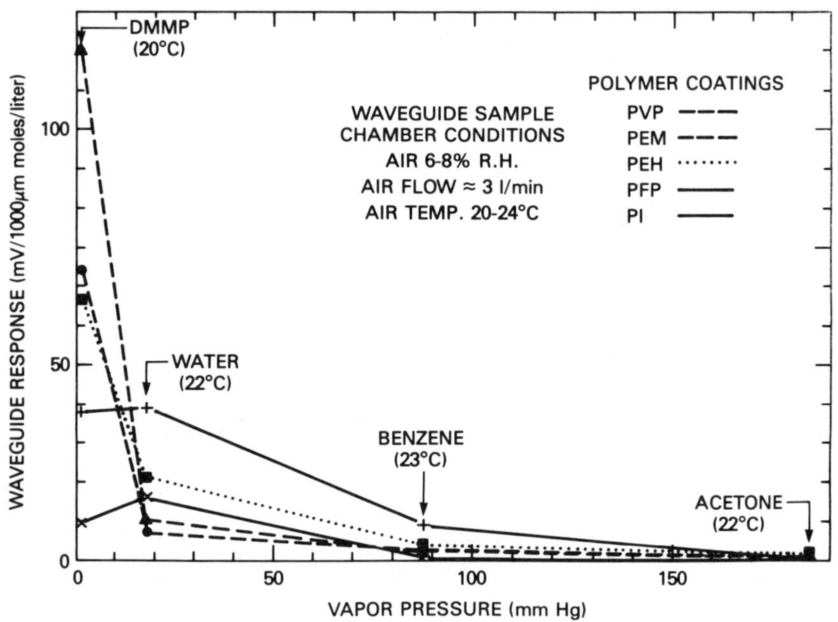

Figure 4. Plot of the optical waveguide response for each polymer coating as a function of the vapor pressure of the vapors to which these caotings are exposed. The scale of the ordinate is taken from the absolute heights shown in the bar chart data shown in Figure 3. It is clearly seen that for all these polymeric films their response increases inversely as the vapor pressure for the particular volatile material tested.

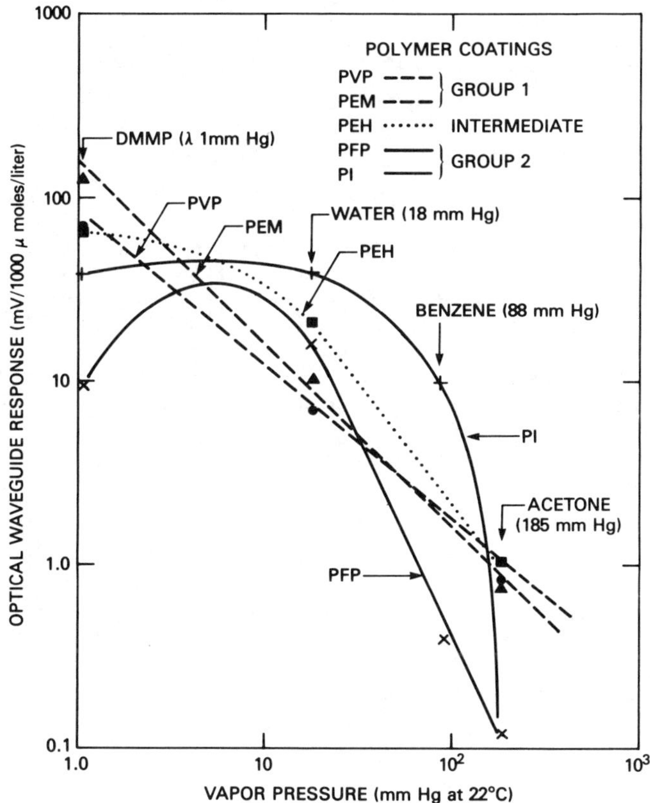

Figure 5.   A replot of Figure 4 on a Log–Log scale clearly suggesting a systemmatic grouping of the polymer films with respect to their responses as a function of vapor pressure for the solvents tested.

number of samplings taken in this study, it is not possible at this
time to fully explain this interesting result. There is a suspicion
however, that the common pendant side groups incorporated into these
polymer structures may have a strong effect on the optical response
differences reported in this work.

     Referring to figure 4, the polymer film PEM appears to produce
the largest response to the CW simulant DMMP, and this film, was
tested further for response to precisely controlled low concentrations of DMMP as a possible device coating. Figure 6 displays a
plot of the percent change in waveguide transmittance at 660 nm with
DMMP vapor concentrations between 1000 ppm and 10 ppm. The data
points indicated by the solid circles and squares represent two
separate simulant/dry air cycles obtained for a single coating. The
dotted line drawn through the data points indicate a reasonable
detection discrimination over this dynamic range.

<u>Conclusions</u>

This study represents the first systemmatic application of the
optical waveguide technique to the study of the response of polymer
film coatings to condensed vapor molecules. These results indicate
that the technique is useful for surveying rapidly potential polymeric films as possible vapor sensor coatings. Moreover, this work
has further substantiated that the vapor pressure is an important
physical property to be taken into account when employing polymeric
films as surface coatings.

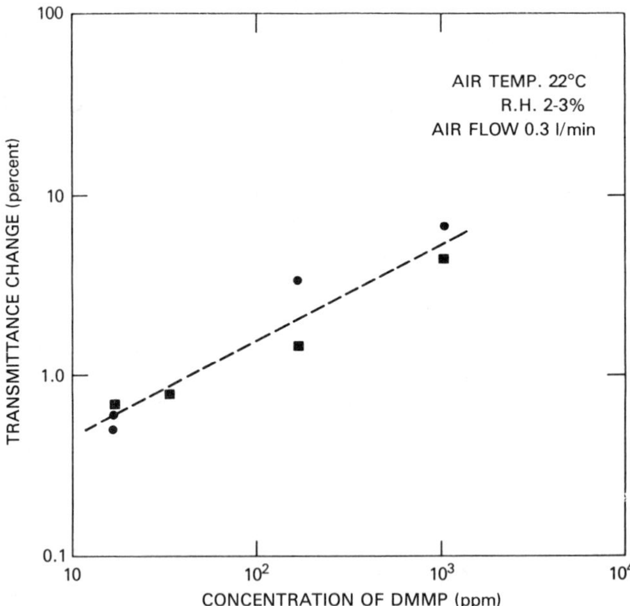

Figure 6. Optical transmittance change at 660 nm as a function of
various DMMP concentrations for the optical waveguide surface
coated with the PEM polymer film.

Acknowledgment

The authors wish to acknowledge the U.S. Air Force, School of
Medicine, Brooks Air Force Base, San Antonio, Texas 78235 for their
support of this work.

Literature Cited

1.   Giuliani, J.F.; Wohltjen, H.; Jarvis, N.L.; Optics Letters 1983,
      8, 54-56.
2.   Giuliani, J.F. (unpublished work).
3.   Snow, A.; Wohltjen, H; Anal. Chem. 1984, 56, 1411-16.
4.   Snow, A.; Wohltjen, H.; Memo Rept. No. 5313, 1984.

RECEIVED February 26, 1986

20

Microbial Sensors for Process and Environmental Control

I. Karube

Research Laboratory of Resources Utilization, Tokyo Institute of Technology, Nagatsuta-cho, Midori-ku, Yokohama 227, Japan

Many organic and inorganic compounds are present in fermentation process and waste waters. Rapid and sensitive on-line monitoring of these compounds is required. Microbial sensors composed of immobilized microorganisms and an electrochemical device are suitable for the determination of these compounds. A microbial sensor for glucose consisted of immobilized whole cells which utilize mainly glucose and an oxygen electrode. The steady state current obtained depended on the concentration of glucose. The microbial sensor could be used for the determination of glucose in fermentation media. Furthermore, alcohols, acetic acid, ammonia, and methane were determined by microbial sensors based on the same principle. On the other hand, potentiometric microbial sensors have been developed for fermentation process. A microbial sensor for glutamic acid consisted of immobilized Escherichia coli having glutamate decarboxylase activity and a carbon dioxide gas-sensing electrode. The concentration of glutamic acid in some fermentation broths were determined by the sensor. Cephalosporines were also measured by a microbial sensor system. Furthermore, microbial sensors were applied to the determination of organic pollution. A microbial sensor for BOD consisted of Trichosporon cutanium and an oxygen electrode. The sensor could be used for a long time for the estimation of BOD. Nitrite and mutagens were also determined by microbial sensor systems.

Many useful compounds such as amino acids, nucleic acids, alcohols, vitamins, antibiotics, foods, etc. are produced in fermentation industries. Furthermore, many organic and inorganic compounds are present in waste waters. The determination of these compounds is required for control of fermentation and environment. Analysis of these compounds can be done by spectrophotometric methods. However, complicated procedures and long reaction times are required.

Furthermore, samples are not usually optically clear.On the other hand,
electrochemical monitoring of these compounds can be done by biosensors. Many reports on enzyme sensors have been published for
clinical and food analyses(1,2). The enzyme sensors consisted of
the immobilized enzyme and an electrochemical device. However,
enzymes are unstable and expensive. Therefore, the enzyme sensors
are not suitable for industrial process and environmental control.
On the other hand, microbial sensors composed of immobilized living
microorganisms and an electrochemical device have been developed
for process and environmental control(3-5). Living microorganisms
are used as a molecular recognition element in this sensor. Assimilation of organic compounds by microorganisms is monitored by the
respiration activity change or the amount of metabolites, produced,
which can be measured directly with various electrodes. The microbial sensors are very stable and can be used for a long time. Microbial sensors for fermentation and environmental control are described
in this chapter.

Microbial Sensors for Fermentation Process

Glucose sensor. The determination of glucose is important for
process control. Assimilation of glucose by microorganisms can
be determined from the respiration activity by using an oxygen electrode. Therefore, a microbial sensor for glucose consisted of immobilized whole cells which utilize mainly glucose and an oxygen elect-
rode(6).
    Immobilized whole cells of Pseudomonas fluorescens were used
for the glucose sensor. The microbial sensor has been applied to
the continuous determination of glucose in molasses.
    A schematic diagram of the microbial sensor is illustrated
in Figure 1. The sensor consisted of double membranes of which
one layer was the bacteria-collagen membrane (thickness 40 μm),
the other an oxygen permeable Teflon membrane (thickness 27 μm),
an alkaline electrolyte, a platinum cathode, and a lead anode.
The double membrane is in direct contract with the platinum cathode
and is tightly secured with rubber rings.
    The microbial sensor was inserted into a sample solution of
glucose in water and the sample solution was saturated with dissolved
oxygen at its partial pressure in air and stirred magnetically while
measurements were taken. The temperature of the sample solution was
maintained at $30\pm0.1°C$. The current was measured by a milliammeter.
    The current at time zero was that obtained in a sample solution
saturated with dissolved oxygen. The bacteria began to utilize
glucose in a sample solution when the sensor was placed in it.
Then, consumption of oxygen by the bacteria in the collagen membranen
began. Consumption of oxygen by the bacteria caused a decrease in dissolved oxygen around the membrane. As a result, the current of
the sensor markedly decreased with time until a steady state was
reached. The steady state indicated that the consumption of oxygen
by the bacteria and the diffusion of oxygen from the solution to
the membrane were in equilibrium. The steady state current was
attained within 10 min at 30°C. The steady state current depended
on the concentration of glucose. 'Current' means the steady state
current hereafter.
    A linear relationship was observed between the current and

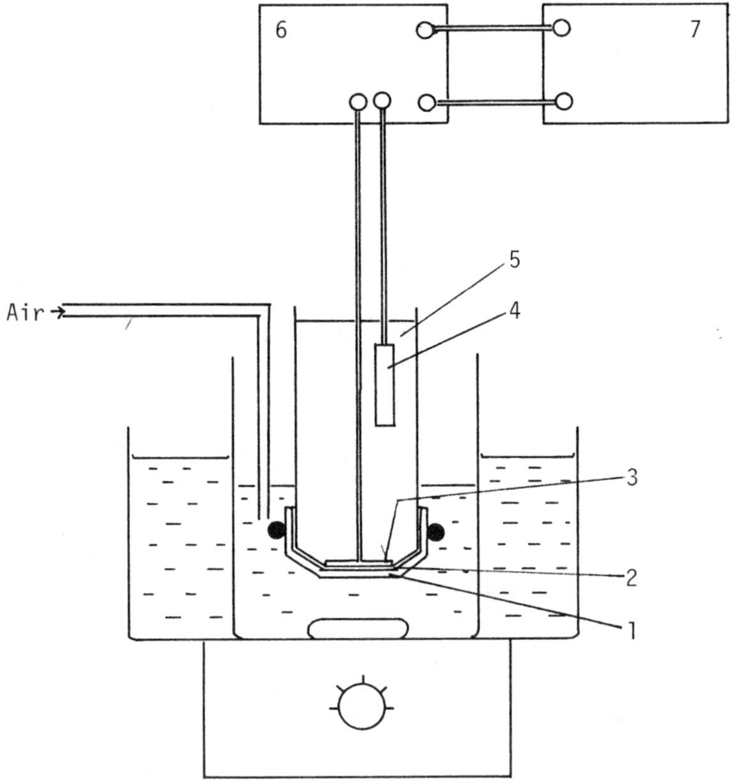

Figure 1. Schematic diagram of the microbial electrode sensor for glucose.
1. Bacteria-collagen membrane; 2. Teflon membrane; 3. Platinum cathode; 4. Lead anode; 5. Electrolyte (KOH); 6. Ammeter; 7. Recorder

the concentration of glucose below 20 mg l^{-1} . The minimum concentration for determination was 2 mg glucose l^{-1} . The current was reproducible within ±6 % of   relative error when a sample solution
containing 10 mg l^{-1} of glucose was employed. The standard deviation was 0.6 mg l^{-1} in 20 experiments. The sensor responded slightly
to fructose, galactose, mannose and saccharose. However, no response
was observed in the case of amino acids.

The microbial sensor for glucose was applied to molasses.
Glucose in molasses was determined with an average relative error
of 10 % by the microbial sensor. No decrease in current output
was observed over a two week period and 150 assays.

The sensitivity of the microbial sensor was almost the same as that
of common enzyme electrodes. Furthermore, the microbial sensor
was more stable than enzyme electrodes when they were applied to
determine glucose in molasses.

Alcohol Sensor. On-line measurements of ethyl alcohol concentration
in culture broth are required in fermentation industries. A microbial electrode consisting of immobilized yeasts or bacteria, a gas
permeable Teflon membrane, and an oxygen electrode was prepared
for the determination of methyl and ethyl alcohols(7).

Unidentified bacterium and Trichosporon brassicae were used
for the methyl and ethyl alcohol sensors, respectively. The microorganisms were adsorbed on a porous acetyl cellulose membrane.

The porous membrane retaining the microorganisms was fixed on the
surface of the Teflon membrane on the electrode. Furthermore, a gas
permeable membrane was placed on the surface of the electrode and
covered with a nylon net.

A linear relationship was observed between the current decrease
and the concentration of ethyl alcohol below 22.5 mg l^{-1} by the
pulse methods(7). The minimum concentration for the determination
was 2 mg ethyl alcohol l^{-1} . The current difference was reproducible
within ±6 % of the relative error when a sample solution containing
16.5 mg l^{-1} of ethyl alcohol was employed. The standard deviation
was 0.5 mg l^{-1} in in 40 experiments. Assay time is within 6 min.

The sensor did not respond to volatile compounds such as methyl
alcohol, formic acid, acetic acid, propionic acid, and other nutrients for microorganisms such as carbohydrates, amino acids, and
ions. The selectivity of the microbial sensor for ethyl alcohol
was satisfactory.

The microbial sensor was applied to fermentation broths of
yeasts. The concentration of ethyl alcohol was determined by the
microbial sensor and by gas chromatography. Satisfactory comparative results were obtained between them. The correlation coefficient was 0.98 with 20 experiments.

The current output of the sensor was almost constant for more
than three weeks and 2100 assays. The microbial sensor can be used
for long time for assay of ethyl alcohol.

A microbial sensor consisting of immobilized methyl alcohol-uti-
lizing bacteria (AJ 3993), a gas permeable membrane, and an oxygen
electrode was applied to the determination of methyl alcohol. A
linear relaionship was also observed between the current decrease
and the concentration of methyl alcohol. Therefore, the sensor
can be also applied to the determination of methyl alcohol.

Acetic Acid. On-line measurement of acetic acid concentrations is required in fermentation processes. The microbial sensor for ethyl alcohol described above could be used for the determination of acetic acid(8).

The pH of the solution had to be kept sufficiently below the pK value for acetic acid (4.75 at 30°C), because acetate ions cannot pass through the gas-permeable membrane.

The calibration graphs obtained showed linear relationships between the current decrease and the concentration of acetic acid up to 72 mg l^{-1}. The minimum concentration for determination was 5 mg of acetic acid l^{-1}. The current difference was reproducible within ±6 % for an acetic acid sample contaning 54 mg l^{-1}. The standard deviation was 1.6 mg l^{-1} in 20 experiments. The sensor did not respond to volatile compounds such as formic acid and methanol or to nonvolatile nutrients such as glucose and phosphate ions. The response to organic compounds depends on the assimilability by the immobilized microorganisms. Trichosporon brassicae, utilized propionic acid, n-butyric acid and ethanol. The measurement can be within 4 min using a flow cell (8).

The microbial sensor for acetic acid was applied to a fermentation broth of glutamic acid. The concentration of acetic acid was determined by the microbial sensor and by a gas chromatographic method. Good agreement was obtained; the regression coefficient was 1.04 for 26 experiments.

The current output (0.29-0.25 μA) of the sensor was constant (within +10 % of the original values) for more than 3 weeks and 1500 assays.

Formic Acid Sensor. Formic acid is found in culture media. Therefore, on-line determination of formic acid is required.

The fuel cell type electrode consisting of a platinum anode and a silver peroxide cathode can be used for measuring hydrogen. Some anaerobic bacteria such as Escherichia coli and Clostiridium butyricum produce hydrogen from formic acid. Therefore, determination of formic acid is possible by using Clostridium butyricum and a fuel cell type electrode(9).

A diagram of the microbial sensor is illustrated in Figure 2. When the sensor was inserted into a sample solution containing formic acid, formic acid permeated through the porous Teflon membrane. Hydrogen, produced from formic acid by C. butyricum, penetrated through the Teflon membrane, and was oxidized on the platinum anode. As a result, the current increased until it reached a steady state. The steady state current depended on the concentration of formic acid. The steady state current was obtained within 20 min.

A linear relationship was obtained between the steady state current and the formic acid concentration below 1,000 mg l^{-1}. The minimum concentration for determination was 10 mg l^{-1}. The current were reproducible with an average relative error of 5 % when a medium containing 200 mg l^{-1} of formic acid was used. The standard deviation was 3.4 mg l^{-1} in 30 experiments.

The sensor did not respond to nonvolatile and other volatile compounds. The microbial sensor was applied to the determination of formic acid in the cultivation medium of Aeromonas formicans. The formic acid concentration was measured by the gas chromatography and by the microbial sensor. Good agreement was obtained between both methods; the regression coefficient was 0.98 for 10 experiments.

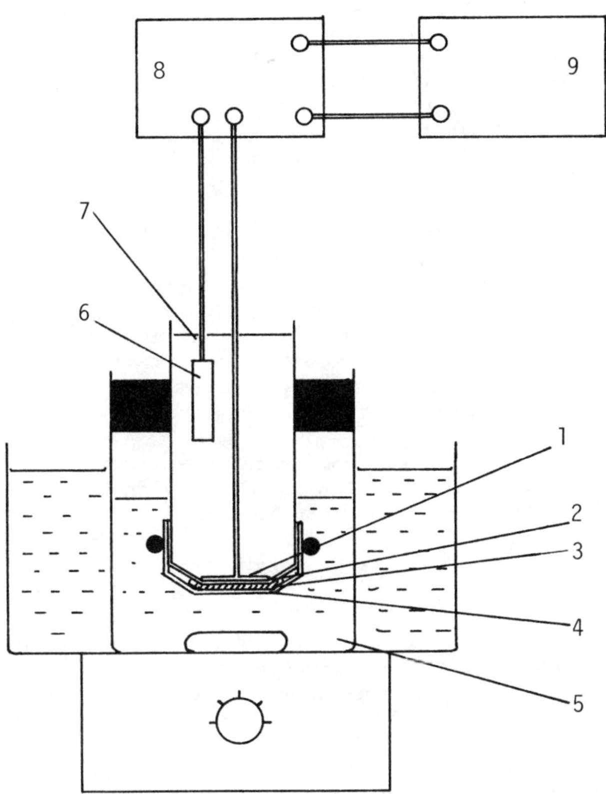

*Figure 2.   Schematic diagram of the microbial sensor for formic acid.
1. Pt anode; 2. Teflon membrane; 3. Immobilized C. butyricum;
4. Porous Teflon membrane; 5. The cell with glycine-HCl buffer (pH
3.1); 6. Ag_2O_2 cathode; 7. Electrolyte (0.1 M phosphate buffer,
pH 7.0); 8. Ammeter; 9. Recorder*

<u>Glutamic Acid Sensor</u>. A rapid automatic measurement of glutamic acid in fermentation media is required in fermentation industries.

Glutamate decarboxylase catalyzes the decarboxylation of glutamic acid, which produces carbon dioxide and amine. Certain microorganisms contain glutamate decarboxylase. Therefore, a microbial sensor for glutamic acid consisted of immobilized <u>Escherichia</u> <u>coli</u> having glutamate decarboxylase activity and a carbon dioxide gas-sensing electrode(<u>10,11</u>).

Preliminary experiments showed that any carbon dioxide produced by the bacteria under anaerobic conditions resulted from the glutamate decarboxylase reaction.

When the sample solution containing glutamic acid was injected into the system, the potential of the carbon dioxide gas-sensing electrode increased with time. The enzyme reaction was carried out at pH 4.4, which was sufficiently below the pK value (6.34 at 25°C) of carbon dioxide.

Figure 3 shows the response of the microbial sensor to various concentrations of glutamic acid. The plot of the maximum potential vs. the logarithm of the glutamic acid concentration was linear over the range 100–800 mg l^{-1}. The slope over this range was approximately Nernstian. When a glutamic acid solution (400 mg l^{-1}) was measured in replicate, the standard deviation was 1.2 mg l^{-1} (20 experiments). The sensor responded to glutamic acid and glutamine and very slightly to some other amino acids. The response to glutamine can be decreased, if necessary, using acetone-treated <u>E. coli</u>.   The measurement time is within 5 min.

The concentrations of glutamic acid in some fermentation broths were determined by the microbial sensor and by the Auto-analyzer method. The results were in good agreement. The response of the sensor was constant for more than 3 weeks and 1500 assays. Thus the microbial sensor appears to be very attractive for the determination of glutamic acid.

<u>Cephalosporine Sensor</u>. Simple, continuous methods for antibiotic determination are required for fermentation industries. <u>Citrobacter</u> <u>freundii</u> produced cephalosporinase, which catalyzes the following reaction of cephalosporin, which liberates hydrogen ions :

Cephalosporin is determined from the proton concentration generated in a medium by using immobilized bacteria. A microbial sensor composed of a bacteria-collagen membrane reactor and a combined glass electrode was applied to the determination of cephalosporins in fermentation media.   The system used for continuous determination of cephalosporins is illustrated in Figure 4.

A linear reltaionship was obtained between the logarithm of the cephalosporin concentration and the potential difference. 7-phenylacetylamidodesacetoxysporanic acid (phenyl-acetyl-7ADCA),

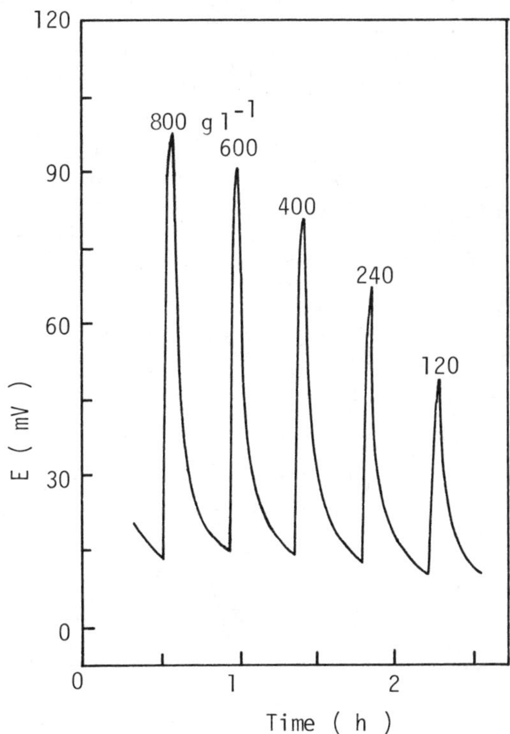

Figure 3.   Responses given by the electrode for glutamic acid solutions of the concentrations stated.   Sample solution (3 ml) was injected for 3 min.   The determination was carried out under the recommended conditions.

cephaloridine, cephalothin and cephalosporin c were determined by
the cephalosporin sensor (Figure 5). Each determination took about
10 min.

The reproducibility was determined with phenylacetyl-7ADCA
solution (125 μg ml^{-1}); the relative standard deviation was 10 %
(2 mV) for 10 experiments.

The reusability of the microbial sensor was examined with a
solution containing 125 μg ml^{-1} of phenylacetyl-7ADCA. The cephalosporin determination was carried out several times a day, and no
change in the potential difference response was observed for a week.

The system was applied to the determination of cephalosporin c
in a broth of <u>Cephalospolium</u> <u>acremonium</u>, and was compared with a
method based on high-pressure liquid chromatography (h.p.l.c.).
The relative error of the determination by the microbial system
was 8 %.

<u>Ammonia Gas Sensor</u>. The determination of ammonia is important in
process analysis. An ammonia gas electrode is usually used for
this purpose. However, volatile compounds such as amines often
interfere with the determination of ammonia. Therefore, a
sensor based on amperometry is desirable for the determination of
ammonia.

Nitrifying bacteria belong to two genera.   One genus
(i.e., <u>Nitrosomonas</u> sp.) of bacteria utilizes ammonia as the sole
source of energy and oxygen is consumed by the respiration as follows :

$$NH_3 + 1.50_2 \xrightarrow{\text{\underline{Nitrosomonas} sp.}} NO_2^- + H_2O + H^+$$

The other genus (i.e., <u>Nitrobacter</u> sp.) of bacteria oxidizes
nitrite to nitrate as follows :

$$NO_2^- + 0.50_2 \xrightarrow{\text{\underline{Nitrobacter} sp.}} NO_3^-$$

Therefore, ammonia is determined by a microbial sensor using immobilized nitrifying bacteria and an oxygen electrode(<u>12-14</u>)(Figure 6).

When the ammonia solution was injected into the buffer solution, ammonium ion changed to ammonia gas. Ammonia gas permeated
through the gas permeable membrane and was assimilated by the immobilized bacteria. Oxygen was then consumed by the bacteria so that
the concentration of dissolved oxygen around the membrane decreased.

The steady-state current depended on the concentration of ammonia. A linear relationship was observed between the current decrease
and the ammonia concentration below 42 mg l^{-1} (current decrease
4.7 μA). The minimum concentration for the determination of ammonia
was 0.1 mg l^{-1} (signal to noise, 20; reproducibility, ±5 %). The
current decrease was reproducible within ±4 % of   relative error
when a sample solution containing 21 mg l^{-1} of ammonium hydroxide
was employed. The standard deviation was 0.7 mg l^{-1} in 20 exper-
ments. The response time of the sensor was within 4 min.

The sensitivity of the microbial sensor was almost at the same
level as that of a ammonia electrode. The sensor did not respond
to volatile compounds such as acetic acid, ethyl alcohol and amines

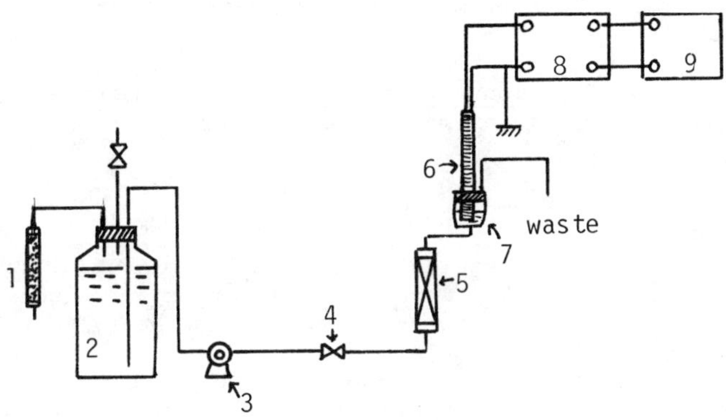

Figure 4.   Immobilized whole cell-based flow-type sensor for cephalosporins. 1. Soda lime; 2. Buffer reservoir; 3. Peristaltic pump; 4. Sample inlet; 5. Immobilized whole cell reactor; 6. Combined glass electrode; 7. Sensing chamber; 8. Amplifier; 9. Recorder

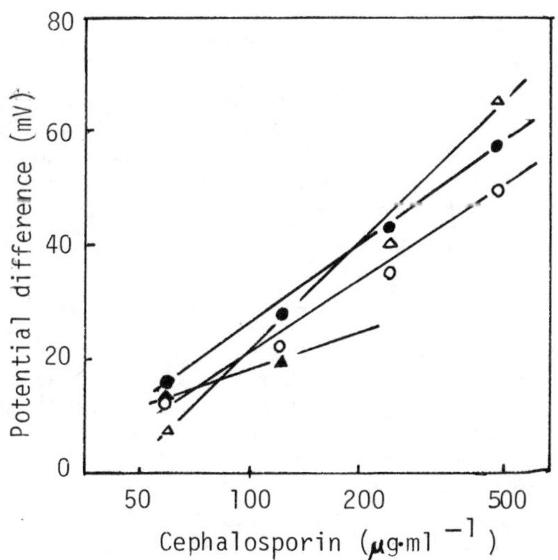

Figure 5.   Calibration curves: (Δ) Phenylacetyl-7ADCA; (▲) Cephalosporin c; (o) Cephalothin; (●) Cephaloridine

(diethylamine, propylamine, and butylamine) or to nonvolatile nutrients such as glucose, amino acids, and metal ions (potassium ion,
calcium ion, and zinc ion). The current output of the sensor was
almost constant for more than 10 days and 200 assays. Therefore,
the microbial sensor can be used for a long time for the assay of
ammonia.

Methane Sensor. Methylomonas flagellata utilizes methane as its
sole source of energy and oxygen is consumed by the respiration
as follows :

$$CH_4 + NADH + H^+ + O_2 \longrightarrow CH_3OH + NAD^+ + H_2O$$

The methane concentration was determined with a microbial sensor
consisting of immobilized M. flagellata and an oxygen elect-
rode(15,16).   M. flagellata were immobilized in acetylcellulose
filters with agar. The microbial sensor system is schematically
illustrated in Figure 7. Methane containing gas introduced into both reactors by a pump at a controlled flow rate. The partial pressure
of oxygen in each line was monitored with an oxygen electrode.
The difference between the output currents of the two electrodes
is related to the amount of methane in the flow lines. The response
time required for the determination of methane gas was 1 min. A
linear relationship was observed between the current difference
of   the electrodes and the concentration of methane (below 6.6
mmol·l^{-1}).The minimum concentration for the determination was 13.1 μmol·l^{-1}
The current decrease was reproducible within $\pm$5 % in 25 experiments
with sample gas containing 0.66 mmol·l^{-1} methane.
        The current output of the sensor system was almost constant
for more than 20 days and 500 assays. The microbial sensor can,
therefore, be used to assay methane over a long period of time.
In the same experiment the concentration of methane was determined
by both the electrochemical sensor and the conventional method (gas
chromatography). A good correlation was obtained between the methane
concentrations determined by the two methods (correlation coefficient 0.97).

Microbial Sensors for Environmental Control

BOD Sensor. The biochemical oxygen demand (BOD) test is one of
the most widely used and important tests in the measurement of organic pollution. Since the BOD test measures biodegradable organic
compounds in waste waters, it requires a long incubation period
(5 days at 20°C). Therefore a simple and reproducible method for
estimation of 5-day BOD is required for pollution control(17).
        Since bacteria are known to utilize organic compounds such
as carbohydrates and proteins, the biofuel cell system using immobilized C. butyricum could be applied to the estimation of the BOD
of waste waters(18).
        C. butyricum was immobilized in polyacrylamide gel membrane
and the immobilized whole cells were fixed on the anode. A linear
relationship was obtained between the steady-state current and the
BOD from 0 to 250 ppm. The steady-state current was reproducible
within $\pm$7 % of   relative error, when the standard solution (50
mg l^{-1} glucose, 50 mg l^{-1} glutamate) was measured repeatedly. The
standard deviation was 2 ppm.

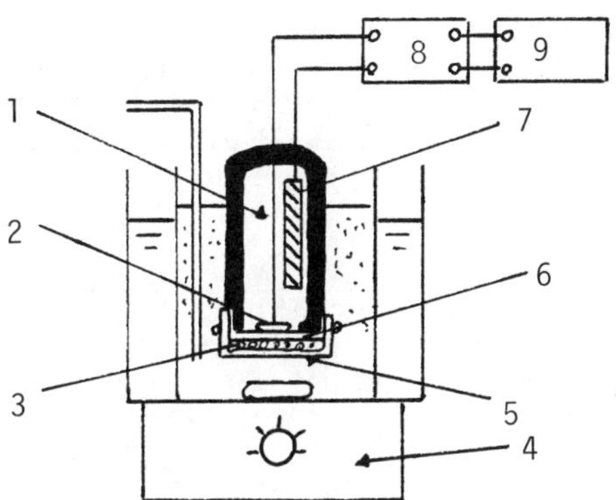

Figure 6. The microbial sensor system for ammonia. 1. Electrolyte (NaOH); 2. Cathode (Pt); 3. Immobilized cells; 4. Magnetic stirrer; 5. Gas permeable Teflon membrane; 6. Teflon membrane; 7. Anode (Pb)

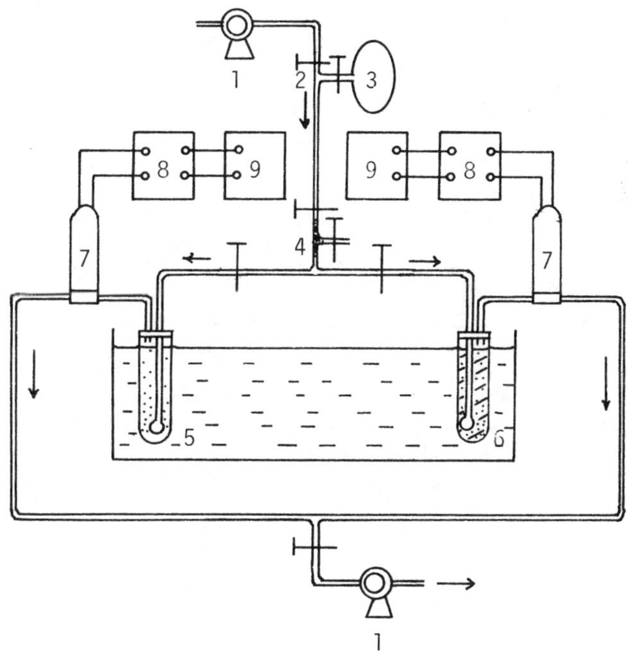

Figure 7. Schematic diagram of the methane gas sensor. 1. Pump; 2. Gas sampler; 3. Sample gas; 4. Cotton filter; 5. Reference reactor; 6. Methane oxidizing bacterial reactor; 7. Oxygen electrode; 8. Amplifier; 9. Recorder

The microbial sensor was applied to the estimation of the BOD
of waste waters. Three kinds of industrial untreated waste water
— slaughter-house, food factory, and alcohol factory — were employed in the experiments. Relative error of BOD estimation of industrial wastewaters was within 10% campared to 5-day BOD test. Assay
could be done within 20 min. No decrease in current output was
observed over a 30-day period.

A second sensor employed _Trichosporon cutaneum_, which was
utilized for wastewater treatment (19).   The steady-state
current depended on the BOD of the sample solution. A
linear relationship was observed between the current difference
(between initial and steady state current) and the 5-day BOD of
the standard solution below 60 mg 1^{-1}. The minimum measureable
BOD was 3 mg 1^{-1}. The current was reproducible within ±6 % of
relative error when BOD of 40 $mg \cdot 1^{-1}$ of the standard solution was employed. The standard deviation was 1.2 $mg \cdot 1^{-1}$ of BOD in 10 experiments.
The current means the current differences hereinafter.
The microbial sensor was applied to estimation of 5-day BOD
for untreated waste waters from a fermentation factory. The 5-day
BOD of the waste waters was determined by the JIS method (Japanese
Industrial Standard Committee, 1974). Good comparative results
were obtained between the BOD estimation by the microbial sensor and
those determined by the JIS method. The regression coefficient was 1.2
in 17 experiments and the ratios (BOD estimated by the microbial
sensor/5-day BOD determined by JIS method) were in the range from
0.85 to 1.36. This variation might have been caused by changes
in composition of organic waste water compounds. Stable responses
to the standard solution (20 $mg \cdot 1^{-1}$ BOD) were observed for more
than 17 days (400 tests). Fluctuations of the current and the base
line (endogeneous level) were within ±20 % and 15 % respectively
for 17 days. The microbial sensor could be used for a long time
for the estimation of BOD. A continuous BOD estimation system is
now commercialized in Japan (Figure 8).

<u>Nitrite Sensor</u>. The principal gaseous oxides of nitrogen of interest
in air pollution sampling and analysis are nitric oxide (NO), and
nitrogen dioxide (NO_2). Nitrogen dioxide is the most reactive of
the gaseous oxides of nitrogen and is a primary absorber of sunlight
in photochemical atmospheric reactions that produce photochemical
smog. Therefore, the determination of nitrogen dioxide is important
in environmental and industrial process analyses. <u>Nitrobacter</u> sp.
utilize nitrite as the sole source of energy and oxygen is consumed
by the respiration as follows :

$$2NO_2^- + O_2 \xrightarrow{\text{Nitrobacter sp.}} 2NO_3^-$$

Therefore, NO_2 generated in the buffer (pH 2.0) can be determined
by the microbial sensor using immobilized <u>Nitrobacter</u> sp. and an
oxygen electrode (20,21). The scheme of the microbial sensor system
is illustrated in Figure 9.
When the sample solution containing sodium nitrite was injected
into the system , nitrogen dioxide was produced in the flow cell

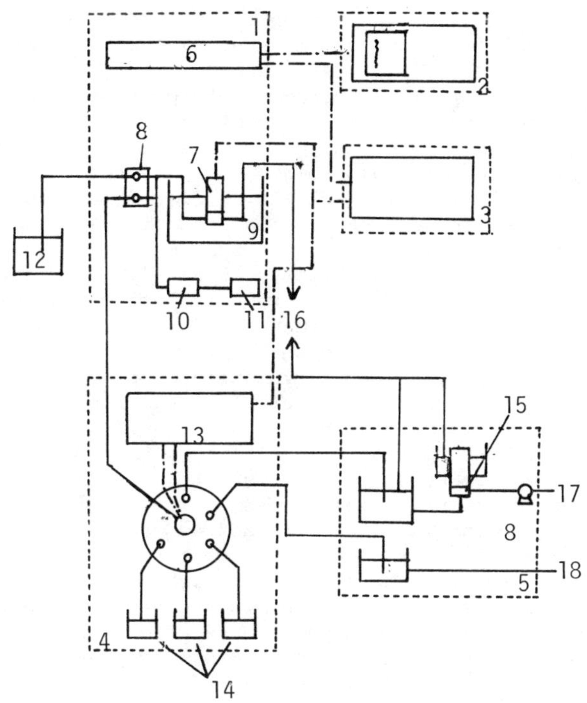

Figure 8. *Schematic diagram of Continual Measuring System for BOD.*
1. Sensor unit; 2. Recorder unit; 3. Data processing unit; 4. Flow
line selector unit; 5. Sampling unit; 6. Amplifier; 7. Microbial
sensor; 8. Pump; 9. Incubator; 10. Flow meter; 11. Air pump;
12. Buffer tank; 13. Selector controle; 14. Standard solution;
15. Filter; 16. Waste

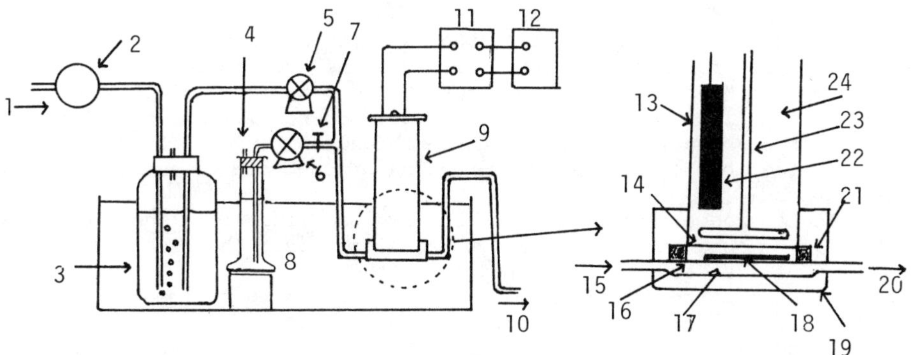

Figure 9. *Schematic diagram of the nitrite sensor system.* 1. Air(280mL/
min); 2. Pump; 3. Buffer (pH 2.0); 4. Sample solution; 5. Peristaltic
pump; 6. Pump; 7. Valve; 8. Incubator (30 C); 9. Microbial electrode;
10. Waste; 11. Amplifier; 12. Recorder; 13. Electrolyte (30% sodium
hydroxide); 14. Teflon membrane; 15. Buffer (pH 2.0) containing NO_2
gas; 16. Immobilized whole cells; 17. Gas permeable membrane;
18. cell space (1.0 mL); 19. Teflon cup; 20. Waste; 21. Rubber rings;
22. Platinum cathode; 23. Lead anode; 24. Insulator

and permeated through the gas permeable membrane. Nitrite was formed in the bacterial layer and assimilated by the immobilized bacteria. The steady-state current is obtained within 10 min. The differences between the initial and steady-state currents were directly
proportional to the concentration of sodium nitrite. A linear relationship was observed between the current decrease and the sodium
nitrite concentration below 0.59 mM (current decrease 0.63 μA).
At greater than 0.65 mM sodium nitrite, a linear relationship was
not observed between the current and concentrations. The minimum
concentration for the determination of sodium nitrite was 0.01 mM
(signal to noise, 20; reproducibility, $\pm$5 %). The current decrease
was reproducible within $\pm$4 % of   relative error and the standard
deviation was 0.01 mM in 25 experiments when a sample solution containing 0.25 mM of sodium nitrite was employed.

The sensor did not respond to volatile compounds such as acetic
acid, ethyl alcohol. and amines (diethylamine, propylamine, and
butylamine) or to nonvolatile nutrients such as glucose, amino acids,
and metal ions (potassium and sodium ions). Therefore, the selectivity of this microbial sensor was satisfactory in the presence of
these different substances. The current output of the sensor was
almost constant for more than 21 days and 400 assays. The microbial
sensor can be used to assay sodium nitrite for a long period. In
the same experiments the concentration of sodium nitrite was determined by both the sensor proposed and the conventional method (dime-
thyl-α-naphtylamine method). A good correlation was obtained between
the sodium nitrite concentrations determined by the two methods
(correlation coefficient 0.99).

<u>Mutagen Sensor</u>. Long-term carcinogenicity tests with laboratory
mammals are not only time-consuming but also demanding of resources.
The mutagenic activity of carcinogens has recently been confirmed
in a great number of cases. The existence of a high correlation
between the mutagenicity and carcinogenicity of chemicals is now
evident. The use of microbial systems is important for a survey
of mutagenic chemicals. Recently, a number of microbial methods
for detecting the various types of mutagens have been developed.
A method named "rec-assay" utilizing <u>Bacillus</u> <u>subtilis</u> has also
been proposed for screening chemical mutagens and carcinogens (22).
However, the "rec-assay" still require a lengthy incubation of bacteria and complicated procedures. An electrode consisting of the
aerobic recombination-deficient bacteria and the oxygen electrode
was applied to the preliminary screening of chemical mutagens and
carcinogens (23-25).

The microbial sensor system is shown in Figure 10. The electrode system consisted of two microbial electrodes: the electrode
of <u>B. subtilis</u> Rec$^-$ (Rec$^-$ electrode) and the electrode of <u>B. subtilis</u>
Rec$^+$ (Rec$^+$ electrode). Each electrode was composed of immobilized
bacteria and an oxygen electrode.

If sufficient nutrients (e.g., 0.3 g l^{-1} glucose) are present
in a sample solution, a   constant current is obtained from the
electrode. The current depends on the total respiration activity
of immobilized cells. Therefore, the total respiration activity
of bacteria, the current, depends on the number of viable cells
immobilized onto the acetylcellulose membrane. The relationship
between the current and the cell numbers on the acetylcellulose

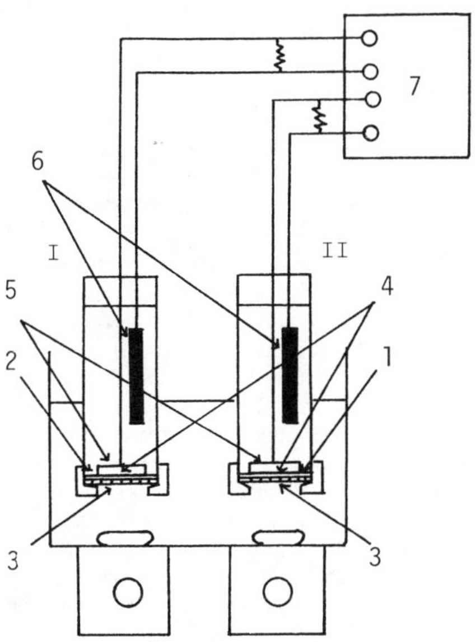

Figure 10. Schematic diagram of the electrode system for rapid detection of chemical mutagen. (I) Rec⁺ electrode; (II) Rec⁻ electrode; 1. Bacillus subtilis Rec⁻; 2. Bacillus subtilis Rec⁺; 3. Membrane filter; 4. Teflon membrane; 5. Pt cathode; 6. Pb anode; 7. Recorder

membrane was linear in the range over $0.1-3.0 \times 10^8$ cells. Consequently, 2.7×10^8 cells of B. subtilis Rec⁻ and Rec⁺ were immobilized, thereafter, on the membrane of the electrode.

When the Rec⁻ and Rec⁺ electrodes were inserted into the glucose-buffer solution (0.3 g.l⁻¹ glucose), steady-state currents were obtained. Then, AF-2, well known mutagen, was added to the solution. After 24-40 min, the current of the Rec⁻ electrode began to increase giving a sigmoidal curve. On the other hand, the current of the Rec⁺ electrode did not increase. The rate of current increase is a measure of the mutagen concentration and is most easily measured as the linear slope at the midpoint of the sigmoidal curve. When chemical mutagens such as AF-2, mitomycin, Captan, 4NQO, N-methyl-N'-nitro-N-nitrosoguanidine, and aflatoxin B₁ were added to the glucose-buffer solution, the rates of the current increase of the Rec⁻ and Rec⁺ electrodes were measured. The current of the Rec⁻ electrode markedly increased when these reagents were added to the system. Therefore, the mutagenicity of chemicals can be estimated with the sensor system. Linear relationships were obtained in the range over $1.6-2.8$ g ml⁻¹ for AF-2 and $2.4-7.3$ g ml⁻¹ for mitomycin.

This microbial sensor system is based on the inhibitory action of the mutagens on the respiration of B. subtilis Rec⁻. B. subtilis M45 (Rec⁻) is genetically deficient in the DNA recombination enzyme system, whereas B. subtilis H17 (Rec⁺) is a wild strain which has the ability to repair   damaged DNA. The subsequent death of Rec bacteria is preceded by the decrease of respiration. As a result, the number of Rec⁻ cells on the surface of the oxygen electrode decreased and the current of the Rec⁻ electrode increased. On the other hand, the damaged DNA of Rec⁺ bacteria is repaired with the recombination system. Therefore, the number of Rec⁺ cells did not change and the current of the Rec⁺ electrode did not increase. Since the respiration of bacterial cells is directly and immediately converted to an electric signal, the preliminary screening of mutagens is possible within 1 h. Moreover, the microbial sensor system employs a homogeneous suspension. Consequently, the sensitivity of the microbial sensor is higher than the "rec-assay". The minimum measurable mutagen concentration is 1.6 μg ml⁻¹ by the microbial sensor and 5.0 μg ml⁻¹ by the "rec-assay" and 10 μg ml⁻¹ by the Ames test (26) for AF 2.

On the other hand, Salmonella typhimurium TA 100 requires histidine for   growth. When the membrane filter-electrode containning various amounts of S. typhimurium revertant was inserted into the glucose-buffer solution saturated with oxygen, glucose was assimilated by the bacteria. However, the revertant of this strain can grow on the histidine-free medium. The electrode response depends on the numbers of viable bacteria retained on the membrane filter. Then, various amounts of AF-2 were added to the medium. The current was measured at 2-h intervals. After 8 h of incubation, the current decreased with increasing incubation time because the revertants grew above the minimum detectable numbers of cells by the electrode. A 10-h incubation gave the greatest sensitivity. On the other hand, there was no decrease in current from the medium in the absence of AF-2. The current decrease became larger with increasing AF-2 concentration. The minimum measurable concentration of AF-2 was 0.001 μg ml⁻¹. The current decrease was reproducible

within 5 % of   relative error when a sample solution containing
0.006 μg ml^{-1} of AF-2 was employed.

     When S. typhimurium was incubated with chemical mutagens such
as N-methyl-N'-nitro-N-nitrosoguanidine, nitrofurazone, methyl methanesulfonate, and ethyl methanesulfonate, for 10 h the current
decrease of the electrode was measured. The response of the electrode increased with increasing concentration of chemical mutagens.
Therefore, the mutagenicity of chemicals can be estimated with the
microbial electrode.

     In this study, a homogeneous bacterial suspension was employed
and mutagens were added to the homogeneous suspension. The complete
medium containing amino acids, vitamins, and mineral salts was used
for experiments. Furthermore, employment of the electrode system
makes possible a large injection of bacterial suspension. As a
result, the time required for mutagen test was shortened to 10 h.
The time is longer than that (1 h) of the micribial electrode system
described above. The sensitivity of the microbial electrode system
was higher than the convetional Ames test. The minimum measurable
mutagen concentration was 0.001 μg ml^{-1} by the microbial electrode,
and 10 μg ml^{-1} by the Ames test for AF-2.

Acknowledgment

The author thank Professor Shuichi Suzuki, The Saitama Institute
of Technology, for his encourage during experiments.

Literature Cited

1. Guilbault, G. G. "Hand-book of Enzymic Analysis"; Dekker: New
    York, 1976.
2. Karube, I.; Suzuki, S. "Ion-Selective Electrode Reviews";
    Pergamon Press: Oxford, 1984; Vol. 6, p. 15.
3. Karube, I.; Suzuki, S. "Annual Reports on Fermentation
    Processes"; Academic Press: New York, 1983; p. 203.
4. Karube, I. "Biotechnology & Genetic Engineering Reviews"; Inter-
    cept Ltd.: Newcastle upon Tyne, 1983; Vol. 2, p. 313.
5. Karube, I. "Biotechnology Handbook"; Butterworth/Ann Arbor
    Sci. Pub. Co.: New Jersey, 1985; p. 134.
6. Karube, I.; Mitsuda, S.; Suzuki, S., Europ. J. Appl. Microbiol.
    Biotechnol. 1979. 7, 343.
7. Hikuma, M.; Kubo, T.; Yasuda, T.; Karube, I.; Suzuki, S.,
    Biotechnol. Bioeng. 1979. 21, 1845.
8. Hikuma, M.; Kubo, T.; Ysauda, T.; Karube, I.; Suzuki, S., Anal.
    Chim. Acta 1979. 109,33.
9. Matsunaga, T.; Karube, I.; Suzuki, S., Europ. J. Appl. Micro-
    biol. Biotechnol. 1980. 10, 235.
10. Hikuma, M.; Obana, H.; Yasuda, T.; Karube, I.; Suzuki S., Anal.
    Chim. Acta 1980. 116, 61.
11. Hikuma, M.; Yasuda, T.; Karube, I.; Suzuki, S., Ann. New York
    Acad. Sci. 1981. 369, 307.
12. Karube, I.; Okada, T.; Suzuki, S., Anal. Chem. 1981. 53, 1852.
13. Okada, T.; Karube, I.; Suzuki, S., Anal. Chim. Acta 1982. 135,
    159.
14. Hikuma, M.; Kubo, T.; Yasuda, T.; Karube, I.; Suzuki, S., Anal.
    Chem. 1980. 52, 1020.

15. Okada, T.; Karube, I.; Suzuki, S., Europ. J. Appl. Microbiol.
    Biotechnol. 1981. 12, 102.
16. Karube, I.; Okada, T.; Suzuki, S., Anal. Chim. Acta 1982. 135,
    61.
17. Karube, I.; Matsunaga, T.; Mitsuda, S.; Suzuki, S., Biotechnol.
    Bioeng. 1977. 19, 1535.
18. Karube, I.; Matsunaga, T.; Suzuki, S., J. Solid phase Biochem.
    1977. 2, 97.
19. Hikuma, M.; Suzuki, H.; Yasuda, T.; Karube, I.; Suzuki, S.,
    Europ. J. Appl. Microbiol. Biotechnol. 1979. 8, 289.
20. Karube, I.; Okada, T.; Suzuki, S., Europ. J. Appl. Microbiol.
    Biotechnol. 1982. 15, 127.
21. Okada, T.; Karube, I.; Suzuki, S., Biotechnol. Bioeng. 1983.
22. Kada, T.; Tutikawa, K.; Sadaie, Y., Mutat. Res. 1972. 16, 165.
23. Karube, I.; Matsunaga, T.; Nakahara, T.; Suzuki, S., Anal.
    Chem. 1981. 53, 1024.
24. Karube, I.; Nakahara, T.; Matsunaga, T.; Suzuki, S., Anal.
    Chem. 1982. 54, 1725.
25. Karube, I., Trends in Analytical Chemistry 1984. 3, 40.
26. Ames, B.N.; Lee, F.D.; Durston, W.E., Proc. Natl. Acad. Sci.
    USA 1972. 70, 782.

RECEIVED February 6, 1986

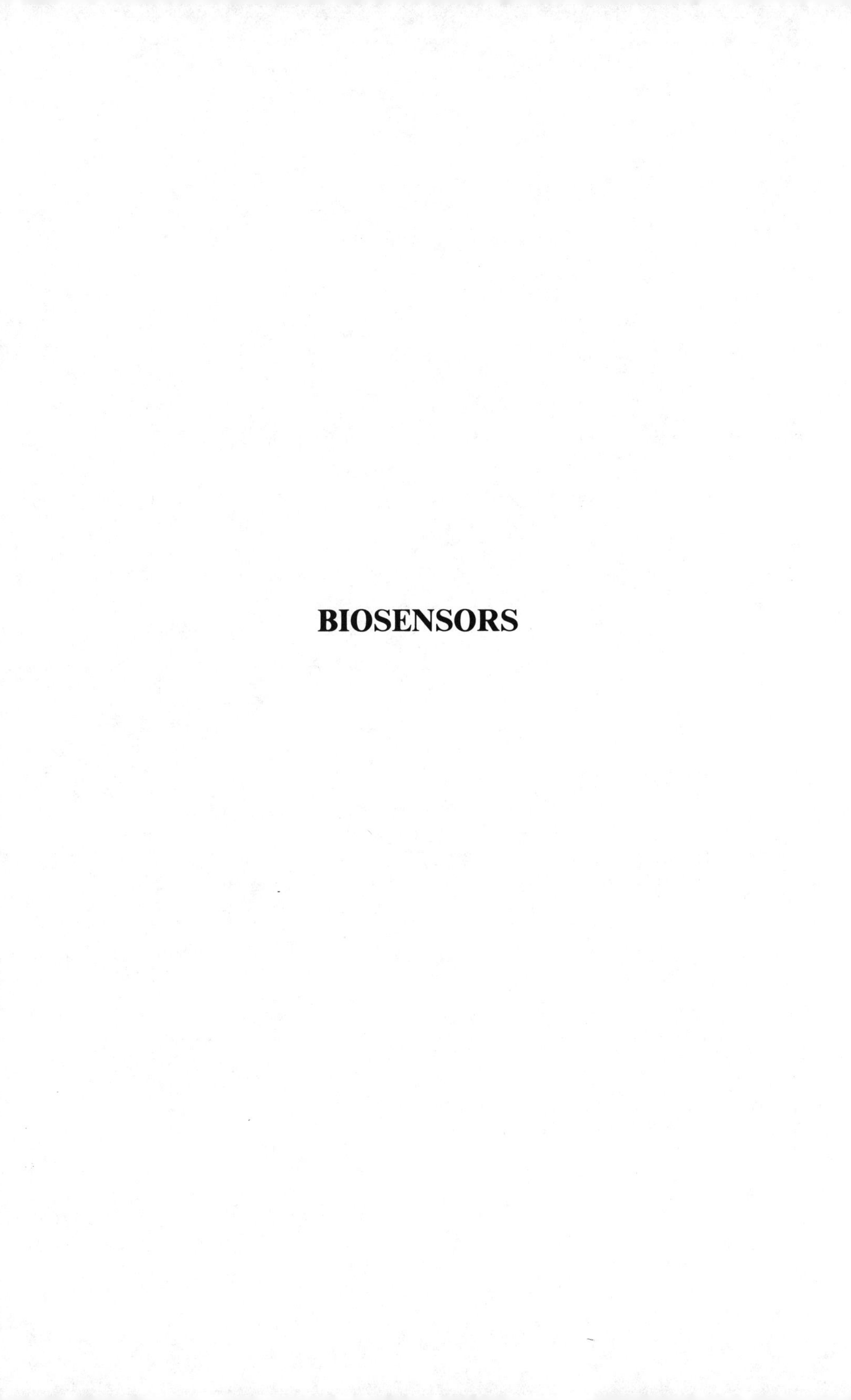

BIOSENSORS

Langmuir–Blodgett Technology and Receptor Action in Stabilized Lipid Membranes

U. J. Krull, M. Thompson, and H. E. Wong

Chemical Sensors Group, Department of Chemistry, University of Toronto, Toronto, Ontario, Canada M5S 1A1

Artificial lipid membranes formed from phospholipid and steroids in both bilayer lipid membrane (BLM) and Langmuir-Blodgett monolayer configurations have been investigated for relationships between BLM Arrhenius energy barrier and intermolecular interaction. The results indicate a large sensitivity of energy barrier to molecular packing on the order of 0.10 eV for 0.01 nm^2 average molecular area change. Stabilized lipid membranes formed on polyacrylamide gel are described with respect to receptor and stabilization requirements.

The selectivity and sensitivity of natural chemoreceptive processes has intrigued scientists for many years, but only recently has its potential for chemical sensor development been fully recognized. Extensive investigation of cellular membrane processes in the biological sciences coupled with successful laboratory models of such membranes has contributed greatly to advancing the understanding of important membrane processes. However the particular chemistry responsible for signal transduction in chemoreceptive tissues remains an unconquered research frontier. Nevertheless valuable models regarding the selective chemical sensor problems can be derived from the structure and mechanisms of natural membrane technology.

The bilayer lipid membrane (BLM) has become one artificial membrane model which has permitted extensive investigation of the chemistry associated with the structural lipids present in natural membranes.(1-2)

Substantial insight has been achieved into the suitability of BLM matrices as the basis for development of a transducer where control of analytical signals is determined by transmembrane ionic conduction variations. The embedding of molecular receptors or complexing agents into BLM has led to the development of selective transducers for organic species, (3-6) whereby such organics would complex with the receptor to alter membrane character and therefore ion current across the membrane.

The complex structure and poorly defined nature of receptors in membranes have been major impediments to the elucidation of the mechanisms of signal generation suitable for analytical measurement. It has been postulated that receptors (7) as well as some anaesthetics (8) may influence transmembrane conductance by altering one or more of the ion current controlling physical parameters listed in Table I. Since these parameters are inherent properties of the BLM matrix, it is possible to investigate the relative impact of their variation on ion current by appropriate modification of lipid chemistry thereby avoiding implementation of receptors. Such modifications can be readily achieved by employment by lipids containing variable: polar headgroups, acyl chain lengths, and acyl chain chemistry including the addition of polar functional groups. Experiments involving the probing of membrane chemistry by chromatography, measurement of molecular packing characteristics by Langmuir-Blodgett monolayer compression experiments, analysis of the temperature dependence of transmembrane ion conduction for Arrhenius energy barrier determination, together with consideration of membrane electrostatic potential leads to an evaluation of the significance of particular membrane parameters.

The results from such experiments indicate that membrane ion current is very dependent on molecular packing with respect to the other membrane parameters of Table I. This factor must be considered in any new design for development of a sensing device to ensure that substantial molecular mobility may be incorporated into stabilized membranes. Such a stabilized, electrochemically active membrane has been produced by lipid monolayer deposition onto hydrated polyacrylamide gel, demonstrating that device fabrication may be possible.

<u>Experimental</u>

The lipids used for BLM and monolayer formation are listed in Table II. The phospholipid was obtained from Avanti Biochemicals, Birmingham, Al and the steroids were supplied by Research Plus, Inc., Bayonne, NJ.

TABLE I.   Summary of BLM Parameters Controlling

Transmembrane Ion Current

BLM Property	Origin	Type of Barrier
Surface Potential	Electrical Double Layer due to surface ions or ionization	Electrostatic
Dipolar Potential	Anisotropy of lipid headgroup dipoles in organized membrane	Electrostatic
Fluidity	Density and mobility of lipids, particularly hydrocarbon chains, determine interstitial spaces.	Steric
Dielectric Constant	Low dielectric in hydro-carbon zone not suitable for ion solvation, Born energy effects	Chemical/Electrostatic
Thickness	Hydrocarbon solvent and lipid acyl chain length, reflects extent of low dielectric zone	Chemical/Steric

Oxidized phospholipid was produced by irradiation of pure phospholipid dissolved in decane at 254 nm in the ambient atmosphere.

The electrochemical equipment and procedures for investigation of BLM's has been described previously in detail. (9-10) The apparatus was based on application of +25 mV DC potential across the membrane between two Ag/AgCl reference electrodes while monitoring ion current with a digital electrometer.

Chemical analysis of lipid composition has been described previously in detail. (11) The technique employs derivatization by methylation of phospholipid acyl chains followed by separation and analysis with capillary gas chromatography.

The Langmuir-Blodgett thin-film trough employed in this work was a Lauda Film Balance Type 1974 (Sybron-Brinkmann, Toronto, Canada).

All experiments using lipid membranes employed equal weight ratios of the phospholipid or oxidized phospholipid, and one of the steroids. For the trough experiments, 2 mg of phospholipid and 2 mg of steroid were dissolved in a total of 5 ml of solvent. Hexane was used as the solvent in most instances, but small quantities of chloroform were necessary as a secondary solvent for complete dissolution of the steroid diol and triol species. Approximately 0.1 ml of solution was slowly spread on the aqueous sub-phase (consisting of pure water or 0.1 M KCl) in the trough, and the solvent was allowed to completely evaporate before compression experiments were initiated. Compression was performed slowly in all cases to allow surface equilibration and each sample solution was investigated at least four times to ensure reproducibility.

Monolayer casting experiments made use of similar experimental conditions to form organized and highly compressed membranes. Such films were deposited on polyacrylamide gel by dipping the polymer through the air-water interface of the trough, with monolayer compression held constant at pressures of 30 to 40 mN.m^{-1}.

The gel was formed by mixing equal volumes of a solution of 20% (w/v) total monomer concentration containing 19:1(w/w) acrylamide/bisacrylamide and a 0.01%(w/v) riboflavin-5'-phosphate solution in 0.1 M pH 7 phosphate buffer, then adding N,N,N',N'-tetramethylethylenediamine to a 1% (v/v) concentration and catalyzing polymerization with UV radiation at 254 nm. These monomeric reagents were mixed in a glass reservoir, and a small volume was transferred to a "well" formed from hydrophobic epoxy resin for in-situ polymerization on the device. The resin coated a Ag/AgCl substrate, the latter acting as one electrode when the monolayer-coated gel was investigated for electrochemical character in an arrangement similar to

that for planar BLM. This device structure is shown in
Figure 1 in an orientation suitable for electrochemical
experimentation, which consisted of solution addition of
the ion carrying probe valinomycin and dipole potential
perturbing probe phloretin (both from Sigma Chemical
Co., St. Louis, MO). Further details of the
construction of this device have been published
elsewhere. (12)

Results and Discussion

Mechanisms of Analytical Signal Generation. The
experiments were designed to test the influence of
various polar functional groups on membrane packing and
dipolar potential in an effort to relate these
parameters to transmembrane ion conduction. Previous
work has indicated that the process of ion conduction
through a membrane can be distinguished by three
distinct processes which are physically related to
compartmentalization of a BLM as follows:
1) The interface between the polar phospholipid
headgroups and the aqueous electrolyte solution provides
a membrane surface which contains weakly selective
cationic binding sites. (10) This generates a reservoir
of cations available for conduction and is apparently
much more important than bulk solution ion content in
the determination of permion and ion current density.
(13)
2) The polar lipid headgroup zone contains the
phospholipid and steroid phosphorus-nitrogen, carbonyl,
hydroxyl and hydration water moieties which combine to
establish a substantial dipolar potential. This
positive potential is of a magnitude of several hundred
millivolts across the membrane headgroup zone. In the
membrane hydrocarbon interior, the electrostatic field
must be at least 450 to 750 mV (14) and controls ion
current across the interior as well as possibly
influencing selective ion adsorption to the membrane
surface.
3) The interior lipid acyl chain hydrocarbon zone is
defined as an area of low dielectric constant ($\varepsilon \approx 2$) so
that Born energy considerations for ion injection into
this region become very significant. (15) The acyl
chains are also densely packed in a compressed situation
such as a BLM where pressures are expected to reach 30
mN.m^{-1}. (16) Ion migration down an electrical potential
gradient through this zone must be influenced by chains
rotation which would sweep out a transient volume
suitable for temporary ion containment. (17) This
feature represents a steric hindrance to ion motion and
is dependent on acyl chain structure and length, and the
temperature of the matrix.
     This work deals with membrane modification to test

the significance of each of these three areas with respect to each other. The incorporation of each of the various steroids into a phospholipid matrix at 0.65 mole fraction concentration allowed observation of the influence of extra polar species on dipolar and steric modifications in the headgroup zone of the membrane. Figure 2 illustrates that a direct relationship exists between BLM Arrhenius energy barrier and the average molecular area occupied by a lipid molecule in a monolayer measured at a low equilibrium pressure of 16 $mN.m^{-1}$. The maximum sensitivity in energy barrier of BLM to differences in molecular area caused by headgroup zone variation was observed to be as high as 0.10 eV per 0.01 nm^2. This result was determined by the structural characteristics of the A and B rings of the steroids and their effects on intermolecular interaction. The results indicate that hydrogen bonding in the headgroup area, as evidenced by the differences between steroid "a" and "g", (Table II) is crucial in determination of headgroup interaction, and this is translated into a dramatic effect on acyl chain interaction.

The influence of surface dipolar potential must also be considered as a factor contributing to the measured Arrhenius energy barrier. Electrostatic fields can certainly moderate ion current, though the significance of this parameter in contrast to molecular packing may be minimal. The molecular dipole moments listed in Table II for steroids "a" and "g", in combination with their measured average molecular areas can be used to calculate on a first approximation the expected surface dipolar potential. (18) Such a calculation indicates that dipolar potentials play a minor role in determination of the Arrhenius barrier and that molecular packing constraints on ion transport dominate. This conclusion still requires experimental substantiation by measurement of monolayer dipolar potentials since no corrections for angular inclination of dipoles, or contributions from hydrogen bonding (including water of hydration) can be readily incorporated into the analysis. Furthermore, a synergic association between molecular packing and dipolar potential must exist. This implies that alterations of molecular packing would necessarily cause changes in dipolar potential, by headgroup realignment and surface dipole reorientation or alteration would cause internal structural order perturbations.

A second important series of experiments dealt with membrane chemistry composed of 5-cholesten-3β-ol (cholesterol) and partially oxidized phospholipid. The ultraviolet radiation induced oxidation led to formation of hydroperoxide moieties at the unsaturated sites on some of the phospholipid acyl chains listed in Table II. The $C_{18:2}$ chain was chosen as a reactive representative

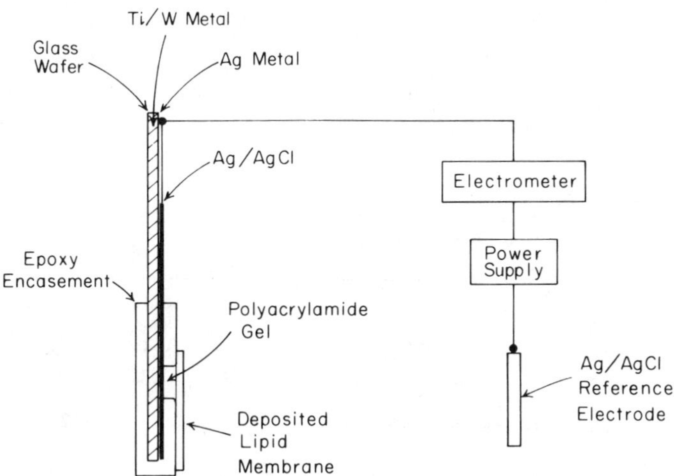

Figure 1.   Schematic of an electrochemical devise based on lipid membrane technology shown in an arrangement suitable for electrochemical experimentation.

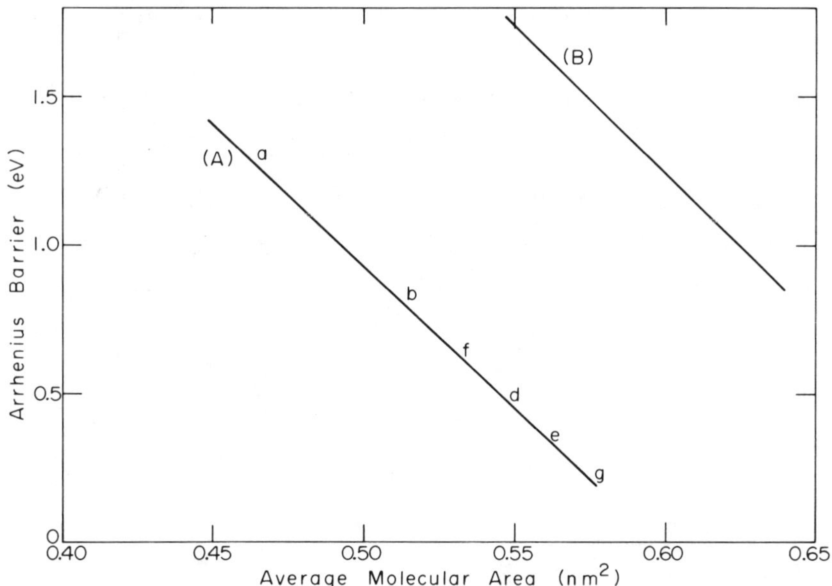

Figure 2.   Relationships between A) phospholipid–steroid BLM Arrhenius energy barrier and monolayer average molecular area measurements as a function of steroids a–g (see Table II), and B) BLM Arrhenius energy barrier and monolayer average molecular area measurements for oxidized phospholipid–cholesterol compositions.

Table II.   Summary of the lipids used in membrane formation

Steroid Designation	Lipid	Dipole Moment(D)
a	5-cholesten-3β-ol	2.01
b	5,7-cholestadiene-3β-ol	–
c	5-cholesten-3β,7α-diol	–
d	5α-cholestan-3β,5α,6β-triol	–
e	5α-cholestan-5α,6α-epoxy-3β-ol	–
f	5-cholesten-3β-ol-7-one	3.79
g	5α-cholestan-3-one	3.01

	Acyl Chain Composition
egg phosphatidyl choline	$C_{16:0}$, 33.0%
	$C_{16:1}$, 2.1%
	$C_{18:0}$, 15.4%
	$C_{18:1}$, 31.7%
	$C_{18:2}$, 17.8%
	$C_{20:4}$, 4.3%
	$C_{22:6}$, 1.7%

chain to monitor and quantitate oxidation. A direct
correlation of the extent of oxidation with average
molecular area from monolayer compression results was
observed.   The average molecular areas were then
contrasted to measured Arrhenius energy barriers to
produce the curve "B" shown in Figure 2.   Again a
correlation of energy barrier to molecular area shows a
sensitivity as high as 0.10 eV for 0.01 nm^2. This
result further supports the significance of the steric
factors in controlling the energy barrier.
Incorporation of polar species into the hydrophobic
portion of a BLM should provide transient binding sites
for ion migration through the membrane interior. This
is related to a reduction of the electrostatic effects
associated with Born Energy considerations. (15)
However the high sensitivity of energy barrier on
molecular packing implies that steric effects dominate
the ion migration mechanism.

<u>Receptor Action in Analytical Membranes</u>.   This work has
identified the importance of some of the relevant
parameters associated with organized lipid matrix ion
conduction.   Receptors embedded in these matrices should
function to perturb one or more of the three
compartmental zones of the BLM.   A number of features
significant to electrochemical receptor operation
follow:
1)   The Arrhenius energy barrier is largely controlled
by molecular interactions and freedom of rotation, and
perturbation of this parameter will cause energy barrier
changes of approximately 0.10 eV for 0.01 nm^2.
2)   In Figure 2, although both curves have similar
slopes, the Arrhenius barriers measured for steroid
modification are significantly lower than for oxidized
phospholipid when comparing similar average molecular
areas.   This implies that molecular packing and/or
electrostatic alterations near the membrane surface may
result in greater signal evolution than similar changes
in the hydrocarbon zone.
3) Synergic action of receptors is preferred for
maximization of an analytical signal.   A concurrent
reduction of dipolar potential and molecular packing
will amplify any one receptor-stimulant complexation in
comparison to singular alteration of either parameter.

Receptors should, therefore, have active sites available
at the membrane/solution interface, and on complexation
with stimulant the following ideal scenario should
occur:
a) The local electrostatic field would adjust to attract
more permions to the membrane surface for transport.
b) The positive dipolar potential present within the
membrane would reduce to allow greater permion flux.

c) A concurrent associated intermolecular association alteration located in the headgroup zone would reduce steric hindrance to ion permeation.
d) A similar concurrent steric hindrance reduction would occur in the hydrophobic depths of the membrane.

To maximize analytical signal generation, the ideal receptor will therefore be chosen or designed by consideration of the electrostatics of the binding site, the location of membrane positioning, intra-membrane protrusion/anchorage, association with adjacent lipids and conformational alteration.

<u>Device Development</u>. A substantial body of work supports the premise that receptor-modified BLM can sensitively and selectively respond to stimulants to provide electrochemical signals. Successful receptor-stimulant combinations include enzyme-substrate, antibody-complement-antigen, hormone-hormone receptor and lectin-saccharide systems. (<u>3-6</u>) However, given that certain receptor systems work, another major difficulty associated with practical device design remains in the technological problem of stabilizing organized lipid membranes for long-term reproducible function. Numerous attempts at incorporation of stabilizing species into BLM have not resulted in structurally practical membranes. (<u>19-20</u>) The re-emergence of Langmuir-Blodgett thin-film technology has resulted in the development of an experimental capability suitable for creation of organized lipid or surfactant monolayers/multilayers, which can be stabilized by membrane-internal polymerization and deposition onto a wide variety of substrates. (<u>21</u>) The first step in the investigation of the feasibility of membrane deposition technology for device fabrication must be the development of an electrochemically active sensing membrane.

The electrochemical experiments which have proven to be successful have to date been almost exclusively based on planar BLM studies. These systems operate by using the BLM as a selectively permeable membrane which physically separates two highly conductive aqueous electrolyte solutions. An analogous arrangement for the testing of deposited membranes indicates that the substrate must not only offer a solid interface, but must also provide conditions suitable for ion conduction. This can be achieved by employment of extensively hydrated polymers such as polyacrylamide gels (PAG). In this work PAG was mounted on a Ag/AgCl electrode surface to provide an aqueous reservoir of ions for electrochemical conduction. This substrate was driven through a compressed phospholipid-cholesterol monolayer at least two times (beginning immersed in the aqueous trough subphase) to deposit the monolayer in-situ onto the gel. Even though the success rate for formation of lipid membranes of character appropriate for ion current modulation was poor, functional systems clearly responded to non-selective stimulants such as the ion carrying antibiotic valinomycin and the dipolar potential reducing agent phloretin. The electrochemical response features to these probe molecules was consistent with BLM deposition onto the gel surface. The deposited membrane could be transported physically throughout the aqueous subphase of the trough without destroying the character of the organized film, however passage of a stabilized membrane through the air-water interface was completely destructive with respect to further electrochemical response.

Concluding Remarks

The experiments reviewed in this paper have addressed two critical areas relating to the development of practical devices based on lipid membrane technology. It has been established that stabilized membranes can be developed, and some insight into receptor design/action has been accrued. Future work must address better methods of membrane stabilization including polymerization, substrate anchorage and fluidity/electrostatic modification by appropriate chemical adjustment. The achievement of reproducible surfaces for device fabrication will open new horizons in the study of receptor action in membranes. The goal of such studies must be directed towards permanent membrane embedding of receptors, and development of technology to induce receptor reversability. It is highly unlikely that the BLM structure and experimental electrochemical arrangement in use today will play any significant role in the practical devices of the future. This niche will be filled by other organized surfactant arrangements and different electrochemical techniques, derived from our knowledge of BLM electrochemistry.

Acknowledgments

We thank the Natural Sciences and Engineering Research Council of Canada, and the Ministry of the Environment, Province of Ontario for support of this work.

Literature Cited

1. Mueller, P.; Rudin, D.O.; Tien, H.T.; Wescott, W.C. Nature (London), 1962, 194, 979.
2. Tien, H.T. "Bilayer Lipid Membranes", Dekker, New York, 1974.
3. Toro-Goyco, E.; Rodriguez, A.; del Castillo, Biochem. Biophys. Res. Comm., 1966, 23, 341.

4. Issaurat, B.; Amblard, G.; Gavach, C. _FEBS Lett._, 1980, 115, 148.
5. Thompson, M.; Krull, U.J.; Venis, M.A. _Biochem. Biophys. Res. Comm._, 1983, 110, 300.
6. Thompson, M.; Krull, U.J.; Bendell-Young, L.I. _Bioelectrochemistry and Bioenergetics_, in press.
7. Thompson, M.; Krull, U.J. _Anal. Chim. Acta_, 1983, 147, 1.
8. Reyes, J.; Latorre, R. _Biophys. J._, 1979, 28, 259.
9. Thompson, M.; Krull, U.J. _Anal. Chim. Acta_, 1982, 141, 33.
10. Thompson, M.; Krull, U.J.; Bendell-Young, L.I.; Lundstrom, I.; Nylander, C. _Anal. Chim. Acta_, in press.
11. Krull, U.J.; Thompson, M.; Arya, A. _Talanta_, 1984, 31, 489.
12. Krull, U.J.; Thompson, M.; Wong, H.E. _Anal. Chim. Acta_, in press.
13. McLaughlin, S. in "Current Topics in Membranes and Transport", Bonner, F.; Kleinzeller, A. eds., Academic, New York, 1977, 9, 71.
14. Haydon, D.A. _N.Y. Acad. Sci. Ann._, 1975, 264, 2.
15. Lundstrom, I.; Nylander, C. _J. Theor. Biol._, 1981, 88, 671.
16. Blume, A. _Biochim. Biophys. Acta_, 1979, 557, 32.
17. Gallay, J.; de Kruyff, B. _FEBS Lett._, 1982, 143, 133.
18. Gaines, G.L. "Insoluble Monolayers at Liquid-Gas Interfaces", Interscience, New York, 1966.
19. Bean, R.C. _Gov't Rep. Announce_, 1973, 73, 76.
20. Shchipunov, Y.A. _Biophysics_, 1978, 23, 162.
21. Thin Solid Films, 1980, 68, (volume dedicated to advances in Langmuir-Blodgett membrane technology).

RECEIVED February 3, 1986

Design of Sensitive Drug Sensors: Principles and Practice

R. P. Buck and V. V. Cosofret

Department of Chemistry, University of North Carolina, Chapel Hill, NC 27514

Design principles of new sensors for ionic drugs
follow from application of potential generation theory
using three ions: M^+, X^- and Y^-, or M^+, N^+ and Y^-.
Studies of liquid/liquid interface transport identify
single-ion free energies of partition as the figures
of merit for selectivity and sensitivity. New liquid
membranes with sensitivities to 10^{-7} mol/l for
bisquaternary muscle relaxants and phenytoin serve as
examples.

The so-called "trapped sites" of classical mobile-site, liquid ion
exchanger electrodes belong to a category of compounds known as ion
association extractants. Examples are long-chain diesters of
phosphoric acid and tricaprylylmethylammonium (Aliquat) ions. The
latter cation was studied extensively by Freiser and co-workers (1-
3) in the design of anion sensors.

In 1970 Higuchi, et al. (4) and Liteanu (5) introduced liquid
membrane electrodes responsive both to organic and inorganic ions.
These electrodes seemed to be neither conventional ion exhanger-
nor neutral carrier-based. At the 1973 IUPAC International
Symposium on Ion-Selective Electrodes, J. R. Cockrell, Jr.
presented an unpublished example of a liquid membrane responsive to
both cations and anions, not unlike responses of silver halide
membranes. Cation or anion response depended on which common ion
of the membrane was in excess, in soluble form, in solution. A
membrane containing a detergent ion pair was responsive to either
cationic or anion detergent species in solution, depending on which
was in excess. At that time, this effect seemed anomalous.

Although the equilibrium principle was available (equality of
electrochemical potential of each ion that reversibly equilibrates
across an immiscible liquid/liquid interface), the elementary
theory and consequences were not explored until recently (6). To
develop an interfacial potential difference (pd) at a liquid
interface, two ions M^+, X^- that partition are required. However,

this condition is necessary but not sufficient, because the pd produced is independent of salt concentration in either phase. To develop a pd dependent solely on a single ion activity, say M^+, three ions are required -- M^+, X^- and Y^-, of which Y^- is very oil soluble; X^- is mainly water soluble and M^+ is soluble in both phases. The salt MY is typically an organic ion pair that may be isolated and dissolved in an organic solvent or prepared _in situ_ by extraction. The anion is typically picrate, tetraphenylborate, triphenylstilbenylborate or tetrabiphenylborate. The salt MX, where X^- = Cl^- preferably, is the sample whose M^+ activity is to be measured at variable values in the aqueous phase. When MX is varied, the interfacial pd is overall S-shaped (mV vs log[MX]), but contains a linear section, the so-called Nernstian region.

It is merely an extension of these ideas to demonstrate the conditions that the same membrane, containing MY, should also be responsive, in a Nernstian fashion, to Y^- activities in solution. These conditions are again a three-ion situation: M^+, Y^- and N^+. The salt NY is the aqueous sample whose Y^- activity is to be measured. N^+ is typically a hydrophilic ion such as Na^+. When aqueous NY activity is varied, the interfacial pd is again S-shaped (mV vs log[NY]). These responses are illustrated from a theoretical calculation in Figure 1. The assumed extraction parameters are given in the legend. The similarity with silver halide membrane electrodes are summarized below.

<u>Saturated Salt AgBr (aq.)</u>	<u>Partitioned Salt MY (very oil soluble)</u>
Anion Responsive	Anion Responsive
$(aq.)Na^+Br^-$ at $C's > K_{sp}^{1/2}$	$(aq.)Na^+Y^-$ at $C's > C_{MY}(aq.)$
Cation Responsive	Cation Responsive
$(aq.)Ag^+NO_3^-$ at $C's > K_{sp}^{1/2}$	$(aq.)M^+Cl^-$ at $C's > C_{MY}(aq.)$

It is the hyperbolic relations $(Ag^+)(Br^-) = K_{sp}$ and $(M^+)(Y^-) = (\overline{M}^+)(\overline{Y})/K^2$ that provides the basic analogy between the two kinds of systems. In the latter, K^2 is the ionic salt partition coefficient relating membrane and bathing solution activities at an equilibrium interface. The latter form can also be derived for insoluble salt membranes. However the salt activities (super bar quantities) are constant and so are hidden in the value of the solubility product K_{sp}.

<u>Equilibrium Theory</u>

Each ion M^+, N^+, X^- and Y^- will generally have different energies in water and in an organic phase: an ester such as dioctyladipate (low dielectric constant). The partition free energy

$$\underline{\triangle}G_i = \overline{\mu}^o_i - \mu^o_i = -RT\ln K_i \tag{1}$$

for the reaction

$$\text{species } i(aq.) = \text{species } i(org.)$$

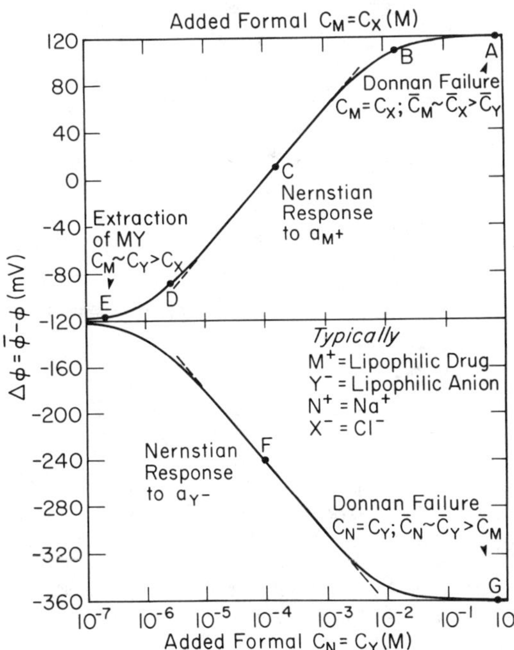

Figure 1. Calculated response curves. Upper curve response of M^+; lower curve response of Y^-. See text for values of the constants K_M, K_Y and K_X.

is a measure of the intrinsic ionic oil-solubility, where K_i is the single-ion partititon coefficient and bars denote organic phase quantities. For oil soluble (hydrophobic) ions $\triangle G$ is more negative and K_i is larger than for water-soluble (hydrophilic) ions. In a typical two phase water/organic system containing equilibrated salts MX and MY, some MX and MY will be present in each phase. This two salt, three ion system provides the necessary conditions to obtain the upper-curve for the M^+-sensor in Figure 1. The concentrations in each phase depend on the salt partition coefficients $K_M K_X$ and $K_M K_Y$, with mass and charge balance equations to provide the additional required relationships. The link between energies and equilibrium concentrations (activities) is the electrochemical potential for each ion at equilibrium across the interface:

$$\tilde{\mu}_i = \mu_i^o + RT\ln a_i + z_i F\phi \tag{2}$$

$$i = M^+,\ X^-,\ Y^- \tag{3}$$

and

$$\tilde{\mu}_i(aq.) = \tilde{\mu}_i(org.) \tag{4}$$

By eliminating ϕ between oppositely charged species one has

$$\frac{(a_{M^+})(a_{X^-})(org.)}{(a_{M^+})(a_{X^-})(aq.)} = K^2 = K_M K_X \tag{5}$$

and

$$\frac{(a_{M^+})(a_{Y^-})(org.)}{(a_{M^+})(a_{Y^-})(aq.)} = K^2 = K_M K_Y \tag{6}$$

Since MY is intentionally more oil soluble than MX, MY maintains virtually constant activity in the organic phase, when MX(aq.) is varied. Only a very small amount of X^- appears in the membrane because M^+ in the membrane is already large and fixed by MY (Donnan Exclusion). However at very high concentrations of MX(aq.), increasing amounts of MX will be extracted into the organic phase. In principle, at high enough MX(aq.) activities, the extracted MX exceeds the MY already present. This is the condition of Donnan Failure because the co-ion X^- enters the membrane at a concentration comparable to or greater than Y^-. A converse argument applies to the fate of M^+ in the aqueous phase as MX(aq.) concentration is decreased. Since MY maintains a very low concentration in water, as MX(aq.) is decreased below the equilibrium value of MY(aq.), M^+(aq.) approaches the constant concentration of MY(aq.). A constant interfacial potential is approached that is independent of further decreases in MX(aq.). In Fig. 1 the single ion partition coefficients were chosen $K_X = 10^{-2}$, $K_M = 10^2$ and $K_Y = 10^6$, with MY(org.) = 10^{-2} M. Consequently the equilibriums aqueous MY concentration is 10^{-6} M.

In the lower part of Figure 1, the corresponding two-phase, two-salt extracting system is MY, NY. The values of partition coefficients selected for illustration were, as above, $K_M = 10^2$, $K_Y = 10^6$ and $K_N = 10^{-2}$. Consequently mirror image potential characteristics result, and Donnan Failure occurs when NY(aq.) exceeds 10^{-2}–10^{-1} M NY(aq.).

Potential Theory

The interfacial potential difference (pd) for the partition equilibrium interface is given by the equality of electrochemical potential in terms of all ions in equilibrium, equation (4).

$$\triangle\phi = \phi(\text{org.}) - \phi(\text{aq.}) = \frac{RT}{z_i F}\ln\frac{K_i a_i}{\overline{a}_i} \tag{7}$$

However, for convenience in calculation, the interfacial pd is determined from those species whose activities are known or easily calculated. For the upper curve in Figure 1, in the linear range MX is predominately in water while MY is predominately in the organic phase since $K_X \ll K_Y$. Consequently M^+ activities are known and used in equation 7, to give

$$\triangle\phi = \frac{RT}{F}\ln\frac{K_M a_M^+}{\overline{a}_M} = \triangle\phi° + \frac{RT}{F}\ln a_M^+ \tag{8}$$

(Nernstian response) over a wide activity range. However, at very low MX activities, the pd becomes insensitive to decreasing M^+ activities and levels off at a value

$$\triangle\phi = \frac{RT}{2F}\ln\frac{K_M}{K_Y} \tag{9}$$

while at very high MX activities (generally only seen when $X^- = I^-$, NO_3^-, ClO_4^-) the pd again levels off since Donnan Exclusion by Y^- is violated in the organic phase and

$$\triangle\phi = \frac{RT}{2F}\ln\frac{K_M}{K_X} \tag{10}$$

These limiting, single-salt pds are derived from the more general expression:

$$\triangle\phi = \frac{RT}{2F}\ln\left[\frac{K_M a_M/\overline{a}_M + K_N a_N/\overline{a}_N}{K_X a_X/\overline{a}_X + K_Y a_Y/\overline{a}_Y}\right] \tag{11}$$

The values of activities to be used are those calculated from equilibrium theory. For one salt, say MX

$$\triangle\phi = \frac{RT}{2F}\ln\left[\frac{K_+a_+/\overline{a}_+}{K_-a_-/\overline{a}_-}\right] \tag{12}$$

This equation is suitable even in the case of unequal
concentrations of M^+ and X^- in each phase. However, if only MX is
present, a simplification is possible because there are equal
concentrations of + and − ions in each phase.

$$\triangle\phi = \frac{RT}{2F}\ln\left[\frac{K_+a_+/\overline{\gamma}_+}{K_-a_-/\overline{\gamma}_-}\right] = \frac{RT}{2F}\ln\left[\frac{K_+\gamma_+\overline{\gamma}_-}{K_-\gamma_-\overline{\gamma}_+}\right] \tag{13}$$

These values are, in some sense derived from corrosion theory,
"mean" or "mixed" potentials because they are determined by
exchange of two charged species. When activity coefficients are
ignored equation 13 reduces to 10. By the same analysis, a single
salt MY partitioned gives a constant potential in equation 9.
These two limiting values are shown in Figure 1 points A and E.
Likewise, Donnan Failure upon addition of excess aq. NY gives a
constant negative limit of

$$\triangle\phi = \frac{RT}{2F}\ln\frac{K_N}{K_Y} \tag{14}$$

Details of the curvature regions in Figure 1 have been given in
reference (6).

Sensitivity and Selectivity

The lower detection limit for ions M^+ or Y^- is often given as the
intersection of the Nernstian region with the limiting potential of
equation 9. This value depends on the membrane loading MY(org.)
and is given by

$$\frac{a_Y(\text{limit})}{\overline{a}_Y} = \frac{a_M(\text{limit})}{\overline{a}_M} = (K_M K_Y)^{-\frac{1}{2}} \tag{15}$$

which is 10^{-6}M M^+ in Figure 1. The total range of available
potentials for M^+ measurements is determined by the two limiting
potentials of equations 9 and 10. This is the so-called "window"
for M^+

$$\triangle\phi(\text{window}) = \frac{RT}{2F}\ln\frac{K_Y}{K_X} \tag{16}$$

There is also a window for Y^- measurement.

The definition of selectivity of the electrode for two ions of the same sign requires consideration of the responses to MX and IX, when the membrane is initially loaded with MY. M^+ is the principal ion and I^+ is the interference.

Defining the fraction f as

$$f = \frac{K_M a_M + \gamma_I}{K_I a_I + \gamma_M} \tag{17}$$

then ion exchange equilibrium requires formation of both MY and IY in the membrane with activities

$$\bar{a}_M + = \bar{\gamma}_M(\bar{Y})f/(1 + f) \tag{18}$$

$$\bar{a}_I + = \bar{\gamma}_I(\bar{Y})/(1 + f) \tag{19}$$

Consequently the interfacial pd is related to both a_M+ and a_I+ according to

$$\triangle\phi = \frac{RT}{F}\ln\frac{K_M}{\bar{\gamma}_M(\bar{Y})}\left[a_M + \frac{K_I\bar{\gamma}_M}{K_M\bar{\gamma}_I}a_I\right] \tag{20}$$

The interfacial pd selectivity coefficient, the factor multiplying a_I is determined by the ratio K_I/K_M, by the activity coefficient ratio, and by the mobility ratio, when the internal diffusion potential contribution is added. Clearly interferences should correlate with the ratio K_I/K_M, which can be determined from salt extraction coefficients K_{I_X}/K_{M_X} for a series of positive drugs, using common anion salts. This result is well documented in the literature (<u>7,8</u>). A curious correlation for N-based drugs studied by us and by Freiser (<u>9</u>) is a trend in selectivity

$$RNH_3^+ < R_2NH_2^+ < R_3NH^+ < R_4N^+$$

in which quarternary drugs of the same carbon number are most sensitively detected.

Omitted from this elementary theory are effects of plasticizer and ion pairing. Ion pair formation constants in the organic phase increase with decreasing dielectric constants of the plasticizer, in the absence of specific bonding effects. In the more general theory the single ion partition coefficients are replaced by the product of partition coefficient and ion pair formation constant.

Experimental: Reagents

All reagents, except TPSB (triphenylstilbenylborate), were of analytical-reagent grade and were used as received. Succinylcholine chloride, hexamethonium bromide and decamethonium bromide were purchased from Sigma (St. Louis, MO). NPOE (2-nitrophenyloctyl ether)(Fluka) and poly(vinylchloride)(Aldrich)

were used. Triphenylstilbenyl borate (potassium salt) was kindly
donated by Dr. D. Daniels (Kodak, Rochester, NY). Injectable
succinylcholine chloride solutions (USP quality) were purchased
from a local drugstore. All cation drug solutions were prepared
with distilled water in a Tris-HCl buffer of pH 7.0.
Succinylcholine solutions contained 0.05% (w/v) methyl-p-
hydroxybenzoate as stabilizer. Phenytoin (sodium salt) and decanol
were supplied by Sigma (St. Louis, MO); other materials were
tricaprylylmethylammonium chloride or Aliquat 336S (General Mills
Chemicals, Inc., Kankakee, IL). Solutions of sodium phenytoin were
prepared by serial dilution while keeping both pH and ionic
strength at constant values, 10 and 0.1 mol/l, respectively. The
selectivity coefficients were determined at pH 10.0 and 0.1 mol/l
ionic strength, both adjusted with borax-NaOH buffer solution of pH
10.0.

Experimental: Electroactive Materials

Bisquaternary, as well as monoquaternary compounds and various
amino derivatives, react with tetraphenylborate and similar
compounds to form stable ion-pair complexes. TPSB was found to be
very suitable as an ion-pairing agent for bisquaternary drugs with
which it forms very insoluble compounds. The TPSB complexes were
prepared in situ, by soaking the TPSB (potassium salt)/PVC
membranes in the appropriate bisquaternary solution. Of the
plasticizers tested, 2-nitrophenyloctyl ether (NPOE),
dioctylphthalate, di-iso-butylphthalate, nitrobenzene, 2-nitro-p-
cymene, NPOE showed the best behavior in terms of response time and
reproducibility. The membrane compositions were 3.2% (w/w) TPSB,
64.5% (w/w) NPOE and 32.3% (w/w) PVC.
       The quaternary ammonium cation, tricaprylylmethylammonium, is
a well known ion-pairing extracting agent and was used to obtain
the ion-pair association complex with 5,5-diphenylhydantoinate
anion. The ion-pair complex was embedded in a PVC matrix,
containing NPOE as plasticizer. The membrane composition was 7.7%
(w/w) electroactive material, 61.5% (w/w) NPOE and 30.8% (w/w) PVC.
       Five grams of Aliquat 336S were mixed with 5.0 g of decanol
and equilibrated with ten separate 15 ml aliquots of 0.1 mol/l
sodium phenytoin solution in 20% (v/v) methanol. The organic phase
was washed twice with distilled water and then centrifuged until a
clear solution was obtained.

Experimental: Construction of Electrodes

The basic principle of the electrode construction has been
described elsewhere (10). The electroactive material (50 mg) was
well mixed with 400 mg plasticizer (NPOE) and later with 200 mg PVC
powder dissolved in 6 ml of tetrahydrofuran. The clear liquid was
poured into a 28 mm i.d. glass ring on as sheet of plate glass. A
pad of filter paper placed on top of the ring was kept in place by
a heavy metallic weight and the assembly left for 48 hrs. to allow
slow solvent evaporation. A disc (0.9 cm diameter) was cut from
the membrane and fixed to the end of a 10 mm Tygon tube by using a
PVC-tetrahydrofuran solution as adhesive. The other end of the
Tygon tube was fitted on to a glass tube to form the electrode

body. A silver/silver chloride wire was then inserted and the
electrode body was filled with typically 0.01 NaCl, saturated with
AgCl and a salt of exchanging ion and a buffer solution. The
internal reference solutions were 10^{-3}M of the respective cation
drug at pH 7.0 (0.1 M Tris-HCl buffer). The TPSB in the polymer
membranes was converted to the bisquaternary-drug form by soaking
the electrodes in the appropriate 10^{-2}M drug for 24 h. When not in
use, the electrodes were stored in the same solution as the
internal solution. For the anion drug, the body was filled with 10^{-3} mol/l sodium phenytoin solution of pH 10.0 (borax-NaOH buffer).

<u>Results</u>

Although this theory has not been succinctly reported until
recently, intuitive applications have been applied earlier.
Ruzicka, et al. (<u>11</u>) increased the oil solubility of phosphate
ester anions to increase the sensitivity of the Ca^{2+} electrode.
Gavach, et al. (<u>12</u>) and Birch, et al. (<u>13</u>) made detergent sensors
for cations and anions.

The electrodes described above show near-Nernstian responses
over a large range on concentrations and very low detection limits.
These electrodes are not affected by pH in the range 2-10. Their
selectivities relative to a number of inorganic ions, amino acids,
neurotransmitters, drugs and various drug-excipients are
outstanding. Response characteristics are given in Tables I and
II. Extensive selectivity tables are given in recent or
forthcoming publications (<u>14</u>,<u>15</u>).

Table I. Response Characteristics for Bisquaternary-drug Membrane
Electrodes

Parameter	Succinylcholine Electrode	Hexamethonium Electrode	Decamethonium Electrode
Slope[a] (mV/log a)	29.05±0.35	28.03±0.44	29.45±0.29
Intercept (mV)	157±1.8[b]	147±2.2	152±2.1
Linear range (M)	10^{-2}—10^{-6}	10^{-2}—2.5×10^{-7}	10^{-2}—2.5×10^{-7}
Usable range (M)	10^{-2}—10^{-7}	10^{-2}—5.0×10^{-6}	10^{-2}—10^{-7}
Detection limit (M)	1.58×10^{-7}	3.16×10^{-8}	1.12×10^{-7}
(ng ml^{-1})	46	6.4	29

[a] Average values calculated for 10^{-3}—10^{-5}M range with standard
 deviation of average slope value for multiple calibration (5-7).

[b] Standard deviation of values recorded during one month.

Source: Reproduced with permission from Ref. 14. Copyright 1984
Elsevier (Amsterdam).

Table II. Response Characteristics for the Phenytoin-Membrane
                          Electrode

Parameter	
Slope(mV/log a)	$56.25 + 0.83^a$
Intercept(mV)	$182 + 2.1^b$
Linear Range(mol/l)	$10^{-1} - 10^{-5}$
Useable Range(mol/l)	$10^{-1} - 10^{-5}$
Detection Limit(mol/l)	1.5×10^{-5}
(µg/ml)	4.1

[a] Standard deviation of average slope value for multiple calibrations in the $10^{-2} - 10^{-4}$ mol/l range.

[b] Standard deviation of values recorded during one month.

Literature Cited

1. Coetzee, C. J.; Freiser, H. Anal. Chem. 1968, 40, 2071.
2. Coetzee, C. J.; Freiser, H. Anal. Chem. 1969, 41, 1128.
3. James, H. J.; Carmack, G. P.; Freiser, H. Anal. Chem. 1972, 44, 853.
4. Higuchi, T.; Illian, C. R.; Tossounian, J. L. Anal. Chem. 1970, 42, 1674.
5. Liteanu, C.; Hopirtean, E. Talanta 1970, 17, 1067.
6. Melroy, O. R.; Buck, R. P. J. Electroanal. Chem. 1973, 143, 23.
7. Koryta, J. "Ion-Selective Electrodes"; Cambridge Univ. Press: Cambridge, 1975, Chapt. 6.
8. Koryta, J.; Stulik, K. "Ion-Selective Electrodes"; Cambridge Univ. Press: Cambridge, 1983, Chapt. 7.
9. Freiser, H., personal communication.
10. Moody, G. J.; Oke, R. B.; Thomas, J. D. R., Analyst 1970, 95, 910.
11. Ruzicka, J.; Hansen, E. H.; Tjell, J. C. Anal. Chim. Acta. 1973, 67, 155.
12. Gavach, C.; Seta, P. Anal. Chim. Acta. 1970, 50, 407.
13. Birch, B. J.; Clarke, D. E. Anal. Chim. Acta. 1973, 67, 387.
14. Cosofret, V.V.; Buck, R. P., Anal. Chim. Acta. 1984, 162, 357.
15. Cosofret, V. V.; Buck, R. P., J. Pharm. Biomed. Anal. 1984 in press.

RECEIVED October 31, 1985

Development of Subcutaneous-Type Glucose Sensors for Implantable or Portable Artificial Pancreas

Kaname Ito[1], Shoichiro Ikeda[1], Kaori Asai[1], Hirotoshi Naruse[1], Kunitoshi Ohkura[2], Hidehito Ichihashi[2], Hideo Kamei[2], and Tatsuhei Kondo[2]

[1]Department of Applied Chemistry, Nagoya Institute of Technology, Gokiso-cho, Showa-ku, Nagoya 466, Japan
[2]Department of Surgery, Nagoya University, School of Medicine, Tsurumai-cho, Showa-ku, Nagoya 466, Japan

Most subcutaneous-type glucose sensors developed to date have been based upon oxygen type enzyme electrodes. A new glucose sensor developed may be characterized by the glucose semipermeable menbrane covered on the enzyme electrodes. The sensor which was prepared from an oxygen permeable PMSP (poly (1-trimethylsilyl-1-propyne)) membrane and a glucose semipermeable AC (acetyl cellulose) membrane, rapidly responded to glucose levels up to 500 mg/dl without the effect of oxygen tension in the range of 5 to 21 %. This sensor appears to hold promise for artificial pancreas applications.

The closed-loop type artificial pancreas (specifically β-cell), which consists of an automatic continuous monitor of blood glucose level (BGL) and an automatic injector of insulin which are coupled with feed-back system, has great potential for prevention of diabetic complication such as micro-angiopathies(1). A large-scale closed-loop type artificial pancreas for bedside use has already been developed and is clinically used at some laboratories and hospitals (2-4). However, this device is limited to only bedside use. On the other hand, the open-loop type artificial pancreas which consists of only a insulin injecting pump without an automatic continuous monitor of BGL, has been developed and is going to be clinically used(5-7). This system, however, can not completely control BGL as well as the bare pancreas in a normal body and often causes lower BGL(8-9).

In order to provide for the complete therapy of diabetic patients, an implantable or portable closed-loop type artificial pancreas must be developed. The key factor in the development of such system is development of a small-size glucose sensor which is able to measure directly up to 500-700 mg/dl of BGL in a blood stream or in a body fluid.

About ten years ago, Bessman et al(10), University of Southern California, developed a glucose sensor of enzyme electrode type with glucose oxidase (GOX) for an artificial pancreas. This sensor had the limitation that the measurable glucose level was up to ca. 150 mg/dl, at maximum.

Shichiri et al(11) (Osaka University in Japan) has developed the micro needle type glucose sensor, which consisted of a hydrogen peroxide electrode and a GOX enzyme immobilized layer. The sensor was clinically used, but it had to be renewed after a few days because of a gradual decline in its output.

In the previous papers(12,13), we reported on the vessel access type, i.e. tubular type, glucose sensor. It consisted of a glucose electrode system with a GOX enzyme immobilized Nylon membrane and a glucose semipermeable membrane, and a reference oxygen electrode system. The sensor could directly measure up to 700 mg/dl of BGL in an arterial blood stream when it was placed into an external A-V shunt. This sensor, however, has some problems such as thrombus during *in vivo* testing without heparin and clinical complexity associated with implanting the sensor in a blood stream.

In the present paper, therefore, we have modified the shape and the performance of the glucose sensor to measure the glucose level in the subcutaneous tissue.

Experimental

Preperation of sensors.   The subcutaneous type glucose sensors developed are essentially an oxygen type enzyme electrode as well as the tubular type sensors previously reported(12,13). Figure 1 shows a schematic diagram of an oxygen electrode system and epoxy resin parts used to construct this type of glucose sensor. The cathode is a 0.5 mm diameter Pt wire sealed with a glass tube, and the anode is a 0.5 mm diameter Ag wire. These electrodes were set inside and outside of epoxy resin ring, respectively, and it was set in an epoxy resin socket.

The oxygen sensing properties of such electrode systems were measured at first. The reproducibility of the output current versus oxygen tension is shown in Figure 2. The coefficient of correlation was 0.987 using eighteen data points.

One pair of oxygen electrodes with similar properties and the epoxy resin parts as shown in Figure 1 were assembled and a subcutaneous type glucose sensor as shown in Figure 3 was prepared by the following process. Initially, both electrode systems were filled with normal saline solution including gelatin. In the case of the glucose electrode system, an oxygen permeable Teflon FEP membrane, a GOX enzyme immobilized Nylon membrane and a glucose semipermeable membrane were set up on the electrode system and were fixed by a waterproof Kapton tape. While the reference oxygen electrode system was prepared by a similar manner as the glucose electrode system, except that a bare Nylon membrane was used in place of an enzyme immobilized membrane.

Measurement.   The measurement apparatus for the *in vitro* test of this sensor is shown in Figure 4. The sensor was dipped into the flask containing 100 ml of phosphate buffered saline solution, pH 7.4, and the output was measured with respect to stepwise changed glucose concentrations of 0 to 2000 mg/dl under the oxygen tensions of 5 to 21%, which prepared by mixing air and N_2 gas.

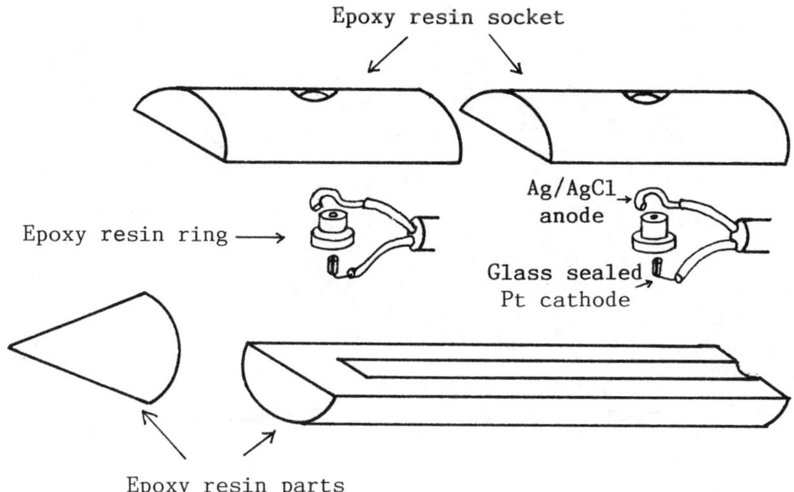

Figure 1.   Schematic diagram of oxygen electrode systems for subcutaenous type glucose sensors.

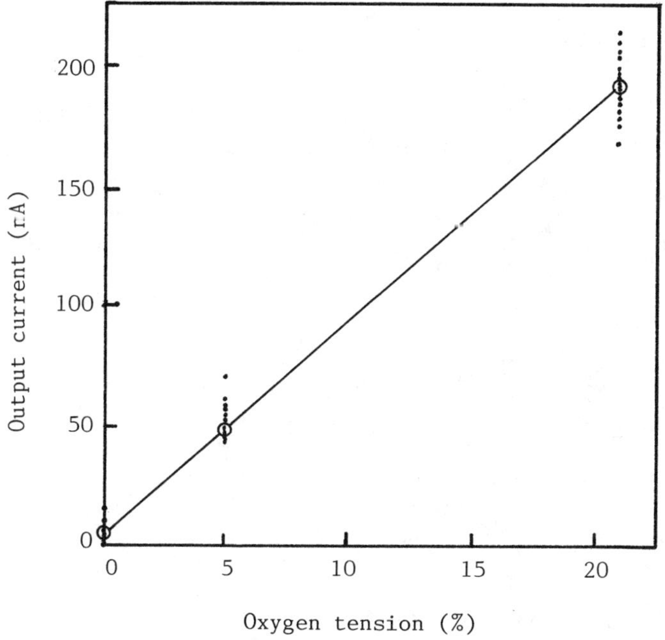

Figure 2.   Reproducibility of output currents vs. oxygen tension of oxygen electrode systems.

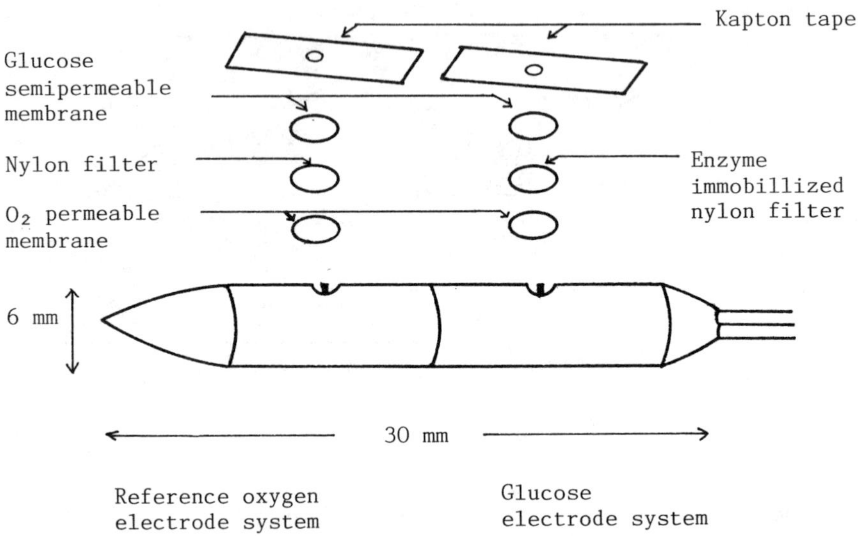

Figure 3. Structure of a subcutaneous type glucose sensor

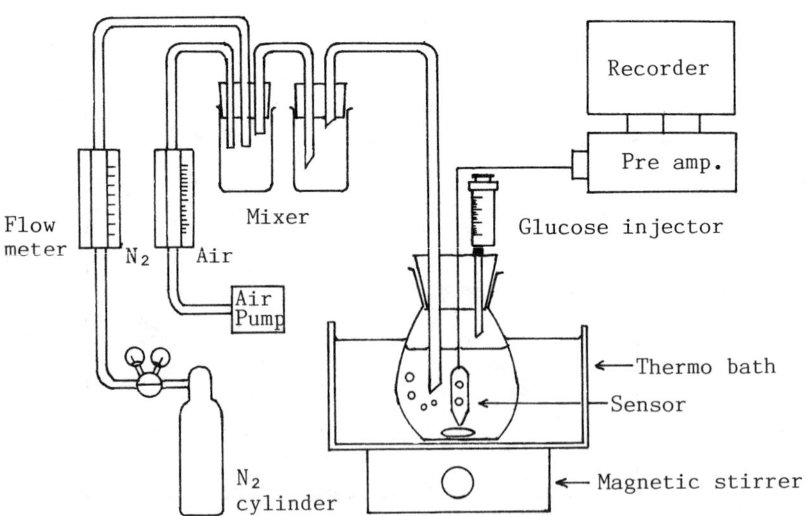

Figure 4. Apparatus for *in vitro* measurement of dynamic response properties of the subcutaneous type glucose sensor

Results and Discussion

In vitro tests.   The oxygen tension in subcutaneous tissue is reported to be about 5 %, and the output of glucose sensors is known to be affected by oxygen tension. Figure 5 illustrates the output currents of the glucose sensor I with respect to the stepwise changed glucose concentrations shown in this figure, under an oxygen tension of 21, 10 or 5 %. The sensor I, which is prepared from the glucose semipermeable Teflon FEP membrane with 25 μm diameter pinhole as shown in Table 1, has rapid response properties for increasing and decreasing of glucose level and 95 % response time is less than 2 min under 21 % oxygen tension. At lower oxygen tensions such as 10 or 5%, the sensor I has a smaller output current and lower response times.

The calibration curves of the glucose sensor I are shown in Figure 6 and the linear range of the curves gradually decrease with decreasing of the oxygen tension from 21 to 5 %.

The glucose oxidation reaction in the enzyme immobilized membrane of the glucose sensors is performed according to the following reaction.

$$\text{Glucose} + O_2 + H_2O \xrightarrow{\text{GOX}} \text{Gluconic acid} + H_2O_2 \qquad (1)$$

where GOX is glucose oxidase and is sufficiantly immobilized on the Nylon membrane by covalent bonding using the glutaraldehyde method. If the glucose diffused into the enzyme membrane is much less than concentration of the dissolved oxygen in it, the reaction rate becomes the first order with respect to glucose concentration. Then, the calibration curve of the glucose sensor will show linear response in the range of the corresponding concentration of glucose. Therefore, in order to obtain the linear response up to a higher glucose level, such as 500 mg/dl, the glucose permeability of the semipermeable membrane, in the case of lower oxygen tension such as 5 %, must be much less than that in oxygen tension of 21 %.

Glucose sensors II and III were prepared from the semipermeable membrane of PMSP, poly(1-trimethylsilyl-1-propyne), which has 4 times the oxygen permeability compared with that of FEP membrane. The response properties of sensor II, using a PMSP membrane with 25 μm diameter pinhole, were almost similer to that of the sensor I, so that their calibration curves were not presented in this paper.

The calibration curves of sensor III , which was prepared using a PMSP semipermiable membrane with smaller pinhole (15 μm diameter) than that of sensors I and II , is shown in Figure 7. The curves under 5, 10 and 21 % of oxygen tension almost agreed with each other in the range of 0 to 500 mg/dl of glucose concentration and the output was hardly affected by oxygen tension in the range of 5 to 21 %.

The calibration curves of the sensor IV , which was prepared with a PMSP membrane as an oxygen permeable membrane and an acetyl celulose(AC) membrane as a glucose semipermeable membrane, were shown in Figure 8. This AC membrane was cast by Manjikian's method and treated at a curing temperature of 85°C. The sensor IV indicated a linear response up to 500 mg/dl of glucose concentration, even under 5 % of oxygen tension, and the output currents were hardly affected by oxygen tension.

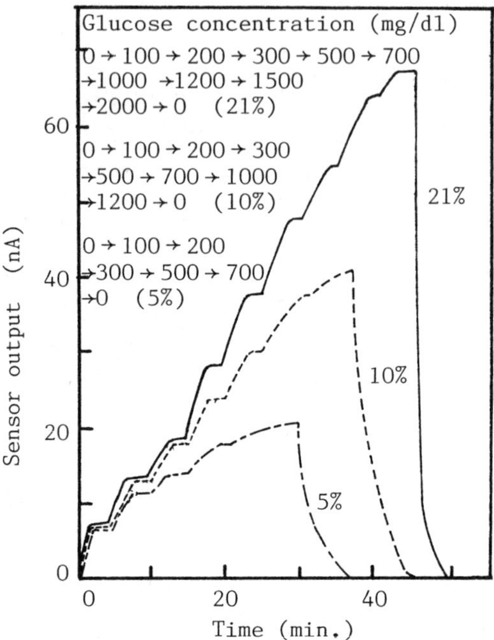

Figure 5. Dynamic response curves of glucose sensor I under various oxygen tensions (%)

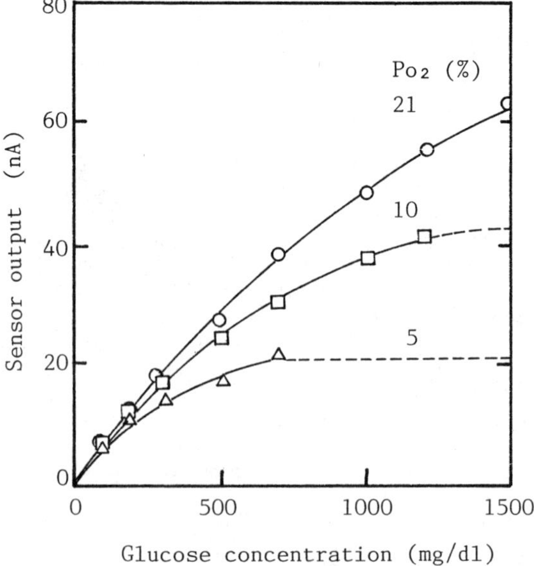

Figure 6. Calibration curves of glucose sensor I having glucose semipermeable FEPp membrane with 25 μm diameter pinhole.

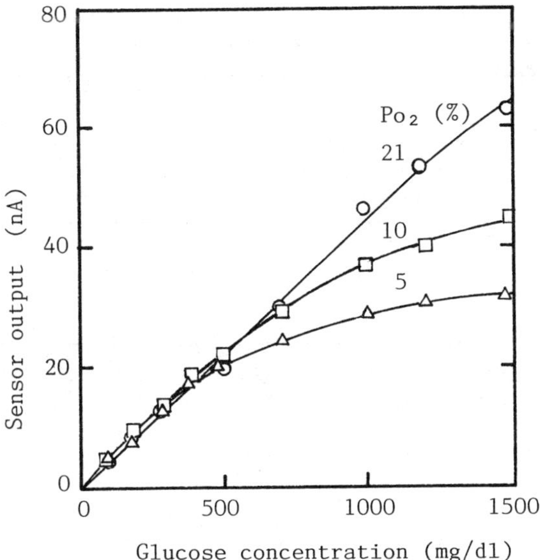

Figure 7. Calibration curves of glucose sensor Ⅲ having glucose semipermeable PMSPp(1) membrane with 15 μm diameter pinhole

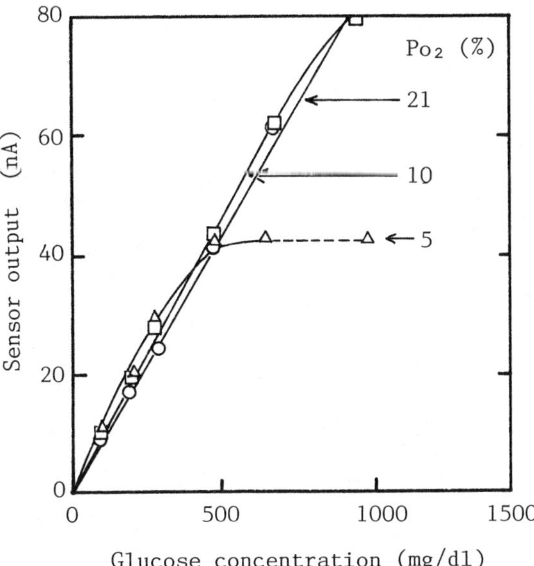

Figure 8. Calibration curves of glucose sensor Ⅳ having a glucose semipermeable AC membrane. The AC membranes were cast using a solution of acetyl cellulose: formamide: aceton = 25:30:45 (wt %) and tempered at 85°C

Table 1   Linear response range and response time of glucose sensors
           with various glucose semipermeable membranes

Sensor No.	O_2 permeable membrane	GOX enzyme immobilyzed membrane	Glucose semipermeable membrane	Max. glucose conc. of liner response range (mg/dl)			95% response (min)
				oxygen tension			
				21%	10%	5%	
I	FEP	Nz	FEPp	500	300	200	3
II	FEP	Nz	PMSPp(I)	700	500	300	2
III	FEP	Nz	PMSPp(II)	2000	700	500	2
IV	PMSP	Nz	AC	2000	1000	500	5

FEP :Teflon FEP membrane   FEPp   :FEP with 25 μm φ pinhole
PMSP:Poly(1-trimethylsilyl-1-   PMSPp(I) :PMSP with 25 μm φ pinhole
     propyne) membrane   PMSPp(II):PMSP with 15 μm φ pinhole
Nz :Enzyme immobilized Nylon   AC   :Acetyl cellulose membrane
     filter

In vivo tests.   The response properties of the glucose sensors I
to Ⅳ are summarized in Table 1. The sensor Ⅲ and Ⅳ have more
desirable responses than the sensor I and Ⅱ as a glucose sensor for
the artificial pancreas. In the present paper, however, the results
of _in vivo_ tests obtained by using the sensor I will be demonstrated
in the later section.
     Before _in vivo_ tests using a dog, the performance of sensor I
was first examined in the human plasma solution, involving similar
substances to that of subcutaneous tissue. The calibration curves of
the sensor I in saline solution before and after plasma test for 10
hours are illustrated in Figure 9, and the output of the sensor I in
human plasma solution was plotted with respect to the plasma glucose
level by using the sign of open triangle. The calibration curves
before and after plasma test in this figure almost agreed with
each other and the open triangle was located on the calibration
curves. Therefore, there seems to be no significant interferences
from substances in human plasma with the glucose sensor performance.
Also the sensor output seems to be able to indicate glucose levels
according to the calibration curves obtained in the saline solution.
     The _in vivo_ tests of glucose sensor I were performed by using a
normal dog under general anesthesia. Several sensors were implanted
in the femoral abdominal or thoracic subcutaneous tissue of a dog,
and the intravenous glucose tolerance tests were carried out.
A typical example of the results is shown in Figure 10. The glucose
levels, which were read from the outputs of the sensor according to
the calibration curve in the right side graph of this figure, are
demonstrated by the solid line in left side graph of this figure. The
BGLs, which were measured with the sampled bloods using a YSI-23A
glucose auto-analyzer, are plotted by the circles. As shown in this
figure, the intravenous glucose tolerance tests of 4 g were performed
twice and the tests of 1 g were done 4 times. In all cases sensor I
responded rapidly to the variation of BGL within 1 min, although it
couldn't respond to rapid peak of BGL after glucose loading
intravenously. The glucose levels indicated from the output of the

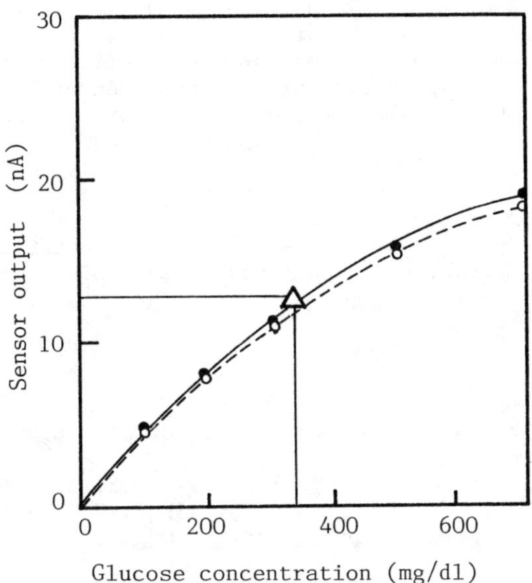

Figure 9.   Calibration curves of glucose sensor I before (—●—) and after (—○—) human plasma test.
△:sensor output (13 nA) in human plasma (GL 333 mg/dl)

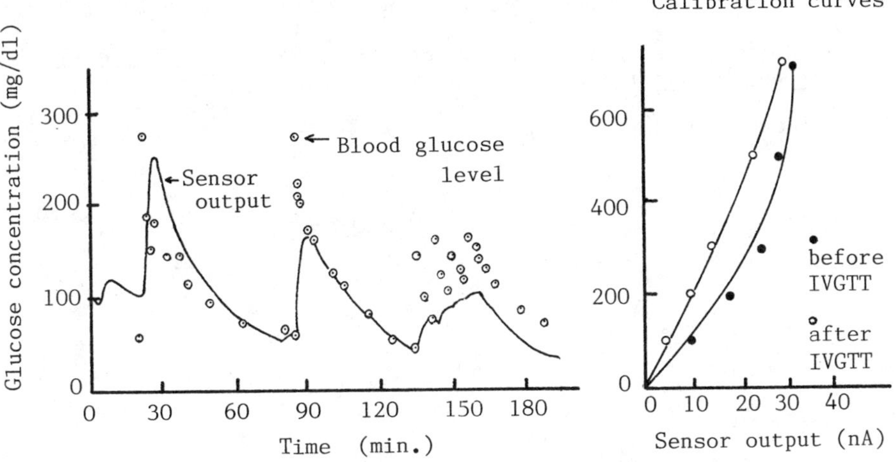

Figure 10.   Results of intravenous glucose tolerance test (IVGTT) using a normal dog in the case of glucose sensor I.

sensor were in fair agreement with the glucose levels of sampled bloods measured by a glucose auto-analyzer.

Glucose sensor III and IV have more desirable properties than that of the sensor I, so that they are promising as a subcutaneous type glucose sensor for an implantable artificial pancreas. Now, life time test and *in vivo* testing of these sensors are going to be performed, and the results of these tests will be reported in the future.

Acknowledgments

This work was supported by Grant-in-Aid for Scientific Research from the Ministry of Education, Science and Culture of Japan.

Literature Cited

1. Cahill, G. F., Jr.; Soeldner, J. S. et al. _Diabetes_ 1972, 21, 703.
2. Albisser, A. M. et al. _Diabetes_ 1974, 23, 389; 1974, 23, 397.
3. Chemene, A. H. et al. _Metab. Rev._ 1977, 7 (Suppl.), 23
4. Shichiri, M. et al. _Jpn. J. Artif. Organs_ 1977, 7, 3 ; 1980, 9, 181
5. Irrigler, K. et al. _Artif. Organs_ 1981, 5 (Suppl.), 754.
6. Pickup J. C. et al. _Br. Mad. J._ 1978, 1, 204.
7. Tamborlane, W. V. _N. Enal. J. Med._ 1979, 300, 573.
8. Unger, K. H. _Diabetes_ 1982, 31, 479.
9. Felig, P.; Bergman, M. _JAMA_ 1983, 250, 1045.
10. Layne, E. C.; Bessman, S.P. et al. _Diabetes_ 1976, 25, 81.
11. Shichiri, M. et al. _Jpn. J. Artif. Organs_ 1982, 11, 683.
12. Ito, K.; Ikeda, S.; Kondo, T. et al. _Artif. Organs_ 1981, 5 (Suppl.), 735.
13. Kondo, T.; Ikeda, S.; Ito, K. et al. _Trans. Am. Soc. Artif. Intern. Organs_ 1981, 27, 250.

RECEIVED October 31, 1985

INDEXES

Author Index

Subject Index

A

Pumping current
  combustible gas-mixture sensor, 152f
  double-cell oxygen sensor, 146f
Pyroelectric-based thermal
    sensors, 21-29
Pyroelectric element
  advantage, 34
  schematic, 26f
Pyroelectric enthalpimetric
    structures, 29,30f
Pyroelectric sensor and attached
    amplifier, equivalent circuit, 24f
Pyroelectricity, discussion, 21-23

R

Rare earth sulfates and silicon
    dioxide, mixed with sodium
    sulfate, 121-134
Receptor action
  analytical membranes, 359-360
  stabilized lipid
      membranes, 351-361
Redox properties of oxides,
    modification, 84-85,86f
Reflectance, definition, 277
Reflectance analysis, near-IR, for
    chemical sensing, 271
Remote readout, near-IR reflectance
    monitoring in a flour
    mill, 285,286f
Resistance
  Pd-doped tin(IV) oxide sensor
    as a function of adsorption
        temperature, 77,79f
    relation to the amount of oxygen
        adsorbed, 77,80f
  perovskite-type oxides, general
      discussion, 87,91-92
  tin oxides, factors affecting, 67
Response(s)
  amperometric gas sensor, 300-301
  bisquaternary-drug membrane
      electrodes, 371t
  glucose sensors, 378f,380t
  sensors coated with different
      phthalocyanines, 162f
  single-cell oxygen sensors, 144
Response and recovery times
  capacitor-type structures, 193-197
  diode-type structures, 186-188
Rich atmosphere, definition, 102
Rocketborne field-widened
    interferometer-spectrometer
    (RBFWI), 233-240

S

Salmonella typhimurium, use in mutagen
    sensors, 346-347
Sample cell, carbon dioxide
    sensor, 223,225,226f

Sample preparation, near-IR
    reflectance analysis, 279-280
Scanning electron micrograph, finished
    tin oxide microsensor, 62f
Schottky-barrier diode gas sensors,
    performance, 177-202
Selective response of polymeric
    film-coated optical waveguide
    devices to water and toxic
    compounds, 320-328
Selectivity, electrodes for ions, 369
Selectivity coefficients
  coated-wire electrodes, 261t
  cyclohexylamines, 268-269t
  divalent cations, 259t
  electrodes, 263t
  interfacial potential
      difference, 369
  liquid-membrane electrodes, 261t
Selectivity measurements, coated-wire
    electrodes, 262-270
Semiconductive metal oxides, relation
    of electrical conductivity to
    oxygen partial pressure, 93,97f
Semiconductor barrier,
    modification, 178-179
Semiconductor gas sensors
  advantages, 59
  applications, 40,42f
  general discussion, 39
Sensing mechanisms
  amperometric sensor, 205-209
  modified amperometric
      sensor, 208,211,212f
  tin oxide microsensors, 63,66f,67
Sensing performance, amperometric
    sensor
  carbon monoxide, 211-213
  hydrogen, 205-207
Sensitive drug sensors,
    design, 363-372
Sensitivity, electrodes for
    ions, 368-369
Sensor(s)
  amperometric, sensitivity to
      hydrogen and carbon
      monoxide, 203-213
  arrays, detection and identifi-
      cation of airborne
      chemicals, 302-305
  based on oxygen pumping,
      designs, 141-152
  bioanalytical, 251-254
  chemical, recent advances, 2-34
  chemically modified
      electrode, 245-254
  coated with different
      phthalocyanines, 162f
  combustible gas
      mixtures, 150-151,152f
  development
    for artificial pancreas, 373-382
    in Japan, 43-54

Production by Hilary Kanter
Indexing by Karen McCeney
Jacket design by Pamela Lewis

Elements typeset by Hot Type Ltd., Washington, DC
Printed and bound by Maple Press Co., York, PA